Henry Seebohm

The Birds of the Japanese Empire

Henry Seebohm

The Birds of the Japanese Empire

ISBN/EAN: 9783337169800

Printed in Europe, USA, Canada, Australia, Japan

Cover: Foto ©berggeist007 / pixelio.de

More available books at **www.hansebooks.com**

The horizontal lines on the sea denote that it is under 100 fathoms in depth. The dotted lines represent the isothermal lines (the mean temperature) of July; the figures the degrees of Fahrenheit.

THE BIRDS

OF THE

JAPANESE EMPIRE.

BY

HENRY SEEBOHM,

AUTHOR OF 'SIBERIA IN EUROPE,' 'SIBERIA IN ASIA,' 'A HISTORY OF
BRITISH BIRDS,' 'CATALOGUE OF THE BIRDS IN THE BRITISH
MUSEUM' (VOL. V.), 'THE GEOGRAPHICAL DISTRIBUTION
OF THE FAMILY CHARADRIIDÆ,' 'CLASSIFICATION
OF BIRDS,' ETC., ETC.

WITH NUMEROUS WOODCUTS.

LONDON:
R. H. PORTER, 18 PRINCES STREET, CAVENDISH SQUARE.
1890.

PRINTED BY TAYLOR AND FRANCIS,
RED LION COURT, FLEET STREET.

PREFACE.

During the last ten years I have been making a collection of Japanese Birds, and have accumulated a carefully selected series of more than two thousand skins, out of the many thousands that have passed through my hands. I am indebted to Captain Blakiston for the most interesting of these specimens, some of which were in the Swinhoe Collection, whilst others were sent to me from time to time during the many years' residence of Captain Blakiston at Hakodadi. I am also indebted to the late Mr. Harry Pryer for many skins of birds from Southern Japan, and at his death, after many years' residence in Yokohama, when his collection came into my hands, I began to prepare my notes on Japanese Birds for the press. I have also to acknowledge my indebtedness to Mr. Heywood Jones for a small collection of skins from Fuji-yama, and to Mr. Ringer, of Nagasaki, for some very rare birds from Kiu-siu. I am much indebted to the Curator of the Norwich Museum for arranging their valuable collection of Japanese skins so that it could be examined without difficulty; and I have to thank Dr. Stejneger of Washington for valuable information respecting the interesting collections made in Southern Japan by Messrs. Jouy and Smith. I also take this opportunity of thanking the Curators of the Museums of Leyden, Bremen, Frankfort, Paris, Vienna, Philadelphia, and Washington for showing me many interesting birds from the Japanese Empire.

The interest attaching to Geographical Distribution is so great that it is not necessary to apologize for the amount of space devoted to the consideration of the breeding-range of each species, and of the sub-species or local races of those that vary in different parts of their range. The interest to be derived from the study of the Geographical Distribution of *genera* has been to a large extent destroyed by the number of unimportant subgenera which have been elevated to generic rank. I have endeavoured to place these pseudogenera in their proper position as much as possible, but much yet remains to be done in this direction. It is, of course, impossible to study the Geographical Distribution of the higher groups until their respective limits have been determined. Of these the subdivisions of the Passeres present the greatest difficulty, and with some slight modifications I have adopted those defined by Mr. Oates (Fauna of British India, Birds, i. p. 8), which seem to me to be a distinct advance upon previous arrangements. In the higher groups I have followed the scheme explained in detail in my 'Classification of Birds,' and have included the diagnoses of the subclasses, orders, and suborders which are not represented in Japan as well as those which are. The absence of a group is often quite as interesting as the presence of others, especially the absence of those whose range almost, but not quite, reaches the district treated of. Further, by making the list of suborders, orders, and subclasses complete, it has given me an opportunity of correcting some inaccuracies in my former work.

I have divided the subject of the Birds of Japan into three sections. The first treats of the Literature relating to Japanese Birds. The second section relates to their Geographical Distribution in the Japanese Empire, and during the breeding-season outside its limits. To this is appended a table of comparison of the number of species in each suborder, and in the case of the Passeres in each subfamily, which have been recorded from the British Islands with those which have been found in the Japanese Empire. I have concluded this section with some remarks on the important fact that there are many cases in which West-European birds more closely resemble East-Asiatic ones than the Siberian races which intervene. The third section, which embraces the larger part of the volume, is devoted to the Classification and Identification of Japanese Birds. I have not occupied the space devoted to each species with useless synonymy or with long descriptions. Those references only are quoted which contain some valuable information, and those details of the plumage only are given which are necessary to distinguish the species from other Japanese birds, or from nearly allied species or races.

I venture to think that the information respecting the Birds of the Japanese Empire, much of it collected together for the first time, will prove to be an interesting and important contribution towards our knowledge of the Geographical Distribution of the Birds of the Palæarctic Region.

CONTENTS.

Literature relating to Japanese Birds.

	Page
PALLAS, Zoographia Rosso-Asiatica	1
TEMMINCK, Nouveau Recueil de Planches Coloriées d'Oiseaux	2
KITTLITZ, Ueber die Vögel des Inselgruppe von Boninsima	2
KITTLITZ, Kupfertafeln zur Naturgeschichte der Vögel	2
TEMMINCK, Manuel d'Ornithologie	2
TEMMINCK and SCHLEGEL, Fauna Japonica	3
CASSIN, Descriptions of New Species of Birds	3
CASSIN, Report on the Birds collected by the Perry Expedition	4
CASSIN, Report on the Birds collected during the cruise of the 'Portsmouth'	4
CASSIN, Report on the Birds collected by the United States North Pacific Surveying and Exploring Expedition	5
BLAKISTON, On the Ornithology of Northern Japan	5
BLAKISTON, Corrections and Additions to the preceding paper	5
SWINHOE, List of the Birds of Japan	6
SWINHOE, Notes on the Birds collected in Japan during the cruise of the 'Portsmouth'	6
WHITELY, Notes on Birds collected near Hakodadi	6
SHARPE, Notes on Birds collected by Mr. R. H. Bergman in Japan	7
SWINHOE, On the White Stork of Japan	7
SWINHOE, On the Rosy Ibis of Japan	7
SWINHOE, Notes on Birds collected near Hakodadi	7
SWINHOE, Further Notes on birds from Hakodadi	7
SWINHOE, Further Notes on birds from Hakodadi	8
SWINHOE, Further Notes on Birds from Hakodadi	8
BLAKISTON and PRYER, Catalogue of the Birds of Japan	9
SEEBOHM, On a Wagtail from Japan	10

	Page
SEEBOHM, Remarks on Messrs. Blakiston and Pryer's 'Catalogue of the Birds of Japan'	11
SEEBOHM, Further Contributions to the Ornithology of Japan	11
BLAKISTON and PRYER, Birds of Japan	11
BLAKISTON, Notes on the Birds of Yezzo	12
SEEBOHM, On the Pied Wagtails of Japan	12
SEEBOHM, On a new Species of Owl from Yezzo	12
BLAKISTON, On the Birds collected by Messrs. Jouy and Smith in Central Japan	12
JOUY, On the Birds of Central Japan	13
BLAKISTON, Notes on the Birds of Yezzo	13
BLAKISTON, Notes on the Birds of the Kurile Islands	13
SEEBOHM, Further Contributions to the Ornithology of Japan	13
BLAKISTON, Amended List of the Birds of Japan	13
HARGITT, On a new Japanese Woodpecker	14
SEEBOHM, Further Contributions to the Ornithology of Japan	14
SEEBOHM, On the Cormorants of Japan	14
SEEBOHM, Further Contributions to the Ornithology of Japan	14
STEJNEGER, On the Woodpeckers of Japan	14
STEJNEGER, On the Tits and Nuthatches of Japan	14
STEJNEGER, On the Rails, Gallinules, and Coots of Japan	15
STEJNEGER, On the Birds of the Central Group of the Loo-Choo Islands	15
BLAKISTON, The Water-Birds of Japan	15
SEEBOHM, On the Bullfinches of Japan	15
SEEBOHM, On the Birds of the Central Group of the Loo-Choo Islands	15
STEJNEGER, On the Thrushes of Japan	16
STEJNEGER, On the Bullfinches of Japan	16
STEJNEGER, On the Birds of the Loo-Choo Islands	16
STEJNEGER, On the Ibises, Storks, and Herons of Japan	16
STEJNEGER, On the Carrion-Crow of Japan	16
STEJNEGER, On the Birds of the Southern Group of the Loo-Choo Islands	17
STEJNEGER, On the Pigeons of Japan	17
STEJNEGER, On the Birds of the Seven Islands	17
STEJNEGER, On a new Species of Fruit-Pigeon from the Central Group of the Loo-Choo Islands	17
STEJNEGER, On the Creepers of Japan	17
JOUY, On Cormorant Fishing in Japan	18
SEEBOHM, On the Birds of the Central Group of the Loo-Choo Islands	18
STEJNEGER, On the Nutcracker of Japan	18
STEJNEGER, On the Wrens of Japan	18
SOUTER, On Japanese Birds	18
SEEBOHM, On the Birds of the Bonin Islands	18

Geographical Distribution of Japanese Birds.

	Page
List of Species, showing their Distribution within the Japanese Empire, and (during the breeding-season) outside its limits	20
Summary of the various ranges (during the breeding-season) of Japanese Birds	30
Distribution of Birds peculiar to the Japanese Empire	32
Summary of the Distribution of Birds within the Japanese Empire	34
Comparative table of British and Japanese Birds	37
Climatic Variations of Palæarctic Birds	38

Classification and Identification of Japanese Birds.

Subclass **PASSERIFORMES.**

Order PICO-PASSERES.

Suborder I. *PASSERES.*

Family PASSERIDÆ.

Subfamily TURDINÆ.

1. *Geocichla varia.*	White's Ground-Thrush	43
2. —— *sibirica.*	Siberian Ground-Thrush	44
3. —— *terrestris.*	Kittlitz's Ground-Thrush	44
4. *Merula cardis.*	Grey Japanese Ouzel	45
5. —— *fuscata.*	Dusky Ouzel	46
6. —— *naumanni.*	Red-tailed Ouzel	47
7. —— *pallida.*	Pale Ouzel	47
8. —— *chrysolaus.*	Brown Japanese Ouzel	48
9. —— *obscura.*	Dusky Ouzel	49
10. —— *cælænops.*	Seven-Island Ouzel	50
11. *Erithacus akahige.*	Japanese Robin	50
12. —— *namiyei.*	Stejneger's Robin	51
13. —— *komadori.*	Temminck's Robin	52
14. —— *calliope.*	Siberian Ruby-throated Robin	52
15. —— *cyaneus.*	Siberian Blue Robin	53
16. *Monticola cyanus solitaria.*	Eastern Blue Rock-Thrush	53
17. *Cinclus pallasi.*	Siberian Black-bellied Dipper	54

18. *Accentor alpinus erythropygius.* Japanese Alpine Accentor	56
19. —— *rubidus.* Japanese Hedge-Sparrow	56
20. *Pratincola maura.* Siberian Stonechat	57
21. *Ruticilla aurorea.* Daurian Redstart	57
22. *Larvivora cyanurus.* Siberian Blue-tail	58
23. *Niltava cyanomelana.* Japanese Blue Flycatcher	59
24. *Siphia luteola.* Mugimaki Flycatcher	60
25. *Xanthopygia narcissina.* Narcissus Flycatcher	61
26. *Muscicapa sibirica.* Siberian Flycatcher	62
27. —— *latirostris.* Brown Flycatcher	62
28. *Terpsiphone princeps.* Japanese Paradise Flycatcher	63

Subfamily CRATEROPODINÆ.

29. *Hypsipetes amaurotis.* Brown-eared Bulbul	64
30. —— *squamiceps.* Bonin-Island Bulbul	65
—— —— *pryeri.* Loo-Choo Bulbul	66
31. *Hapalopteron familiare.* Bonin White-eyed Warbler	66
32. *Zosterops palpebrosa nicobarica.* Loo-Choo White-eye	67
33. —— *japonica.* Japanese White-eye	68

Subfamily SYLVIINÆ.

34. *Phylloscopus coronatus.* Temminck's Crowned Willow-Warbler	68
35. —— *borealis.* Arctic Willow-Warbler	69
36. —— *xanthodryas.* Swinhoe's Willow-Warbler	70
37. —— *tenellipes.* Pale-legged Willow-Warbler	70
38. *Acrocephalus orientalis.* Chinese Great Reed-Warbler	71
39. —— *bistrigiceps.* Schrenck's Reed-Warbler	71
40. *Locustella fasciolata.* Gray's Grasshopper-Warbler	72
41. —— *ochotensis.* Middendorff's Grasshopper-Warbler	73
42. —— *lanceolata.* Temminck's Grasshopper-Warbler	73
43. *Cettia squamiceps.* Swinhoe's Bush-Warbler	74
44. —— *cantans.* Large Japanese Bush-Warbler	74
45. —— *cantillans.* Small Japanese Bush-Warbler	76
46. —— *diphone.* Bonin Bush-Warbler	77
47. *Cisticola cisticola brunneiceps.* Fan-tailed Warbler	77
48. *Lusciniola pryeri.* Pryer's Grass-Warbler	79

Subfamily PARINÆ.

49. *Regulus cristatus orientalis.* Eastern Goldcrest	80
50. *Parus palustris japonicus.* Japanese Marsh-Tit	81

			Page
51.	*Parus ater pekinensis.*	Eastern Cole Tit	82
52.	—— *atriceps minor.*	Manchurian Great Tit	83
53.	—— *varius.*	Japanese Tit	85
	—— —— *castaneiventris.*	Formosan Tit	86
54.	*Acredula caudata.*	Continental Long-tailed Tit	87
55.	—— *trivirgata.*	Japanese Long-tailed Tit	87
56.	*Ægithalus consobrinus.*	Swinhoe's Penduline Tit	88
57.	*Troglodytes fumigatus.*	Japanese Wren	89
	—— —— *kurilensis.*	Kurile-Island Wren	90
58.	*Certhia familiaris.*	Common Creeper	91
59.	*Sitta cæsia amurensis.*	Daurian Nuthatch	92
	—— —— *uralensis.*	Siberian Nuthatch	93
	—— —— *albifrons.*	Kamtschatkan Nuthatch	93

Subfamily CORVINÆ.

60.	*Corvus corax.*	Raven	94
61.	—— *macrorhynchus japonensis.*	Japanese Oriental Raven	94
	—— —— *levaillanti.*	Small Oriental Raven	95
62.	—— *corone.*	Carrion-Crow	96
63.	—— *dauricus.*	Pallas's Jackdaw	97
64.	—— *neglectus.*	Swinhoe's Jackdaw	97
65.	—— *pastinator.*	Eastern Rook	98
66.	*Nucifraga caryocatactes.*	Nutcracker	99
67.	*Cyanopolius cyanus.*	Eastern Blue Magpie	99
68.	*Garrulus brandti.*	Brandt's Jay	100
69.	—— *japonicus.*	Japanese Jay	101
70.	—— *sinensis.*	Chinese Jay	101
71.	*Pica caudata.*	Magpie	102
	—— *caudata kamtschatkensis.*	Kamtschatkan Magpie	103

Subfamily LANIINÆ.

72.	*Lanius major.*	Pallas's Grey Shrike	103
73.	—— *magnirostris.*	Thick-billed Shrike	104
74.	—— *superciliosus.*	Japanese Red-tailed Shrike	104
75.	—— *lucionensis.*	Chinese Red-tailed Shrike	105
76.	—— *bucephalus.*	Bull-headed Shrike	106
77.	*Pericrocotus cinereus.*	Siberian Minivet	106
78.	—— *tegimæ.*	Loo-Choo Minivet	107

Subfamily STURNINÆ.

	Page
79. *Sturnus cineraceus.* Grey Starling	107
80. *Sturnia pyrrhogenys.* Red-cheeked Starling	108
81. *Ampelis garrulus.* Bohemian Waxwing	110
82. —— *japonicus.* Japanese Waxwing	110

Subfamily MOTACILLINÆ.

83. *Motacilla lugens.* Kamtschatkan Wagtail	111
84. —— *japonica.* Japanese Wagtail	112
85. —— *boarula melanope.* Eastern Grey Wagtail	114
86. —— *flava.* Blue-headed Wagtail	114
87. *Anthus maculatus.* Eastern Tree-Pipit	115
88. —— *spinoletta japonicus.* Japanese Alpine Pipit	116
89. —— *cervinus.* Red-throated Pipit	117

Subfamily ALAUDINÆ.

90. *Alauda arvensis pekinensis.* Large Japanese Sky-Lark	118
—— *arvensis japonica.* Small Japanese Sky-Lark	118
91. —— *alpestris.* Shore-Lark	119

Subfamily FRINGILLINÆ.

92. *Coccothraustes vulgaris.* Common Hawfinch	120
93. —— *personatus.* Japanese Hawfinch	121
94. *Loxia curvirostra.* Common Crossbill	121
95. *Chaunoproctus ferreirostris.* Bonin Grosbeak	122
96. *Pinicola enucleator.* Pine-Grosbeak	122
97. *Carpodacus roseus.* Rose-Finch	123
98. —— *erythrinus.* Scarlet Rose-Finch	123
99. —— *sanguinolentus.* Japanese Rose-Finch	124
100. *Fringilla spinus.* Siskin	125
101. —— *linaria.* Mealy Redpole	125
102. —— *montifringilla.* Brambling	126
103. —— *sinica.* Chinese Greenfinch	127
104. —— *kawarahiba.* Japanese Greenfinch	127
105. —— *kitlitzi.* Bonin-Island Greenfinch	128
106. *Montifringilla brunneinucha.* Japanese Snow-Finch	128
107. *Pyrrhula griseiventris.* Oriental Bulfinch	129
—— *griseiventris kurilensis.* Kurile-Island Bulfinch	129
108. *Passer montanus.* Tree-Sparrow	130
109. —— *rutilans.* Russet-Sparrow	131

		Page
110.	*Emberiza ciopsis.* Bonaparte's Japanese Bunting	131
111.	—— *yessoensis.* Swinhoe's Japanese Bunting	132
112.	—— *schœniclus palustris.* Eastern Reed-Bunting	133
113.	—— *rustica.* Rustic Bunting	134
114.	—— *fucata.* Grey-headed Bunting	134
115.	—— *sulphurata.* Siebold's Bunting	135
116.	—— *personata.* Temminck's Japanese Bunting	136
117.	—— *spodocephala.* Black-faced Bunting	137
118.	—— *elegans.* Temminck's Yellow-browed Bunting	137
119.	—— *rutila.* Ruddy Bunting	138
120.	—— *aureola.* Yellow-breasted Bunting	138
121.	—— *variabilis.* Grey Bunting	139
122.	—— *nivalis.* Snow-Bunting	140
123.	—— *lapponica.* Lapland Bunting	140

Subfamily HIRUNDININÆ.

124.	*Hirundo rustica gutturalis.* Eastern Chimney-Swallow	141
125.	—— *javanica namiyei.* Loo-Choo Bungalow-Swallow	142
126.	—— *alpestris nipalensis.* Nepalese Mosque-Swallow	142
127.	*Chelidon dasypus.* Black-chinned Martin	144
128.	*Cotyle riparia.* Sand-Martin	144

Suborder II. *EURYLÆMI.*
Suborder III. *TROCHILI.*
Suborder IV. *SCANSORES.*

129.	*Gecinus awokera.* Japanese Green Woodpecker	147
130.	—— *canus.* Grey-headed Green Woodpecker	148
131.	*Picus martius.* Great Black Woodpecker	149
132.	—— *richardsi.* Tristram's Woodpecker	149
133.	—— *noguchii.* Pryer's Woodpecker	151
134.	—— *leuconotus.* White-backed Woodpecker	152
	—— —— *subcirris.* Japanese White-backed Woodpecker	152
135.	—— *namiyei.* Stejneger's Woodpecker	153
136.	—— *major japonicus.* Japanese Great Spotted Woodpecker	153
137.	—— *minor.* Lesser Spotted Woodpecker	155
138.	*Iyngipicus kisuki.* Temminck's Pigmy Woodpecker	156
	—— —— *seebohmi.* Hargitt's Pigmy Woodpecker	156
	—— —— *nigrescens.* Loo-Choo Pigmy Woodpecker	156
139.	*Iynx torquilla.* Wryneck	157

Suborder V. UPUPÆ.

140. *Upupa epops.* Hoopoe 159

Order TROGONES.

Suborder VI. *TROGONES.*

Order COLUMBÆ.

Suborder VII. *COLUMBÆ.*

141. *Columba livia.* Blue Rock-Pigeon 160
142. *Turtur orientalis.* Eastern Turtle-Dove 160
143. —— *risorius.* Common Indian Dove 162
144. —— *humilis.* Chinese Red Dove 162
145. *Treron sieboldi.* Japanese Green Pigeon 163
146. —— *permagna.* Loo-Choo Green Pigeon 164
147. *Carpophaga ianthina.* Japanese Fruit-Pigeon 165
148. —— *versicolor.* Bonin Fruit-Pigeon 166
149. —— *jouyi.* Loo-Choo Fruit-Pigeon 167

Order COCCYGES.

Suborder VIII. *MUSOPHAGI.*

Suborder IX. *CUCULI.*

150. *Cuculus canorus.* Common Cuckoo 169
151. —— *intermedius.* Himalayan Cuckoo 169
152. —— *poliocephalus.* Little Cuckoo 171
153. *Hierococcyx hyperythrus.* Amoor Cuckoo 171

Subclass CORACIIFORMES.

Order PICARIÆ.

Suborder X. *HALCYONES.*

154. *Halcyon coromanda.* Ruddy Kingfisher 173
155. *Ceryle guttata.* Oriental Spotted Kingfisher 174
156. *Alcedo ispida bengalensis.* Eastern Common Kingfisher 175

Suborder XI. CORACIÆ.

		Page
157. *Cypselus pacificus.*	White-rumped Swift	177
158. *Chætura caudacuta.*	Needle-tailed Swift	178
159. *Caprimulgus jotaka.*	Japanese Goatsucker	178
160. *Eurystomus orientalis.*	Broad-billed Roller	179

Suborder XII. BUCEROTES.

Order MIMOGYPES.

Suborder XIII. MIMOGYPES.

Subclass FALCONIFORMES.

Order PSITTACI.

Suborder XIV. PSITTACI.

Order RAPTORES.

Suborder XV. STRIGES.

161. *Bubo maximus.*	Eagle-Owl	183
162. —— *blakistoni.*	Blakiston's Eagle-Owl	184
163. *Surnia nyctea.*	Snowy Owl	185
164. *Strix uralensis.*	Ural Owl	185
—— —— *fuscescens.*	Kiu-siu Ural Owl	185
165. —— *otus.*	Long-eared Owl	186
166. —— *brachyotus.*	Short-eared Owl	187
167. *Ninox scutulata.*	Brown Owlet	187
168. *Scops semitorques.*	Feathered-toed Scops Owl	188
169. —— *elegans.*	Cassin's Scops Owl	188
170. —— *scops.*	Scops Owl	189
171. —— *pryeri.*	Pryer's Scops Owl	190

Suborder XVI. ACCIPITRES.

172. *Falco gyrfalco.*	Jer-Falcon	192
173. —— *peregrinus.*	Peregrine Falcon	192

	Page
174. *Falco subbuteo*. Hobby	193
175. —— *æsalon*. Merlin	193
176. —— *tinnunculus japonicus*. Japanese Kestrel	194
177. *Pandion haliaetus*. Osprey	195
178. *Butaster indicus*. Javan Buzzard	196
179. *Pernis apivorus*. Honey-Buzzard	197
180. *Milvus ater melanotis*. Siberian Black Kite	197
181. *Haliaetus albicilla*. White-tailed Eagle	198
182. —— *pelagicus*. Steller's Sea-Eagle	199
183. *Aquila chrysaetos*. Golden Eagle	199
184. —— *lagopus*. Rough-legged Buzzard-Eagle	200
185. *Spizaetus nipalensis*. Indian Crested Eagle	200
186. *Buteo hemilasius*. Siberian Buzzard	201
187. —— *vulgaris plumipes*. Eastern Buzzard	201
188. *Circus cyaneus*. Hen-Harrier	202
189. —— *æruginosus*. Marsh-Harrier	203
—— —— *spilonotus*. Eastern Marsh-Harrier	203
190. *Accipiter palumbarius*. Goshawk	204
191. —— *nisus*. Common Sparrow-Hawk	204
192. —— *gularis*. Chinese Sparrow-Hawk	205

Suborder XVII. *SERPENTARII*.

Subclass **ANSERIFORMES**.

Order PELECANO-HERODIONES.

Suborder XVIII. *STEGANOPODES*.

193. *Phalacrocorax carbo*.	Common Cormorant	208
194. —— *capillatus*.	Temminck's Cormorant	209
195. —— *pelagicus*.	Resplendent Shag	210
196. —— *bicristatus*.	Bare-faced Shag	211
197. *Sula leucogastra*.	Booby Gannet	212
198. —— *piscatrix*.	Red-footed Booby	213
199. *Phaeton rubricauda*.	Red-tailed Tropic-bird	213
200. *Fregata minor*.	Lesser Frigate-bird	214

Suborder XIX. *HERODIONES.*

201. *Ardea cinerea.*	Common Heron	215
202. —— *alba.*	Great White Egret	216
—— *alba modesta.*	Eastern Great White Egret	216
203. —— *intermedia.*	Plumed Egret	217
204. —— *garzetta.*	Little Egret	218
205. —— *coromanda.*	Eastern Buff-backed Heron	219
206. —— *jugularis.*	Eastern Reef-Heron	220
207. *Nycticorax nycticorax.*	Night-Heron	222
208. —— *crassirostris.*	Bonin Night-Heron	222
209. —— *goisagi.*	Japanese Night-Heron	223
210. —— *javanicus stagnatilis.*	Australian Mangrove-Heron	224
211. —— *prasinosceles.*	Chinese Squacco-Heron	225
212. *Botaurus stellaris.*	Bittern	226
213. —— *sinensis.*	Oriental Little Bittern	227
214. —— *eurhythma.*	Schrenck's Little Bittern	227
215. *Ciconia boyciana.*	Japanese Stork	228

Suborder XX. *PLATALEÆ.*

216. *Platalea leucorodia.*	Common Spoonbill	229
217. —— *minor.*	Swinhoe's Black-faced Spoonbill	231
218. *Ibis nippon.*	Japanese Crested Ibis	232
219. —— *melanocephala.*	White Ibis	232

Order LAMELLIROSTRES.

Suborder XXI. *PHŒNICOPTERI.*

Suborder XXII. *ANSERES.*

220. *Cygnus musicus.*	Hooper Swan	234
221. —— *bewicki.*	Bewick's Swan	235
222. *Anser cygnoides.*	Chinese Goose	235
223. —— *segetum serrirostris.*	Eastern Bean-Goose	236
224. —— *albifrons.*	White-fronted Goose	237
225. —— *minutus.*	Lesser White-fronted Goose	238
226. —— *hyperboreus.*	Snow-Goose	238
—— —— *nivalis.*	Greater Snow-Goose	238
227. —— *hutchinsi.*	Hutchins' Bernacle Goose	239
228. —— *nigricans.*	Pacific Brent Goose	240
229. *Dendrocygna javanica.*	Indian Whistling Teal	240

			Page
230.	*Tadorna cornuta.*	Common Sheldrake	241
231.	—— *rutila.*	Ruddy Sheldrake	241
232.	*Anas strepera.*	Gadwall	242
233.	—— *clypeata.*	Shoveller	242
234.	—— *boschas.*	Mallard	243
235.	—— *zonorhyncha.*	Dusky Mallard	243
236.	—— *crecca.*	Common Teal	244
237.	—— *formosa.*	Spectacled Teal	244
238.	—— *falcata.*	Falcated Teal	245
239.	—— *circia.*	Garganey	246
240.	—— *acuta.*	Pintail	246
241.	—— *penelope.*	Wigeon	247
242.	—— *galericulata.*	Mandarin Duck	248
243.	*Fuligula americana.*	American Black Scoter	248
244.	—— *fusca stejnegeri.*	Asiatic Velvet Scoter	251
245.	—— *glacialis.*	Long-tailed Duck	252
246.	—— *clangula.*	Golden-eye	253
247.	—— *histrionica.*	Harlequin Duck	253
248.	—— *baeri.*	Siberian White-eyed Duck	254
249.	—— *ferina.*	Pochard	254
250.	—— *cristata.*	Tufted Duck	255
251.	—— *marila.*	Scaup	256
252.	*Somateria spectabilis.*	King Eider	256
253.	—— *stelleri.*	Steller's Eider	257
254.	*Mergus merganser.*	Goosander	257
255.	—— *serrator.*	Red-breasted Merganser	258
256.	—— *albellus.*	Smew	258

Suborder XXIII. *PALAMEDEÆ.*

Subclass GALLIFORMES.

Order TUBINARES.

Suborder XXIV. *TUBINARES.*

257.	*Diomedea albatrus.*	Steller's Albatross	261
258.	—— *nigripes.*	Audubon's Albatross	263
259.	*Puffinus leucomelas.*	Siebold's Shearwater	264
260.	—— *tenuirostris.*	Pink-footed Shearwater	265
261.	—— *griseus.*	Sooty Shearwater	266
262.	—— *tenuirostris.*	Slender-billed Shearwater	267

		Page
263.	*Fulmarus glacialis.* Fulmar	268
264.	*Œstrelata hypoleuca.* Bonin-Island Shearwater	269
265.	*Procellaria leachi.* Leach's Fork-tailed Petrel	270
266.	—— *melania.* Black Petrel	270
267.	—— *furcata.* Grey Fork-tailed Petrel	271

Order IMPENNES.

Suborder XXV. *IMPENNES.*

Order GALLO-GRALLÆ.

Suborder XXVI. *GAVLÆ.*

268.	*Alca troile arra.* Pallas's Guillemot	273
269.	—— *carbo.* Sooty Guillemot	274
270.	—— *columba.* Pigeon-Guillemot	275
271.	—— *antiqua.* Bering's Guillemot	276
272.	—— *wumizusume.* Temminck's Guillemot	277
273.	—— *marmorata.* Marbled Guillemot	278
274.	—— *brevirostris.* Kittlitz's Guillemot	279
275.	*Fratercula corniculata.* Horn-eyed Puffin	280
276.	—— *cirrhata.* Tufted Puffin	281
277.	—— *monocerata.* Horn-billed Puffin	283
278.	—— *psittacula.* Parrot-billed Puffin	284
279.	—— *cristatella.* Crested Puffin	285
280.	—— *pygmœa.* Whiskered Puffin	286
281.	—— *pusilla.* Least Puffin	287
282.	*Stercorarius richardsoni.* Richardson's Skua	288
283.	—— *buffoni.* Buffon's Skua	289
284.	—— *pomarinus.* Pomarine Skua	289
285.	*Larus glaucus.* Glaucous Gull	290
286.	—— *glaucescens.* Glaucous-winged Gull	290
287.	—— *marinus schistisagus.* Eastern Great Black-backed Gull	291
288.	—— *cachinnans.* Pallas's Herring-Gull	291
289.	—— *leucopterus.* Iceland Gull	292
290.	—— *crassirostris.* Temminck's Gull	293
291.	—— *canus.* Common Gull	293
292.	—— *tridactylus.* Kittiwake	294
293.	—— *ridibundus.* Black-headed Gull	295
294.	*Sterna dougalli.* Roseate Tern	295

	Page
295. *Sterna longipennis.* Daurian Tern	296
296. —— *melanauchen.* Black-naped Tern	297
297. —— *sinensis.* Oriental Lesser Tern	298
298. —— *aleutica.* Aleutian Tern	299
299. —— *bergii.* Rüppell's Tern	299
300. —— *stolida.* Noddy Tern	300
301. —— *anæstheta.* Bridled Tern	301
302. —— *fuliginosa.* Sooty Tern	302

Suborder XXVII. *LIMICOLÆ.*

303. *Charadrius fulvus.* Asiatic Golden Plover	303
304. —— *helveticus.* Grey Plover	304
305. —— *morinellus.* Common Dotterel	305
306. —— *minor.* Little Ringed Plover	306
307. —— *placidus.* Hodgson's Ringed Plover	307
308. —— *mongolicus.* Mongolian Sand-Plover	308
309. —— *cantianus.* Kentish Plover	309
310. —— *geoffroyi.* Geoffroy's Sand-Plover	310
311. *Lobivanellus cinereus.* Grey-headed Wattled Lapwing	311
312. *Vanellus cristatus.* Common Lapwing	312
313. *Hæmatopus osculans.* Japanese Oystercatcher	313
314. —— *niger.* North-American Black Oystercatcher	313
315. *Numenius arquatus lineatus.* Eastern Curlew	314
316. —— *cyanopus.* Australian Curlew	315
317. —— *phæopus variegatus.* Eastern Whimbrel	316
318. —— *minutus.* Least Whimbrel	317
319. *Phalaropus fulicarius.* Grey Phalarope	318
320. —— *hyperboreus.* Red-necked Phalarope	318
321. *Totanus fuscus.* Dusky Redshank	319
322. —— *calidris.* Common Redshank	320
323. —— *glottis.* Greenshank	321
324. —— *stagnatilis.* Marsh-Sandpiper	322
325. —— *incanus.* Asiatic Wandering Tattler	323
—— —— *brevipes.* American Wandering Tattler	323
326. —— *glareola.* Wood-Sandpiper	324
327. —— *ochropus.* Green Sandpiper	325
328. —— *terekius.* Terek Sandpiper	326
329. —— *hypoleucus.* Common Sandpiper	326
330. —— *pugnax.* Ruff	327
331. *Limosa rufa uropygialis.* Eastern Bar-tailed Godwit	328
332. —— *melanura melanuroides.* Eastern Black-tailed Godwit	329

		Page
333.	*Macrorhamphus griseus scolopaceus.* Alaskan Snipe-billed Sandpiper	330
334.	*Strepsilas interpres.* Turnstone	331
335.	*Tringa crassirostris.* Japanese Knot	332
336.	—— *canutus.* Knot	333
337.	—— *alpina pacifica.* Pacific Dunlin	334
338.	—— *maritima.* Purple Sandpiper	335
339.	—— *arenaria.* Sanderling	336
340.	—— *platyrhyncha.* Broad-billed Sandpiper	337
341.	—— *minuta ruficollis.* Red-throated Stint	337
342.	—— *subminuta.* Middendorff's Stint	338
343.	—— *pygmæa.* Spoon-billed Sandpiper	338
344.	—— *acuminata.* Siberian Pectoral Sandpiper	339
345.	*Rhynchæa capensis.* Painted Snipe	340
346.	*Scolopax australis.* Latham's Snipe	342
347.	—— *solitaria japonica.* Japanese Solitary Snipe	342
348.	—— *megala.* Swinhoe's Snipe	343
349.	—— *gallinula.* Jack Snipe	344
350.	—— *stenura.* Pintail Snipe	345
351.	—— *gallinago.* Common Snipe	346
352.	—— *rusticola.* Woodcock	347

Suborder XXVIII. *GRALLÆ.*

353.	*Grus cinerea.* Common Crane	348
354.	—— *leucogeranus.* Siberian White Crane	349
355.	—— *japonensis.* Sacred Crane	351
356.	—— *leucauchen.* White-naped Crane	352
357.	—— *monachus.* White-headed Crane	353
358.	*Turnix blakistoni.* Blakiston's Hemipode	354

Suborder XXIX. *FULICARIÆ.*

359.	*Otis dybowskii.* Eastern Great Bustard	355
360.	*Crex pusilla.* Pallas's Crake	356
361.	—— *fusca erythrothorax.* Siberian Ruddy Crake	357
362.	—— *undulata.* Swinhoe's Crake	358
363.	—— *sepiaria.* Loo-Choo Crake	358
364.	*Rallus aquaticus indicus.* Eastern Water-Rail	359
365.	*Gallicrex cinereus.* Water-Cock	360
366.	*Fulica atra.* Common Coot	360
367.	*Gallinula chloropus.* Water-Hen	360

Suborder XXX. *PYGOPODES*.

	Page
368. *Colymbus adamsi.* White-billed Diver	362
369. —— *arcticus.* Black-throated Diver	363
370. —— *septentrionalis.* Red-throated Diver	364
371. *Podiceps rubricollis major.* Eastern Red-necked Grebe	364
372. —— *nigricollis.* Black-necked Grebe	366
373. —— *cornutus.* Sclavonian Grebe	367
374. —— *minor.* Little Grebe	367

Suborder XXXI. *GALLINÆ*.

375. *Phasianus torquatus.* Chinese Ring-necked Pheasant	369
376. —— *versicolor.* Japanese Green Pheasant	370
377. —— *sœmmeringi.* Copper Pheasant	370
378. —— *scintillans.* Hondo Copper Pheasant	371
379. *Tetrao mutus.* Common Ptarmigan	372
380. —— *bonasia.* Hazel-Grouse	373
381. *Coturnix communis.* Common Quail	373
—— —— *japonica.* Eastern Common Quail	373

Suborder XXXII. *CRYPTURI*.

Subclass **STRUTHIONIFORMES**.

Order APTERYGES.

Suborder XXXIII. *APTERYGES*.

Order RATITÆ.

Suborder XXXIV. *RHEÆ*.

Suborder XXXV. *CASUARII*.

Suborder XXXVI. *STRUTHIONES*.

LIST OF WOODCUTS.

	Page
Map of the Japanese Empire	Frontispiece
Deep plantar tendons of *Patagona gigas*	145
——— ——— of *Picus martius*	146
Picus richardsi	150
——— *noguchii*	151
Pterylosis of *Upupa epops*	158
Deep plantar tendons of *Trogon*	159
——— ——— ——— of *Catharista atratus*	172
Sternum of *Upupa epops*, of *Buceros albirostris*, and of *Merops apiaster*	176
Deep plantar tendons of *Cathartes aura*	180
Foot of *Bubo maximus*	183
Deep plantar tendons of *Pandion haliaetus*	195
Foot of *Serpentarius secretarius*	206
Head of *Phalacrocorax carbo*	208
——— ——— ——— *capillatus*	209
——— ——— ——— *pelagicus*	210
——— ——— ——— *bicristatus*	211
Pterylosis of neck of *Ardea cinerea*	214
Side of head and throat of *Platalea leucorodia*	230
——— ——— ——— ——— *minor*	231
Skull of *Anas boschas*	234
Head of *Fuligula americana*	249
——— ——— *nigra*	249
——— ——— *fusca stejnegeri*	250
——— ——— ——— *velvetina*	251
——— ——— *fusca*	251
——— *Diomedea albatrus*	261
——— ——— *nigripes*	263
——— *Puffinus leucomelas*	264
——— ——— *carneipes*	265
——— ——— *griseus*	266
——— ——— *tenuirostris*	267
——— *Fulmarus glacialis*	268
——— *Procellaria leachi*	270
——— ——— *furcata*	271
——— *Fratercula corniculata*	281
——— ——— *cirrhata*	282
——— ——— *monocerata*	283
——— ——— *psittacula*	284
——— ——— *cristatella*	285
——— ——— *pygmaea*	287
——— ——— *pusilla*	288

	Page
Head of *Sterna dougalli*	296
———— ———— *longipennis*	296
———— ———— *melanauchen*	297
———— ———— *sinensis*	298
———— ———— *bergii*	299
———— ———— *stolida*	300
———— ———— *anaestheta*	301
———— ———— *fuliginosa*	302
———— *Charadrius helveticus*	304
Charadrius morinellus	305
Head of *Charadrius minor*	306
———— ———— *placidus*	307
———— ———— *mongolicus*	308
Charadrius cantianus	309
Head of *Charadrius geoffroyi*	310
Vanellus cristatus	312
Bill of *Hæmatopus niger*	314
Numenius arquatus	315
———— *phæopus*	316
Phalaropus fulicarius	318
———— *hyperboreus*	319
Totanus fuscus	320
———— *calidris*	321
———— *glottis*	322
———— *glareola*	324
———— *ochropus*	325
———— *hypoleucus*	327
Limosa rufa	328
———— *melanura*	330
Macrorhamphus scolopaceus and *griseus*	331
Strepsilas interpres	332
Tringa canutus	333
———— *alpina*	334
———— *maritima*	335
———— *arenaria*	336
———— *pygmaea*	339
Rectrices of *Tringa acuminata*	340
Scolopax gallinula	344
Rectrices of *Scolopax stenura*	345
Scolopax gallinago	346
———— *rusticola*	347
Grus cinerea	349
———— *leucogeranus*	350
———— *japonensis*	351
———— *leucauchen*	352
———— *monachus*	353
Femur and tibia of *Colymbus glacialis*	361
———— ———— ———— *Podiceps rubricollis*	362
Sternum of *Crax carunculata*, *Lophophorus impeyanus*, and *Megapodius rubripes*	368
Deep plantar tendons of *Gallus domesticus*	369

BIRDS

OF THE

JAPANESE EMPIRE.

LITERATURE RELATING TO JAPANESE BIRDS.

IN the following brief notices of the most important books and papers in various periodicals which treat of the Birds of Japan, an attempt has been made to trace the gradual growth of our knowledge of the subject during the present century. They are arranged in the order of the date of publication.

PALLAS. Zoographia Rosso-Asiatica. Printed in 1809, but not published until 1826.

This important work, the value of which can scarcely be overestimated, embodies the results of thirty years' work upon the Zoology of Siberia and the adjacent Islands. Very little information regarding the birds of Japan is to be found in it, but the occurrence of 50 species on the Kurile Islands is recorded. Most of these are given on the authority of Steller, whose manuscripts were placed at the disposal of Pallas; others were sent to him by his friend Captain Billings, and a few by Dr. Merk.

TEMMINCK. Nouveau Recueil de Planches Coloriées d'Oiseaux. 1827-1836.

The discoveries of Dr. Siebold during his stay in Japan were of so much importance that many of the birds sent by him to Leyden were figured by Temminck in the 'Planches Coloriées' from time to time. No fewer than 35 species were described and most of them figured in this publication from 1827 to 1836.

KITTLITZ. Ueber die Vögel des Inselgruppe von Bouinsima.— Mémoires présentés à l'Académie Impériale des Sciences de St. Pétersbourg par divers savans. 1830, pp. 231-248.

This short but imperfect, though important, paper is a record of the birds obtained by Kittlitz during a fortnight's visit from the 1st to the 14th of May, 1828, to the three larger islands of the Bonin group. Three new species of birds are described:—

Hapalopteron familiare (placed by Kittlitz in the genus *Ixos*).
Cettia diphone (placed by Kittlitz in the genus *Sylvia*).
Geocichla terrestris (placed by Kittlitz in the genus *Turdus*).

KITTLITZ. Kupfertafeln zur Naturgeschichte der Vögel. 1832.

In this little volume some of the birds found by F. H. von Kittlitz on the Bonin Islands are figured:—*Columba ianthina*, *Columba versicolor*, *Fringilla papa* (*Chaunoprocta ferreirostris*), *Galgulus amaurotis* (*Hypsipetes squamiceps*), and *Ardea caledonica* (*Nycticorax crassirostris*).

TEMMINCK. Manuel d'Ornithologie. Second edition. Vol. iii., 1835; Vol. iv., 1840.

In these two volumes, which form a Supplement to the first and second volumes of Temminck's important work on European birds, many references to Japan are added to the geographical distribution of the various species which range across the Palæarctic Region. It might have been a very valuable addition to the knowledge of Japanese Ornithology, but unfortunately he mentions so many birds as occurring in Japan that have never been found there by any recent collector that very little importance can be attached to these statements. For example, he says of *Strix flammea* (Man. d'Orn.

iii. p. 18), " l'espèce est exactement la même au Japon ;" of *Sylvia atricapilla* (Man. d'Orn. iii. p. 132), " Habite jusqu'au Japon, où elle est absolument la même qu'en Europe ;" of *Parus cœruleus* (Man. d'Orn. iii. p. 210), " Se trouve aussi en Morée et au Japon ;" of *Perdix rubra* (Man. d'Orn. iv. p. 333), " On trouve cette espèce au Japon, sans qu'elle y ait éprouvé la moindre différence dans les formes ou la coloration du plumage ;" of *Pelecanus onocrotalus* (Man. d'Orn. iv. p. 560), " Les sujets reçus du Japon ne diffèrent point de ceux d'Europe," &c., &c.

It seems probable that Temminck must have been imposed upon by some fraudulent dealer, or that by some unfortunate accident in the management of the Leyden Museum a number of European skins were mixed with the Japanese collections.

TEMMINCK & SCHLEGEL. Fauna Japonica. Aves. 1845–1850.

This book is the standard work upon the birds of Japan. It comprises all the species that were obtained by Dr. Siebold during his residence in Southern Japan from 1823 to 1830, some of which had already been described in the 'Planches Coloriées.' Unfortunately, no information as to the exact locality where each species was obtained is given, and scarcely a word is said as to the habits of any of the birds. The number of species enumerated in the 'Fauna Japonica' is 200; but after eliminating one or two obvious errors, and discarding those which were introduced solely on the authority of Japanese pictures, which may or may not have been drawn from native birds, the number of species known to inhabit Japan at the date of the publication of this important work is reduced to 175. This does not include the birds mentioned by Pallas as found on the Kurile Islands, or those discovered by Kittlitz on the Bonin Islands.

This work was published in numbers. Parts 1 to 3, containing the Raptores and Striges, were issued in 1845 (Engelmann, Bibl. Hist. Nat. p. 342), a statement confirmed by the fact that plates 8 and 9 and page 25 are quoted in September 1845 (Gray, Genera of Birds, i. p. 38), and plate 10 in October 1845 (*tom. cit.* p. 39).

CASSIN. Proc. Ac. Nat. Sc. Philadelphia, vi. pp. 184–188. Descriptions of New Species of Birds, specimens of which are in the Collection of the Academy of Natural Sciences of Philadelphia. 1852.

In this paper nine supposed new species of birds from various localities are described, amongst which is an Owl (*Ephialtes elegans*), which was obtained by Dr. Wilson from M. J. P. Verreaux, of Paris, labelled " Eu Mer, côtes du Japon, lat. 29° 47′ N., long. 126° 13′ 30″ E." Unfortunately the collector's name is not added.

CASSIN. Exp. Amer. Squad. China Seas and Japan. ii. pp. 219-248 (1856).

This paper is an important addition to the history of Japanese birds. It is a report of a collection made by Mr. Heine, the artist of the Perry Expedition, during the years 1852-1854. It principally relates to birds obtained at Hakodadi, which was then almost virgin ground. Of the species obtained at Nagasaki by the Siebold Expedition, 18 were found by the Perry Expedition at Hakodadi, 6 others at Simoda near Yokohama, and 2 on the Loo-Choo Islands. Nine species were added to the Japanese fauna, of which the first mentioned had been recorded by Pallas from the Kurile Islands. Two were obtained at Simoda:—

> *Fratercula mystacea.*
> *Larus ridibundus.*

Two were procured on the Loo-Choo Islands:—

> *Gallinula chloropus.*
> *Sterna sinensis.*

And the remaining five were collected at Hakodadi:—

> *Picus major japonicus.*
> *Scolopax stenura.*
> *Phalaropus hyperboreus.*
> *Fratercula monocerata.*
> *Numenius phœopus variegatus.*

Some interesting notes on the habits of the birds, as observed by Mr. Heine, are added.

CASSIN. Proc. Acad. Nat. Sc. Philad. 1858, pp. 191-196.

This is a catalogue of a small collection of birds made by Dr. Henderson during the cruise of the 'Portsmouth' in the year 1857, but it adds something to our knowledge of Japanese birds. All the examples were obtained at Hakodadi.

Fifteen species included in the 'Fauna Japonica' are added to the list of Yezzo birds, and 6 new species are added to the Japanese list:—

    Parus palustris.
    Locustella ochotensis.
        ,,   *lanceolata.*
    Sitta europæa.
    Charadrius morinellus.
    Totanus glottis.

CASSIN. Proc. Acad. Nat. Sc. Philad. 1862, pp. 312–327.—Catalogue of Birds collected by the United States North Pacific Surveying and Exploring Expedition in command of Capt. John Rodgers, United States Navy, with notes and descriptions of new species. (Apparently from 1853 to 1855.)

The chief interest attaching to this paper is the addition of seven species to the list of Loo-Choo birds, of which one, *Ardea jugularis*, was new to the Japanese fauna. Three species are also added to the list of Bonin birds; and one species, *Alca carbo*, is added to the birds of Japan.

BLAKISTON. On the Ornithology of Northern Japan. Ibis, 1862, pp. 309–333.

BLAKISTON. Corrections and Additions to Captain Blakiston's Paper on the Ornithology of Northern Japan. Ibis, 1863, pp. 97–100.

These important contributions to the ornithology of Japan are the result of a visit of three months (August, September, and October) in 1861 to Hakodadi. The number of species added to the list of Yezzo birds was at least 40, of which the following 10 were new to Japan:—

    Chelidon dasypus.
    Parus ater.
    Picus leuconotus.
       ,,   *martius.*
    Gecinus canus.
    Garrulus brandti.
    Nucifraga caryocatactes.
    Tetrao bonasia.
    Charadrius cantianus dealbatus.
    Tringa minuta ruficollis.

SWINHOE. Catalogue of the Birds of China, with remarks principally on their geographical distribution. Proc. Zool. Soc. 1863, pp. 332–338.

In these pages, at the close of his article on the Birds of China, Swinhoe adds a comparative list of the Birds of Amoorland, of Japan, and of Formosa. The list of the Birds of Japan possesses no special interest.

SWINHOE. Notes on the Ornithology of Northern Japan. Ibis, 1863, pp. 442–415.

This paper is little more than an introduction to British ornithologists of the information in regard to Japanese birds contained in Cassin's account of the species obtained by Dr. Henderson at Hakodadi, during the cruise of the 'Portsmouth.'

WHITELY. Notes on Birds collected near Hakodadi in Northern Japan. Ibis, 1867, pp. 193–211.

This paper is a list of birds procured during the residence of the writer for a year or more (1864–1865) at Hakodadi, to which is added the briefest possible notes on their habits. It forms a very important addition to our knowledge of Japanese birds. At least 40 more species were added to the list of Yezzo birds, of which the following 14 were new to Japan:—

> *Certhia familiaris.*
> *Montifringilla brunneinucha.*
> *Tringa acuminata.*
> *Scolopax gallinula.*
> *Fuligula marila.*
> „ *glacialis.*
> „ *fusca stejnegeri.*
> *Colymbus septentrionalis.*
> *Podiceps cornutus.*
> *Larus marinus schistasagus.*
> „ *canus.*
> *Fratercula cristatella.*
> *Strix otus.*
> „ *brachyotus.*

Sharpe. Ann. & Mag. Nat. Hist. 1870, vi. p. 157.

This paper is a catalogue of birds procured by Mr. R. H. Bergman in China and Japan. Fourteen species are enumerated from the latter country, but the only interesting point is the occurrence of *Fuligula marila* at Nagasaki.

Swinhoe. On the White Stork of Japan. Proc. Zool. Soc. 1873, pp. 512–514.

The Japanese Stork is described for the first time, under the name of *Ciconia boyciana*.

Swinhoe. On the Rosy Ibis of China and Japan (*Ibis nippon*). Ibis, 1873, pp. 249–253.

This is a very interesting paper on the breeding-habits and changes of plumage of the Japanese Crested Ibis.

Swinhoe. On some Birds from Hakodadi, in Northern Japan. Ibis, 1874, pp. 150–166.

This paper is Swinhoe's report upon a collection of birds sent to him by Captain Blakiston, who had procured them near Hakodadi. It adds 16 species to the list of Yezzo birds, of which the following 9 were new to Japan :—

> *Acrocephalus bistrigiceps.*
> *Cettia squameiceps.*
> *Acredula caudata.*
> *Emberiza yezzoensis.*
> *Charadrius placidus.*
> *Scolopax australis.*
> *Larus glaucus.*
> „ *glaucescens.*
> *Alca marmorata.*

Of these *Emberiza yezzoensis* had not previously been described.

Swinhoe. On the contents of a second Box of Birds from Hakodadi, in Northern Japan. Ibis, 1875, pp. 447–458.

This paper is Swinhoe's report upon another collection of birds from Hakodadi, sent to him by Captain Blakiston. It adds 20 more species to the list of Yezzo birds, of which the following were new to Japan :—

Falco subbuteo.
Chætura caudacuta.
Lanius superciliosus.
Emberiza aureola.
Charadrius minor.
Totanus fuscus.
Tringa arenaria.
  ,, *subminuta.*
  ,, *pygmæa.*
Anser segetum serrirostris.
  ,, *brachyrhynchus.*
Macrorhamphus griseus scolopaceus.
Podiceps minor.

SWINHOE. On the contents of a third Box of Birds from Hakodadi, in Northern Japan. Ibis, 1876, pp. 330-335.

This paper is Swinhoe's report upon another collection of birds sent to him from Hakodadi by Captain Blakiston. It adds 8 species to the list of Japanese birds :—

Phylloscopus xanthodryas.
Emberiza schœnicola palustris.
Botaurus eurythma.
Crex undulata.
Locustella fasciolata.
Numenius cyanopus.
Cypselus pacificus.
Turtur risorius.

Of these the last-named is from "Yedo" (Yokohama); and in addition to these he adds two species to the list of Yezzo birds.

SWINHOE. On the contents of a fourth Box of Birds from Hakodadi, in Northern Japan. Ibis, 1877, pp. 144-147.

This paper is Swinhoe's report upon another collection of birds from Hakodadi sent by Captain Blakiston. It adds two species to the list of Yezzo birds, both of which were new to Japan :—

Colymbus adamsi.
Fuligula histrionica.

It also adds four other species to the Japanese list, three from examples obtained near Yokohama, and one from the extreme north of Hondo:—

    Geocichla varia.
    Carpodacus roseus.
    Falco æsalon.
    Circus æruginosus.

BLAKISTON & PRYER. A Catalogue of the Birds of Japan. Ibis, 1878, pp. 209-250.

This is by far the most important contribution to the ornithology of the Japanese Islands that has appeared; inasmuch as for the first time an attempt is made to collect information as to the habits and distribution of the birds in the Japanese Islands themselves. For this purpose the authors were specially adapted—Captain Blakiston having resided some years in Hakodadi, and Mr. Harry Pryer in Yokohama. To the 174 species of birds which were known to inhabit Japan when the 'Fauna Japonica' was published, 17 species had been added by the collections which passed through Cassin's hands, and 14 had been added by Whitely, making in all 205 species. This number Captain Blakiston had succeeded in increasing by 46, so that the number of Japanese birds known previously to the publication of this paper was 251. After visiting the various museums in Japan, and comparing their collections, the united labours of Messrs. Blakiston and Pryer greatly increased this number. After eliminating the doubtful species there still remained an addition of the following 44 species, raising the total number of Japanese birds from the 174 of the 'Fauna Japonica' to 295 :—

    Fratercula cirrhata.
     ,, *pusilla.*
    Alca troile arra.
    Cygnus bewicki.
    Anser minutus.
     ,, *canadensis hutchinsi.*
     ,, *brenta nigricans.*
    Anas circia.
    Fuligula baeri.
     ,, *americana.*
    Somateria stelleri.

Sterna longipennis.
Larus tridactylus.
Fulmarus glacialis.
Procellaria furcata.
  ,,   *leucorrhoa.*
Charadrius mongolicus.
Hæmatopus osculans.
Totanus pugnax.
Ibis propinqua.
Botaurus sinensis.
Grus japonensis.
Crex pusilla.
Otis dybowskii.
Tetrao bonasia.
  ,,   *mutus.*
Columba livia.
Cuculus poliocephalus.
Upupa epops.
Cotyle riparia.
Corvus corax.
Lanius major.
Muscicapa sibirica.
Pericrocotus cinereus.
Accentor alpinus erythropygius.
Anthus cervinus.
Erithacus cyaneus.
Emberiza aureola.
  ,,   *nivalis.*
Loxia enucleator.
Aquila lagopus.
Accipiter palumbarius.

SEEBOHM. Contributions to the Ornithology of Siberia. Ibis, 1878, p. 345.

In this paper an apparently new species of Wagtail is described under the name of *Motacilla amurensis*, from the valley of the Amoor and Japan. It was afterwards proved by Captain Blakiston to be the first summer plumage of *Motacilla lugens* of Kittlitz, originally described from Kamtschatka; but as this species is quite distinct from the *Motacilla lugens* of Temminck and Schlegel, it forms an addition to the Japanese fauna.

SEEBOHM. Remarks on Messrs. Blakiston and Pryer's Catalogue of the Birds of Japan. Ibis, 1879, pp. 18-43.

This paper is a report on the skins sent for identification by Messrs. Blakiston and Pryer, the results of which are included in the list of new species added to the Japanese fauna appended to the remarks on the previous paper. Three species are added on the authority of examples sent by Captain Blakiston whilst the paper was in progress:—

Ægithalus consobrinus.
Picus minor.
Phylloscopus borealis.

SEEBOHM. Further Contributions to the Ornithology of Japan. Ibis, 1882, pp. 368-371.

This paper adds but little to the number of Japanese species, but it clears up one or two doubtful points, the most important being that the race of Canada Goose which occurs in Japan is *Anser canadensis hutchinsi*. A new species of Bullfinch is described, but subsequent investigations have thrown some doubt upon its validity.

BLAKISTON & PRYER. Birds of Japan. Trans. Asiatic Society Japan, 1882, pp. 84-186.

This is a revised list of the Birds of Japan, including the additions made during the four years which had elapsed since the previous list was published, correcting the identification and nomenclature of the previous list, adding much information respecting the distribution of Japanese birds, and enumerating the following species, which had not previously been recorded from Japan:—

Fratercula corniculata.
,, *psittacula.*
Alca columba.
Sterna stolida.
Larus cachinnans.
Stercorarius richardsoni.
,, *buffoni.*
,, *pomarinus.*
Attagen minor.
Puffinus fuliginosus.
Totanus stagnatilis.

Tringa canutus.
    ,, *platyrhyncha.*
Phalaropus fulicarius.
Gallicrex cristatus.
Phasianus torquatus.
Turtur humilis.
Eurystomus orientalis.
Garrulus sinensis.
Lanius magnirostris.
Motacilla flava.
Surnia scandiaca.

BLAKISTON. The Chrysanthemum, ii. pp. 124-128; pp. 471-475; pp. 521-525. Ornithological Notes. 1882.

These three papers appeared in the 'Chrysanthemum,' a periodical published in Yokohama, and consist of a series of interesting notes on the various species of birds observed by the writer on the southeast coast of Yezzo during a trip which he made in May and June.

SEEBOHM. Observations on the Pied Wagtails of Japan. Ibis, 1883, pp. 90-92.

This paper describes an apparently new species of Wagtail under the name of *Motacilla blakistoni*, which has since been proved to be the fully adult, in the second year, of *Motacilla lugens* of Kittlitz.

SEEBOHM. Exhibition of a new species of Owl from Yezzo. Proc. Zool. Soc. 1883, p. 166.

This notice refers to a new species of Owl sent from Japan by Captain Blakiston, and named *Bubo blakistoni*.

BLAKISTON. The Chrysanthemum, iii. pp. 76-81. Ornithological Notes. 1883.

This paper contains much very interesting information respecting the collections of birds made by Mr. P. L. Jouy, of the Smithsonian Institution at Washington, and Mr. A. J. M. Smith in the neighbourhood of Fuji-yama and Tate-yama, two mountain-ranges in the largest of the Japanese Islands. One bird is added to the Japanese list:—

Emberiza spodocephala.

JOUY. Ornithological Notes on Collections made in Japan from June to December 1882. Proc. United States Nat. Mus. 1883, pp. 273–318.

Although this paper does not add any new species of bird to the Japanese list, it gives much interesting information respecting the breeding of many species of birds on the mountains of Central Japan.

BLAKISTON. The Chrysanthemum, iii. pp. 26–36. Ornithological Notes. 1883.

This paper is entitled "Autumn collecting at Sapporo, Yezo," and contains a detailed account of the various species observed during an ornithological expedition in the months of September and October.

BLAKISTON. The Chrysanthemum, iii. pp. 172–174. Ornithological Notes. 1883.

This paper chiefly refers to birds obtained by the collectors of Messrs. Owston Snow & Co., on the Kurile Islands. One bird is added to the Japanese list :—

Ardea alba.

SEEBOHM. Further Contributions to the Ornithology of Japan. Ibis, 1884, pp. 30–43.

This paper is a report on a small collection of birds from Japan sent by Captain Blakiston. Many species which are included with some doubt in the second list of Japanese birds issued by Messrs. Blakiston and Pryer are identified for the first time. The following species are added to the Japanese list :—

Puffinus griseus.
Ardea prasinosceles.
Lusciniola pryeri.

BLAKISTON. Amended List of the Birds of Japan. 1884.

This pamphlet corrects some errors in the previous lists, and arranges the Birds of Japan in four groups:—A, B. Species common to Yezo and Southern Japan. C. Species not found in Yezo or the Kuriles. D. Species not found south of Yezo. E. Species found on the Kuriles, but not in Japan proper. The following species are added to the Japanese list :—

Diomedea nigripes.
Puffinus carneipes.

HARGITT. Notes on Woodpeckers.—No. V. On a new Japanese Woodpecker. Ibis, 1884, p. 100.

In this paper the Pygmy Woodpecker of Yezzo and Hondo is separated from that of Kiusiu under the name of *Iyngipicus seebohmi*.

SEEBOHM. Further Contributions to the Ornithology of Japan. Ibis, 1884, pp. 174-183.

This paper is principally a record of the identification of skins sent by Captain Blakiston from Japan, and published in his Amended List.

SEEBOHM. On the Cormorants of Japan and China. Ibis, 1885, pp. 270-271.

In this paper the differences between *Phalacrocorax carbo* and *P. capillatus* are pointed out; the two species were correctly separated by Temminck, and incorrectly united by Schlegel.

SEEBOHM. Further Contributions to the Ornithology of Japan. Ibis, 1885, pp. 363-364.

This short paper records the result of the examination of a few birds sent for determination by the Japanese Government through Mr. Harry Pryer. The reoccurrence of three species which had previously only been known to have been once obtained in Japan is recorded, and one species new to the Japanese fauna is added to the list, *Totanus calidris*.

STEJNEGER. Review of Japanese Birds.—I. The Woodpeckers. Proc. United States Nat. Mus. 1886, pp. 99-124.

This is a very important paper, and is principally founded upon the collections brought from Japan by Mr. Jouy. Two subspecies of *Gecinus canus* are described, *G. canus jessoensis* from Yezzo, and *G. canus perpallidus* from Manchuria. The first will scarcely be maintained, but it is possible that the second may have some claim to be recognized. *Picus leuconotus subcirris* appears to be a fairly good subspecies. *Picus namiyei* appears to be new, but very closely allied to *Picus insularis*.

STEJNEGER. Review of Japanese Birds.—II. Tits and Nuthatches. Proc. United States Nat. Mus. 1886, pp. 374-394.

This paper adds but little to our knowledge of Japanese birds. A new subspecies of Nuthatch from Yezzo is described as *Sitta amurensis clara*, but it seems hardly worthy of recognition.

STEJNEGER. Review of Japanese Birds.—III. Rails, Gallinules, and Coots. Proc. United States Nat. Mus. 1886, pp. 395-408.

This paper adds little to the previous knowledge of the subject.

STEJNEGER. On a Collection of Birds made by Mr. M. Namiye in the Liu-kiu Islands, Japan, with descriptions of New Species. Proc. United States Nat. Mus. 1886, pp. 634-651.

This is a most important paper. The occurrence of a *Turnix* on the Loo-Choo Islands is most interesting. *Treron permagna* is a new species of Fruit-Pigeon very closely allied to *Treron formosæ*. *Scops elegans*, previously known only from one or two examples, is established as a good species. *Hypsipetes pryeri* is a small race of *Hypsipetes squamiceps*. *Erithacus namiyei* is a species of Robin very closely allied to *Erithacus komodori*. *Hirundo namiyei* is probably a subspecies of *Hirundo javanica*. *Pericrocotus tegimæ* is a new species of Minivet allied to *Pericrocotus cinereus*. The occurrence of *Parus castaneoventris* on the Loo-Choo Islands is very interesting.

BLAKISTON. Water-Birds of Japan. Proc. United States Nat. Mus. 1886, pp. 612-660.

This paper is an analysis of the Water-Birds of Japan, which are divided into four groups: those which are circumpolar, those which range across the Palæarctic Region, those which are confined to the eastern half of Asia, and those which are found on both shores of the Pacific.

SEEBOHM. On the Bullfinches of Siberia and Japan. Ibis, 1887, pp. 100-103.

In this paper an attempt is made to fix the respective ranges of the various species and subspecies of the genus *Pyrrhula* which occur in Siberia and Japan.

SEEBOHM. Notes on the Birds of the Loo-Choo Islands. Ibis, 1887, pp. 173-182.

This is a very important paper, the greater part of it being written from information supplied by Mr. Pryer, who visited the Loo-Choo

Islands, and sent a small collection of birds obtained at Naha, the capital of Okinawa-sima, the largest island of the central group. *Iyngipicus kizuki nigrescens* is described as a small dark race of Temminck's Pigmy Woodpecker. *Picus noguchii* is described as a new species, possibly allied to the genus *Blythipicus*. Other species new to the Japanese fauna are

Sterna melanauchen.
„ *dougalli.*
Dendrocygna javanica.

STEJNEGER. Review of Japanese Birds.—IV. Synopsis of the Genus *Turdus*. Proc. United States Nat. Mus. 1887, pp. 4-5.

In this paper a supposed new species of Thrush is described from the main island of Japan under the name of *Turdus jouyi*.

STEJNEGER. Notes on the Northern Palæarctic Bullfinches. Proc. United States Nat. Mus. 1887, pp. 103-110.

In this paper various points relating to the genus *Pyrrhula* are discussed, amongst others the complete intergradation of *Pyrrhula orientalis* and *Pyrrhula rosacea*.

STEJNEGER. Zeitschr. gesammte Ornith. 1887, pp. 166-176.—A List of the Birds hitherto reported as occurring in the Liu-kiu Islands, Japan.

This paper combines the information respecting the birds of the Loo-Choo Islands contained in the writer's article on that subject with that furnished by Mr. Pryer, to which are added the species previously recorded from this locality.

STEJNEGER. Review of Japanese Birds.—V. Ibises, Storks, and Herons. Proc. United States Nat. Mus. 1887, pp. 271-319.

This paper contains much interesting matter; the claim of *Platalea minor* to be regarded as a good species is substantiated. A supposed new species of Reef-Heron is described under the name of *Demiegretta ringeri* from the island of Tsu-sima.

STEJNEGER. On the systematic name of the Kamtschatkan and Japanese Carrion-Crow. Proc. United States Nat. Mus. 1887, pp. 320-321.

In this paper the writer comes to the conclusion that the Japanese Crow ought to bear the name of *Corvus corone orientalis*.

STEJNEGER. Further contributions to the Avifauna of the Liu-Kiu Islands, Japan, with descriptions of new species. Proc. United States Nat. Mus. 1887, pp. 391–415.

This paper contains important information respecting a collection of birds made by Mr. Uishi on the Yaye-yama Islands, the most southerly group of the Loo-Choo chain. An alleged new species of Rail is described as *Porzana phæopyga*, which seems to be an immature example of *Crex fusca*. A new species of *Crex* allied to *Crex mandarina* is described as *Euryzona sepiaria*. An alleged new species of Turtle-Dove is described as *Turtur stimpsoni*, apparently an example of *Turtur orientalis*. The true home of *Erithacus komadori* has been at last discovered.

STEJNEGER. Review of Japanese Birds.—VI. The Pigeons. Proc. United States Nat. Mus. 1887, pp. 416–429.

This paper is valuable, inasmuch as it contains a careful and minute description of the type of *Columba versicolor* described by Kittlitz from the Bonin Islands, and also of an example of *Carpophaga ianthina*, also obtained by Kittlitz on the Bonin Islands—both specimens being preserved in the Museum of the Imperial Academy of Science in St. Petersburg. The latter is made the type of a supposed new species, *Ianthœnas nitens*.

STEJNEGER. On a Collection of Birds made by Mr. M. Namiye in the Islands of Idzu, Japan. Proc. United States Nat. Mus. 1887, pp. 482–487.

This short paper contains two very important statements. An entirely new species of *Merula* is described as *Turdus celænops*; and the breeding-grounds of the very rare Guillemot *Alca wumizusume* are pointed out.

STEJNEGER. Description of a New Species of Fruit-Pigeon (*Ianthœnas jouyi*) from the Liu-Kiu Islands, Japan. The American Naturalist, 1887, pp. 583–584.

The species described in this paper appears to be a very well-marked one.

STEJNEGER. Review of Japanese Birds.—VII. The Creepers. Proc. United States Nat. Mus. 1887, pp. 606–611.

In this paper the two races of the Common Creeper found in the Japanese Islands are discussed.

Jouy. On Cormorant Fishing in Japan. The American Naturalist, 1888, pp. 1–3.

This short paper contains some very interesting information respecting the capture of fish in the rapid rivers of Japan by Cormorants especially trained for the work.

Seebohm. Further notes on the Birds of the Loo-Choo Islands. Ibis, 1888, pp. 232–236.

This paper is principally a correction of a few inaccurate identifications on the part of Mr. Pryer in his paper on the Loo-Choo Islands; based upon the information contained in Dr. Stejneger's article, and confirmed by a small collection of skins. One species, *Zosterops simplex*, is added to the Japanese list.

Stejneger. Review of Japanese Birds.—VIII. The Nutcracker. Proc. United States Nat. Mus. 1888, pp. 425–432.

This paper is an attempt to prove that the Japanese Nutcrackers are more nearly allied to the slender-billed Siberian race than to the thick-billed European race of the species.

Stejneger. Review of Japanese Birds.—IX. The Wrens. Proc. United States Nat. Mus. 1888, pp. 547–548.

In this paper the Wren inhabiting the Kurile Islands is described as *Troglodytes fumigatus kurilensis*.

Jules Soller. Archives des Missions Scientifiques. 3rd Series, vol. xv. pp. 269–280. 1889.

This paper contains some interesting particulars respecting the birds of Japan and their migrations. Mons. Soller was surgeon on board a French steamship which navigated the Japanese Seas in 1885, 1886, and 1887.

Seebohm. On the Birds of the Bonin Islands. Ibis, 1890, pp. 95–108.

This paper contains notes upon an important collection of birds made in 1889 by Mr. P. A. Holst on the Bonin Islands.

GEOGRAPHICAL DISTRIBUTION OF JAPANESE BIRDS.

In the following Table the species which have been recorded from the Japanese Empire are arranged in systematic order. The columns on the right hand represent the distribution within the Japanese Empire; whilst that on the left represents the distribution during the breeding-season outside its limits. C. means that the species is Circumpolar; P. that it ranges across the Palæarctic Region, but is not found in the Nearctic Region; E.P. that it ranges across the Palæarctic Region, but that there are two races, an Eastern Race and a Western Race, which intergrade with each other, and are therefore regarded as only subspecifically distinct; S. that it ranges across Siberia into Eastern, but not into Western Europe; E.S. that the breeding-range of the species is confined to East Siberia, and does not extend to West Siberia or to Europe; P.O. that the species breeds on both the Asiatic and American shores of the North Pacific Ocean; A. that the species breeds in America, but is not known to do so in Asia, though probably such is generally the case; T. that the species is a Tropical one, and breeds chiefly in the Oriental Region, or in a few instances in the Australian Region; J. that although it has been found as a winter migrant, or as an accidental visitor on the mainland, or on more southerly islands, it is not known to breed beyond the limits of the Japanese Empire; and J.J. that the species is believed to be peculiar to the Japanese Empire.

In this Table the subspecific names are omitted, but full particulars of the variations, if any, from the typical form, and their geographical distribution where known, will be found amongst the notes devoted to the species which are represented in the Japanese Empire by closely allied and intergrading races.

	Systematic List of Species.	Kurile Isles.	Yezzo.	Hondo.	Kiu-siu.	Loo-Choo Isles.	Seven Isles.	Bonin Isles.
	PASSERES.							
	Turdinæ.							
E.S.	1. Geocichla varia			*	*	*		
E.S.	2. „ sibirica			*	*			
J.J.	3. „ terrestris							*
J.	4. Merula cardis..............			*	*	*		
E.S.	5. „ fuscata			*	*	*		
E.S.	6. „ naumanni			*	*		*	
E.S.	7. „ pallida			*	*	*	*	
E.S.	8. „ chrysolaus			*	*	*	*	
E.S.	9. „ obscura.............				*	*		
J.J.	10. „ eckenops						*	
E.S.	11. Erithacus akahige			*	*			
J.J.	12. „ namiyei					*		
J.J.	13. „ komadori					*		
E.S.	14. „ calliope	*	*	*	*	*		
E.S.	15. „ cyaneus...........			*	*	*		
E.S.	16. Monticola cyanus			*	*	*	*	*
E.S.	17. Cinclus pallasi			*	*	*		
E.P.	18. Accentor alpinus			*				
J.J.	19. „ rubidus			*	*	*		
S.	20. Pratincola maura	*	*	*	*	*		
E.S.	21. Ruticilla aurorea			*	*	*		
S.	22. Tarsiger cyanurus............			*	*	*	*	
E.S.	23. Niltava cyanomelæna			*	*	*		
E.S.	24. Siphia luteola			*	*	*		
E.S.	25. Xanthopygia narcissina			*	*	*		
E.S.	26. Muscicapa sibirica...........			*	*			
E.S.	27. „ latirostris...........	*	*	*	*			
T.	28. Terpsiphone princeps				*	*	*	
	Crateropodinæ.							
J.	29. Hypsipetes amaurotis			*	*	*	*	
J.J.	30. „ squamiceps..........						*	*
J.J.	31. Hapalopteron familiare							*
T.	32. Zosterops palpebrosa					*		
J.J.	33. „ japonica			*	*	*	*	
	Sylviinæ.							
E.S.	34. Phylloscopus coronatus			*	*	*	*	
S.	35. „ borealis	*	*	*	*	*		
J.	36. „ xanthodryas	*	*	*	*			
J.	37. „ tenellipes			*				
E.S.	38. Acrocephalus orientalis			*	*			
E.S.	39. „ bistrigiceps			*	*	*		
E.S.	40. Locustella fasciolata...........			*				

	Systematic List of Species.	Kurile Isles.	Yezzo.	Hondo.	Kiu-siu.	Loo-Choo Isles.	Seven Isles.	Bonin Isles.
	Sylviinæ (continued).							
E.S.	41. Locustella ochotensis	*	*					
S.	42. „ lanceolata		*					
J.	43. Cettia squamiceps			*				
E.S.	44. „ cantans			*	*	*	*	
E.S.	45. „ cantillans			*	*	*	*	
J.J.	46. „ diphone							*
E.P.	47. Cisticola cisticola			*	*	*		
J.J.	48. Lusciniola pryeri			*				
	Parinæ.							
E.P.	49. Regulus cristatus		*	*				
E.P.	50. Parus palustris	*	*	*	*			
P.	51. „ ater		*	*		*		
E.S.	52. „ atriceps		*	*	*	*	*	
J.	53. „ varius		*	*		*	*	
P.	54. Acredula caudata		*	*				
J.	55. „ trivirgata			*	*			
E.S.	56. Ægithalus consobrinus				*			
E.S.	57. Troglodytes fumigatus	*	*	*	*			
P.	58. Certhia familiaris	*	*	*				
E.P.	59. Sitta cæsia	*	*	*				
	Corvinæ.							
C.	60. Corvus corax	*						
T.	61. „ macrorhynchus	*	*	*	*	*	*	*
P.	62. „ corone	*	*	*	*			
E.S.	63. „ dauricus				*			
E.S.	64. „ neglectus				*			
E.S.	65. „ pastinator			*	*			
P.	66. Nucifraga caryocatactes		*	*				
E.S.	67. Cyanopolius cyanus				*			
S.	68. Garrulus brandti		*					
J.J.	69. „ japonicus			*	*			
T.	70. „ sinensis				?			
P.	71. Pica caudata	*			*			
	Laniinæ.							
E.S.	72. Lanius major		*					
E.S.	73. „ magnirostris			*				
E.S.	74. „ superciliosus		*	*				
T.	75. „ lucionensis					*		
E.S.	76. „ bucephalus		*	*	*		*	
E.S.	77. Pericrocotus cinereus			*				
J.J.	78. „ tegimæ					*		

	Systematic List of Species.	Kurile Isles.	Yezzo.	Hondo.	Kiu-siu.	Loo-Choo Isles.	Seven Isles.	Bonin Isles.
	Sturninæ.							
E.S.	79. Sturnus cineraceus		*	*	*			
J.	80. Sturnia pyrrhogenys		*	*	*	*		
C.	81. Ampelis garrulus		*	*	*			
E.S.	82. " japonicus		*	*	*	*		
	Motacillinæ.							
E.S.	83. Motacilla lugens		*	*	*	*		
E.S.	84. " japonica		*	*	*	*		
E.P.	85. " boarula	*	*	*	*	*		
P.	86. " flava	*						
E.S.	87. Anthus maculatus		*	*	*	*		
S.	88. " spinoletta		*	*	*			
S.	89. " cervinus		*					
	Alaudinæ.							
P.	90. Alauda arvensis	*	*	*				
C.	91. " alpestris	*						
	Fringillinæ.							
P.	92. Coccothraustes vulgaris		*	*	*			
E.S.	93. " personatus		*	*	*			
C.	94. Loxia curvirostra	*	*	*				
J.J.	95. Chaunoproctus ferreirostris							*
C.	96. Pinicola enucleator		*					
E.S.	97. Carpodacus roseus		*	*	*			
S.	98. " erythrinus			*				
E.S	99. " sanguinolentus	*	*	*				
P.	100. Fringilla spinus		*	*	*	*		
P.	101. " linaria		*	*	*			
P.	102. " montifringilla		*	*	*	*		
E.S.	103. " sinica		*	*	*			
J.J.	104. " kawarahiba	*	*	*	*		*	
J.J.	105. " kittlitzi							*
E.S.	106. Montifringilla brunneinucha		*	*	*			
E.S.	107. Pyrrhula griseiventris	*	*	*	*			
P.	108. Passer montanus		*	*	*	*	*	
T.	109. " rutilans		*	*	*			
J.J.	110. Emberiza ciopsis		*	*	*			
J.J.	111. " yessoensis		*	*	*			
P.	112. " schoeniclus		*	*	*			
S.	113. " rustica		*	*	*			
E.S.	114. " fucata		*	*	*			
J.	115. " sulphurata		*	*	*			
J.J.	116. " personata	*	*	*	*			
E.S.	117. " spodocephala		*	*	*			
E.S.	118. " elegans		*	*	*			

	Systematic List of Species.	Kurile Isles.	Yezzo.	Hondo.	Kiu-siu.	Loo-Choo Isles.	Seven Isles.	Bonin Isles.
	Fringillinæ (continued).							
E.S.	119. Emberiza rutila	..	..	..	*			
S.	120. „ aureola	..	*	*				
E.S.	121. „ variabilis	..	*	*	*			
C.	122. „ nivalis	*	*					
P.	123. „ lapponica	*						
	Hirundininæ.							
E.P.	124. Hirundo rustica	..	*	*	*	..	..	*
T.	125. „ javanica	..	..	..	..	*		
E.P.	126. „ alpestris	..	..	*				
J.	127. Chelidon dasypus	..	*	*				
C.	128. Cotyle riparia	..	*	*				
	SCANSORES.							
J.J.	129. Gecinus awokera	..	..	*	*			
P.	130. „ canus	..	*					
P.	131. Picus martius	..	*					
J.J.	132. „ richardsi	..	..	†				
J.J.	133. „ noguchii	..	..	..	*			
E.P.	134. „ leuconotus	..	*	*				
J.J.	135. „ namiyei	..		*				
E.P.	136. „ major	..	*	*				
S.	137. „ minor	..	*					
E.S.	138. Iyngipicus kisuki	*	*	*	*	*	*	
P.	139. Iynx torquilla	..	*	*	*			
	UPUPÆ.							
P.	140. Upupa epops	..	*					
	COLUMBÆ.							
P.	141. Columba livia	..	..	*	*	*		
E.S.	142. Turtur orientalis	*	*	*	*	*	*	
T.	143. „ risorius	..	..	*				
T.	144. „ humilis	..	..	*				
J.J.	145. Treron sieboldi	..	*	*	*			
J.J.	146. „ permagna	..	..	..				
J.J.	147. Carpophaga ianthina	..	..	*	*	*	*	*
J.J.	148. „ versicolor	..	..	..	..	..	..	*
J.J.	149. „ jouyi	..	..	..	*			

† Known only from Tsu-sima.

24　　　　　　　　BIRDS OF THE JAPANESE EMPIRE.

	Systematic List of Species.	Kurile Isles.	Yezzo.	Hondo.	Kiu-siu.	Loo-Choo Isles.	Seven Isles.	Bonin Isles.
	CUCULI.							
P.	150. Cuculus canorus	*	*	*				
E.S.	151. „ intermedius		*	*				
T.	152. „ poliocephalus			*	*			
E.S.	153. Hierococcyx hyperythrus		*	*				
	HALCYONES.							
T.	154. Halcyon coromanda		*	*	*	*		
T.	155. Ceryle guttata		*	*	*			
E.P.	156. Alcedo ispida	*	*	*	*	*		
	CORACIÆ.							
E.S.	157. Cypselus pacificus	*	*	*				
E.S.	158. Chætura caudacuta		*	*				
E.S.	159. Caprimulgus jotaka		*	*	*			
T.	160. Eurystomus orientalis			*	*			
	STRIGES.							
P.	161. Bubo maximus						†	
J.J.	162. „ blakistoni		*					
C.	163. Surnia nyctea							
S.	164. Strix uralensis		*					
P.	165. „ otus			*				
C.	166. „ brachyotus			*				
T.	167. Ninox scutulata			*	*	*		
J.J.	168. Scops semitorques		*	*	*	*		
J.J.	169. „ elegans					*		
E.P.	170. „ scops			*	*	*		
J.J.	171. „ pryeri						*	
	ACCIPITRES.							
C.	172. Falco gyrfalco		*					
C.	173. „ peregrinus	*	*	*	*			
P.	174. „ subbuteo			*				
P.	175. „ æsalon			*				
E.P.	176. „ tinnunculus			*	*			
C.	177. Pandion haliaetus		*	*	*			
T.	178. Butaster indicus		*	*	*			
P.	179. Pernis apivorus			*	*			
E.P.	180. Milvus ater	*	*	*	*			
P.	181. Haliaetus albicilla		*	*	*			
E.S.	182. „ pelagicus		*					

† Known only in the Japanese Empire from the Goto Isles.

GEOGRAPHICAL DISTRIBUTION.

	Systematic List of Species.	Kurile Isles.	Yezzo.	Houdo.	Kiu-siu.	Loo-Choo Isles.	Seven Isles.	Bonin Isles.
	ACCIPITRES (continued).							
P.	183. Aquila chrysaetus	..	*	*				
P.	184. „ lagopus	..	*					
T.	185. Spizaetus nipalensis	..	*	*	*			
E.S.	186. Buteo hemilasius	..	..	..	*			
E.P.	187. „ vulgaris	..	*	*	*	..	..	*
P.	188. Circus cyaneus	*	*	*	*			
E.P.	189. „ aeruginosus	..	*	*				
P.	190. Accipiter palumbarius	..	*	*				
P.	191. „ nisus	..	*	*	*			
T.	192. „ gularis	..	*	*	*			
	STEGANOPODES.							
P.	193. Phalacrocorax carbo	..	..	*				
J.	194. „ capillatus	..	*	*	*			
P.O.	195. „ pelagicus	*	*	*	*			
P.O.	196. „ bicristatus	*						
T.	197. Sula leucogastra	..	..	..	*	*	..	*
T.	198. „ piscatrix	..	..	*				
T.	199. Phaeton rubricauda	..	..	..		..	..	*
T.	200. Fregata minor	..	*					
	HERODIONES.							
P.	201. Ardea cinerea	..	*	*	*	*		
E.P.	202. „ alba	..	*	*	*	*		
T.	203. „ intermedia	..	*	*	*			
P.	204. „ garzetta	..	..	*	*			
T.	205. „ coromanda	..	..	*	*			
T.	206. „ jugularis	..	..	..	*			
C.	207. Nycticorax nycticorax	..	..	*	*	*		
J.J.	208. „ crassirostris	..	..	..	..	..	..	*
J.	209. „ goisagi	..	*	*				
T.	210. „ javanicus	..	*	..	*			
T.	211. „ prasinoscoles	..	*					
P.	212. Botaurus stellaris	..	*	*	*			
T.	213. „ sinensis	..	*	*	*			
E.S.	214. „ eurhythma	..	*	*				
E.S.	215. Ciconia boyciana	..	..	*				
	PLATALEÆ.							
P.	216. Platalea leucorodia	..	*	*				
T.	217. „ minor	..	..	..	*			
T.	218. Ibis nippon	..	*	*	*			
T.	219. „ melanocephala	..	..	*				

BIRDS OF THE JAPANESE EMPIRE.

	Systematic List of Species.	Kurile Isles.	Yezzo	Hondo.	Kiusiu.	Loo-Choo Isles.	Seven Isles.	Bonin Isles.
	ANSERES.							
P.	220. Cygnus musicus		•	•	•			
P.	221. ,, bewicki			•				
E.S.	222. Anser cygnoides	•		•				
E.S.	223. ,, segetum		•	•				
C.	224. ,, albifrons		•	•	•			
S.	225. ,, minutus		•	•				
A.	226. ,, hyperboreus			•	•			
A.	227. ,, hutchinsi	•	•	•				
A.	228. ,, nigricans		•	•	•			
T.	229. Dendrocygna javanica						•	
P.	230. Tadorna cornuta			•	•			
P.	231. ,, rutila				•			
C.	232. Anas strepera			•	•			
C.	233. ,, clypeata		•	•	•			
C.	234. ,, boschas	•	•	•	•			
E.S.	235. ,, zonorhyncha	•	•	•	•			
P.	236. ,, crecca	•	•	•	•	•		
E.S.	237. ,, formosa			•	•			
E.S.	238. ,, falcata			•	•			
P.	239. ,, circia		•	•	•			
C.	240. ,, acuta	•	•	•	•			
P.	241. ,, penelope		•	•	•			
E.S.	242. ,, galericulata		•	•				
A.	243. Fuligula americana	•	•	•	•			
P.	244. ,, fusca	•	•	•	•			
C.	245. ,, glacialis	•	•					
C.	246. ,, clangula	•	•	•	•			
C.	247. ,, histrionica	•	•					
E.S.	248. ,, baeri		•	•				
P.	249. ,, ferina		•	•				
P.	250. ,, cristata	•	•	•	•			
C.	251. ,, marila		•	•	•	•		
C.	252. Somateria spectabilis	•						
S.	253. ,, stelleri	•						
P.	254. Mergus merganser	•	•	•	•			
C.	255. ,, serrator	•	•	•	•			
P.	256. ,, albellus		•	•	•			
	TUBINARES.							
P.O.	257. Diomedea albatrus		•	•	•			•
T.	258. ,, nigripes		•	•				
T.	259. Puffinus leucomelas		•	•	•			
T.	260. ,, carneipes		•					
T.	261. ,, griseus	•						

GEOGRAPHICAL DISTRIBUTION.

	Systematic List of Species.	Kurile Isles.	Yezzo.	Hondo.	Kiu-siu.	Loo-Choo Isles.	Seven Isles.	Bonin Isles.
	TUBINARES (*continued*).							
T.	262. Puffinus tenuirostris			*	*			
C.	263. Fulmarus glacialis	*						
P.O.	264. Œstrelata hypoleuca							*
C.	265. Procellaria leachi	*						
P.O.	266. ,, melania			*				
P.O.	267. ,, furcata	*						
	GAVIÆ.							
C.	268. Alca troile	*	*					
P.O.	269. ,, carbo	*	*	*				
P.O.	270. ,, columba	*	*					
P.O.	271. ,, antiqua	*	*	*	*			
P.O.	272. ,, wumizusume				*		*	
P.O.	273. ,, marmorata	*	*	*				
P.O.	274. ,, brevirostris	*						
P.O.	275. Fratercula corniculata	*						
P.O.	276. ,, cirrhata	*	*					
P.O.	277. ,, monocerata	*	*	*				
P.O.	278. ,, psittacula	*						
P.O.	279. ,, cristatella	*	*	*				
P.O.	280. ,, pygmæa	*	*	*				
P.O.	281. ,, pusilla		*	*				
C.	282. Stercorarius richardsoni	*						
C.	283. ,, buffoni	*						
C.	284. ,, pomarinus			*				
C.	285. Larus glaucus		*	*				
A.	286. ,, glaucescens		*	*				
E.P.	287. ,, marinus	*	*					
P.	288. ,, cachinnans		*	*	*			
S.	289. ,, leucopterus		*					
E.S.	290. ,, crassirostris		*	*	*		*	
P.	291. ,, canus	*	*	*				
C.	292. ,, tridactylus	*	*					
P.	293. ,, ridibundus		*	*	*			
C.	294. Sterna dougalli					*		
E.S.	295. ,, longipennis	*	*	*				
T.	296. ,, melanauchen					*		
T.	297. ,, sinensis		*	*		*		
A.	298. ,, aleutica			*				
T.	299. ,, bergii					*		
T.	300. ,, stolida			*		*		
T.	301. ,, anæstheta		*	*				
T.	302. ,, fuliginosa				*			

		Systematic List of Species	Kurile Isles.	Yezzo.	Hondo.	Kiu-siu.	Loochoo Isles.	Seven Isles.	Bonin Isles.
		LIMICOLÆ.							
C.	303.	Charadrius fulvus............	*	*	*		*		*
C.	304.	„ helveticus...........	*	*	*	*	*		
P.	305.	„ morinellus	..	*	*				
P.	306.	„ minor	..	*	*				
P.	307.	„ placidus	..	*	*				
S.	308.	„ mongolicus	*	*	*				
E.P.	309.	„ cantianus...........	..	*	*				
T.	310.	„ geoffroyi	..			*			
E.S.	311.	Lobivanellus cinereus........	..			*			
P.	312.	Vanellus cristatus	..	*	*	*			
E.S.	313.	Hæmatopus osculans	*	*	*				
A.	314.	„ niger	*						
E.P.	315.	Numenius arquatus	*	*	*	..	*		
E.S.	316.	„ cyanopus........	..	*	*				
E.P.	317.	„ phæopus........	*	*	*	*			
E.S.	318.	„ minutus	..	*	*	*			
C.	319.	Phalaropus fulicarius.........	*	*					
C.	320.	„ hyperboreus	*	*	*				
P.	321.	Totanus fuscus	*	*	*				
P.	322.	„ calidris............	..	*	*				
P.	323.	„ glottis.............	..	*	*				
P.	324.	„ stagnatilis..........	..	*	*				
E.S.	325.	„ incanus............	*	*	*	..	*	..	*
P.	326.	„ glareola...........	*	*	*	*			
P.	327.	„ ochropus	..	*	*	*			
S.	328.	„ terekius............	..	*	*				
P.	329.	„ hypoleucus	*	*	*	*			*
P.	330.	„ pugnax	..	*	*				
E.P.	331.	Limosa rufa.................	*	*	*	*			
E.P.	332.	„ melanura	..	*	*	*			
A.	333.	Macrorhamphus griseus.......	..	*	*				
C.	334.	Strepsilas interpres..........	*	*	*				
E.S.	335.	Tringa crassirostris	*	*	*				
C.	336.	„ canutus	..	*	*				
C.	337.	„ alpina..............	*	*	*	*			
C.	338.	„ maritima............	*	*	*				
C.	339.	„ arenaria	..	*	*				
P.	340.	„ platyrhyncha	..	*	*				
E.P.	341.	„ minuta	*	*	*				
E.S.	342.	„ subminuta	*	*	*				
E.S.	343.	„ pygmæa	..	*	*				
E.S.	344.	„ acuminata	..	*	*	*			
T.	345.	Rhynchæa capensis	..	*	*				
J.	346.	Scolopax australis	..	*	*				
E.S.	347.	„ solitaria	..	*	*				

GEOGRAPHICAL DISTRIBUTION.

	Systematic List of Species.	Kurile Isles.	Yezzo.	Hondo.	Kiu-siu.	Loo-Choo Isles.	Seven Isles.	Bonin Isles.
	LIMICOLÆ (*continued*).							
E.S.	348. Scolopax megala..............				*			
P.	349. ,, gallinula		*	*				
E.S.	350. ,, stenura		*					
C.	351. ,, gallinago	*	*	*	*			
P.	352. ,, rusticola		*	*	*			
	GRALLÆ.							
P.	353. Grus cinerea				*			
E.S.	354. ,, leucogeranus				*			
E.S.	355. ,, japonensis				*			
E.S.	356. ,, leucauchen		*	*	*			
E.S.	357. ,, monachus		*	*	*			
T.	358. Turnix blakistoni					*		
	FULICARIÆ.							
E.S.	359. Otis dybowskii		*		*			
E.S.	360. Crex pusilla		*	*	*			
E.S.	361. ,, fusca		*	*	*			
E.S.	362. ,, undulata...............		*	*				
J.J.	363. ,, sepiaria					*		
E.P.	364. Rallus aquaticus.............		*	*	*			
T.	365. Gallicrex cinereus				*			
P.	366. Fulica atra		*	*	*	*		
C.	367. Gallinula chloropus		*	*	*	*		
	PYGOPODES.							
C.	368. Colymbus adamsi	*	*		*			
C.	369. ,, arcticus		*	*	*			
C.	370. ,, septentrionalis	*	*	*				
E.P.	371. Podiceps rubricollis		*	*	*			
S.	372. ,, nigricollis...........		*	*	*			
C.	373. ,, cornutus		*	*	*			
P.	374. ,, minor		*	*	*			
	GALLINÆ.							
E.S.	375. Phasianus torquatus				†			
J.J.	376. ,, versicolor			*				
J.J.	377. ,, sœmmeringi				*			
J.J.	378. ,, scintillans			*				
C.	379. Tetrao mutus	*		*				
P.	380. ,, bonasia..............		*					
P.	381. Coturnix communis		*	*	*			

† Known only in the Japanese Empire from Tsu-sima.

The geographical distribution of the Birds found in the Japanese Empire presents several points of interest. The avifauna of Japan is typically Palæarctic. If we consider the Birds of Japan with regard to their distribution during the breeding-season, we shall find that about 75 per cent. are Palæarctic species, of which 39 per cent. range across the Palæarctic Region, and 36 per cent., though breeding in Eastern Siberia, are not found in Western Europe. The remaining 25 per cent. consist of 12 per cent. of tropical species, and 13 per cent. of species not known to breed outside the Japanese Empire. The percentage of Oriental and Australian species which invade the southern portions of the Eastern Palæarctic Region is probably about the same as that of Ethiopian species which invade the southern portions of the Western Palæarctic Region.

The species of birds known to have occurred in the Japanese Empire, if classified according to the range of their distribution during the breeding-season, may be summarized as follows:—

Circumpolar species	49
Palæarctic species	71
Eastern races of Palæarctic species	26
Palæarctic species	146
Siberian and East-European species	17
East-Siberian species	95
Both shores of the Pacific	27
East Palæarctic species	139
Tropical species	47
Only known to breed in the Japanese Empire	49
Total	381

This geographical distribution of Japanese Birds can only be regarded as typically Palæarctic. This is all the more remarkable, because not only the Flora of Japan and the Lepidoptera of Japan, which may be more or less connected, but also the Reptiles and Batrachians of Japan show an affinity to the Eastern Nearctic species. It seems impossible to imagine any connection between Japan and the Eastern States of North America, to the exclusion of the

Western States. It must, however, be admitted that similar difficulties present themselves in other parts of the world; for example, the Reptilia of Madagascar are allied to those of South America and not to those of Africa. With the single exception of the oft-repeated story of the Blue Magpie of Japan, a species which reappears in a slightly modified form in Spain, where it may have been introduced by human agency, the Birds of Japan offer no insoluble problems to the student of geographical distribution.

Japan is part of the eastern subtropical or Manchurian Subregion of the Palæarctic Region, and is closely connected with the mainland of that subregion by the islands in the Straits of Corea. Its connection with the eastern arctic or Siberian Subregion of the Palæarctic Region is equally close. The island of Sakhalien forms a bridge from Yezzo to the mouth of the Amoor, whilst the long chain of the Kurile Islands forms a second bridge to Kamtschatka. To the south the chain of the Loo-Choo Islands connects it with Formosa and the Indo-Chinese Subregion of the Oriental Region.

Thus the geographical position of the Japanese Empire fully explains the character of its avifauna, so far as regards the Palæarctic and Oriental species which have been found within its limits. Some further explanation may, however, be given respecting the species which are not known to breed elsewhere.

So far as is known there are 49 species of birds which do not breed beyond the limits of the Japanese Empire, but some of these are migratory birds and wander southwards in autumn.

The following species have been recorded in winter from Formosa or South China:—*Merula cardis*, *Phylloscopus xanthodryas*, *Phylloscopus tenellipes*, *Cettia squamiceps*, *Parus varius*, *Emberiza sulphurata*, *Nycticorax goisagi*, and *Phalacrocorax capillatus*.

In addition to these, *Hypsipetes amaurotis* and *Acredula trivirgata* have occurred on the Corean peninsula, whilst *Scolopax australis* is a regular winter visitor to Australia. *Sturnia pyrrhogenys* and *Chelidon dasypus* winter in the Malay Archipelago. Some of these species may hereafter be found breeding on the mainland, in which case they must be struck off the roll of birds only known to breed in the Japanese Empire.

There still remain no fewer than 36 species of birds which have never been found beyond the limits of the Japanese Empire. Some of these may possibly be found hereafter on the mainland, but when we consider how few of the smaller Japanese Islands have been explored, there can be little doubt that many more peculiar species remain to be discovered. Those already known may be classified as follows:—

Species peculiar to Yezzo :—*Bubo blakistoni*	1
Species common to Yezzo and Southern Japan :—*Accentor rubidus, Zosterops japonicus, Fringilla kawarahiba, Emberiza ciopsis, Emberiza yessoensis, Emberiza personata, Treron sieboldi, Scops semitorques*	8
Species found in Southern Japan but not in Yezzo :—*Lusciniola pryeri, Garrulus japonicus, Gecinus awokera, Picus namiyei, Carpophaga ianthina, Phasianus versicolor, Phasianus sœmmeringi, Phasianus scintillans*	8
Species peculiar to Tsu-sima :—*Picus richardsi* . . .	1
Species peculiar to the Seven Islands :—*Merula celænops*	1
Species peculiar to the Loo-Choo Islands :—*Erithacus namiyei, Erithacus komadori, Pericrocotus tegimæ, Picus noguchii, Treron permagna, Carpophaga jouyi, Scops elegans, Scops pryeri, Crex sepiaria* .	9
Species found only in the Loo-Choo and Bonin Islands :—*Hypsipetes squamiceps*	1
Species peculiar to the Bonin Islands :—*Geocichla terrestris, Hapalopteron familiare, Cettia diphone, Chaunoproctus ferreirostris, Fringilla kittlitzi, Carpophaga versicolor, Nycticorax crassirostris* .	7
Species peculiar to the Japanese Empire . .	36

The number of peculiar species here, as elsewhere, is in direct proportion to the number of opportunities presented by the configuration of the land for geographical isolation. Species may change in the course of ages by natural selection and many other causes, but they can only be multiplied by Isolation. It is impossible to overestimate the importance of Geographical Isolation in studying the Origin of Species. Darwin most justly observes ('Life and

Letters of Charles Darwin,' iii. p. 159), "I do not believe that one species will give birth to two or more new species as long as they are mingled together in the same district." And the importance of Isolation as a factor in the multiplication of species is over and over again recognized by Wallace in his 'Island Life,' as the following quotations prove:—" We " (p. 243) " have every reason to believe that special modifications would soon become established in any animals completely isolated under such conditions;" and again (p. 258), " however long they may have inhabited the islands, there has been no chance for them to have acquired any distinctive characters through isolation."

It is not known that any species of bird is peculiar to the Kurile Islands; but a local race of the Japanese Wren, *Troglodytes fumigatus kurilensis*, has not been obtained elsewhere. Like the local race of the European Wren found on the Faroe Islands, it is remarkable for its long bill. The winters are very cold in the Kurile Islands; consequently few birds are resident, and the opportunities for isolation are very small.

Picus major japonicus is a local race confined to the three main islands, but connected with the typical race by intermediate forms in Sakhalien and the valley of the Amoor. *Picus leuconotus subcirris* is a local race principally confined to Hondo, but occasionally occurring in Yezzo, though most examples from that island are almost typical. *Parus palustris japonicus* is a local race found in Kiu-siu and Hondo, and represented by intermediate forms in Yezzo.

One species only is supposed to be peculiar to Kiu-siu, *Phasianus sœmmeringi*; but *Strix uralensis fuscescens* and *Iyngipicus kisuki* are two local races peculiar to that island.

Very little is known of the Seven Islands, only twenty-two species having as yet been obtained there, but one of these, *Merula celænops*, is peculiar to the islands.

Still less is known of the island of Tsu-sima; but one species, *Picus richardsi*, has not been found elsewhere.

Of the Loo-Choo Islands, the large island of the central group and an island of the southern group are the only ones that have been partially explored, with the result that several new species and three local races have been discovered. *Iyngipicus kisuki nigrescens* is a local race peculiar to the islands; *Hirundo javanica namiyei* is another peculiar local race; and *Hypsipetes squamiceps pryeri* is a local race only differing from the typical form in being on an average slightly smaller.

The southern group of the Loo-Choo Islands has so far produced only two peculiar species, *Erithacus komadori* and *Crex sepiaria*.

The Bonin Islands are remarkable for several peculiar species.

If we consider the birds recorded from the Japanese Empire with regard to their distribution within its limits, the list may be analyzed as follows:—

Resident and migratory species not found south of the Kurile Islands	20
Species not found south of Yezzo; residents 7, winter visitors 25	32
Species common to Yezzo and Southern Japan; residents 108, winter visitors 83, summer visitors 47	238
Species found in Southern Japan, but not in Yezzo; residents 25, summer visitors 36	61
Additional species from	
Loo-Choo Islands	17
Bonin Islands	9
Islands in Corean Straits	3
Seven Islands	1
Total	381

These figures are, of course, approximate. Many of the migratory species which have only been recorded from the Kurile Islands or from Yezzo may occasionally wander further south in winter; and it is not at all improbable that some of the residents and summer visitors which have hitherto been only known from Southern Japan may hereafter be found to occur in Yezzo. In either case the number of species common to Yezzo and Southern Japan (which already amounts to 62 per cent. of the whole) would be increased. If the list be restricted to the birds of Yezzo and Southern Japan, the number of species common to both is raised to 72 per cent.; but if the winter visitors be excluded, it is only raised to 69 per cent.

A somewhat anomalous fact in the distribution of Japanese birds is the occurrence of East-Siberian species in Hondo which for some reason do not visit Yezzo.

Cyanopolius cyanus inhabits the valley of the Amoor, and has been recorded from Lake Baikal; nevertheless it is not known to have occurred in Yezzo, though it is not uncommon in Hondo.

Aquila chrysaetus, *Pernis apivorus*, *Butaster indicus*, *Falco tinnunculus*, *Emberiza elegans*, *Emberiza spodocephala*, *Merula obscura*,

Geocichla sibirica, Accentor alpinus, Pericrocotus cinereus, Lanius magnirostris, Corvus pastinator, Corvus dauricus, Hirundo daurica, Grus cinerea, Grus leucogeranus, Grus monachus, Ciconia boyciana, Numenius minutus, Tringa canutus, Totanus terekius, Totanus calidris, Stercorarius pomarinus, Anas strepera, Anas formosa, Tadorna cornuta, Anser hyperboreus, Cygnus bewicki, and possibly one or two other species have a somewhat similar distribution.

The explanation of these at first sight rather startling facts is not difficult to find. In the first place, about half of the species enumerated above are winter visitors to Japan, and migrate every spring and autumn along the coasts of Yezzo to and from their winter-quarters. It is not surprising that they have escaped detection in Yezzo, because they only pass through on migration and do not winter there; nor is it surprising that they do not winter there, because the mean winter temperature of Yezzo is so much lower than that of Southern Japan. According to the 'Physikalische Atlas' of Berghaus the mean temperature of Hakodadi during January is 4 degrees (Cent., or 7¼ degrees Fahr.) below freezing, whilst at Yokohama it is as much above it. In the second place, the remaining half of these species breed in Southern Japan, and many of them may not breed in Yezzo because of the difference in the mean summer temperature. According to the same authority, the mean temperature of the valley of the Amoor and its tributaries during July ranges from about 63° (Fahr.) in the north to about 73° (Fahr.) in the south. The mean temperature of Hakodadi for the same period is below the lowest of these figures, whilst that of Yokohama is above the highest. The mean temperature appears to be a much more potent factor in the distribution of Japanese birds than the distance from the land or the depth of the intervening ocean. The reason why the Tsugaru Straits, or Blakiston's Line, is an important one in the distribution of birds is not because it represents deep sea as Wallace's Line does, but because it happens to coincide with certain Isothermal Lines which bound both the breeding-grounds and the winter-quarters of so many species.

Besides the 30 species that have been recorded from Yezzo but not from Hondo, there are at least 50 species of birds which have been recorded from Hondo but not from Yezzo, and there are very many more that have been recorded from Yezzo but not from the Kurile Islands. Most of the former are species which breed in the Arctic regions and seldom migrate so far south as Japan; but many of them are species that migrate further south than Japan, and it is only an accident that they have been recorded from Yezzo but not from Hondo. Precisely the same remark applies to many of the latter species; they must have passed Yezzo, and they may have passed the Kuriles in order to get to Hondo. There are, however, amongst the resident birds, three remarkable instances of species inhabiting Siberia and Yezzo which are represented in Hondo by allied but different species:—

Gecinus canus is represented in Hondo by *Gecinus awokera*.
Garrulus brandti is represented in Hondo by *Garrulus japonicus*.
Acredula caudata is represented in Hondo by *Acredula trivirgata*.

A third point of view from which the birds recorded from the Japanese Empire may be regarded, is in relation to those recorded from the British Islands. About 130 species in each list are absolutely identical, or so closely allied that they are not regarded as more than subspecifically distinct. An analysis of the species belonging to each suborder, and, in the case of the Passeres, of those belonging to each family, represented in the two districts, shows a remarkable similarity between the two faunas, which is all the more remarkable when the relative position of the two groups of islands to the mainland is taken into consideration.

The Japanese Empire consists of a range of islands extending from Kamtschatka, in latitude $53\frac{1}{2}°$, southwards to Formosa, in latitude $23\frac{1}{2}°$, a range of thirty degrees. A similar range on the Atlantic coast of the Palæarctic Region would extend from Yorkshire to the Canary Islands. The parallels of latitude have, however, little to do with the distribution of birds, which appears to be governed by the Isothermal Lines. The January isothermals of the Japanese Empire transferred to the European coast would range from Cherry Island to Gibraltar; those of July from John o' Groat's to the Cape Verdes; whilst those of mean annual temperature would range from Iceland to the Canaries. Japan proper, from the north of Yezzo to the south of Kiu-siu, is much less extensive, and only ranges from $44\frac{1}{2}°$ to $31°$, or only thirteen degrees and a half. A similar range on the map of Europe would extend from Bordeaux to Morocco. The corresponding January isothermals would range from Jan Mayen to Lisbon, those of July from London to the Canaries, whilst those of mean annual temperature would range from the Orkneys to Gibraltar.

With a climatic range of so much greater extent than is possessed by the British Islands, it would be reasonable to expect that the number of species found in the Japanese Empire should much exceed those of its Atlantic rival, were it not for other considerations.

The ornithology of the British Islands has been studied for a century or more by a succession of students in every part of the country, who have vied with each other in detecting every rare or accidental visitor to our shores. On the other hand, the ornithology of the Japanese Empire has only been studied during the last half-century, by a dozen visitors who have spent a month or two in a few isolated spots, and by two or three residents who have occupied the leisure of a busy life in the study of Zoology, of which the collection of birds has only formed a branch. It is therefore fair to assume that there are many of the rarer residents, or of the irregular visitors, which have hitherto escaped detection in Japan; and there can be scarcely any doubt that new species remain to be discovered on the islands which have not yet been explored.

	British.	Japanese.
PASSERES—Turdinæ	27	28
Crateropodinæ	0	5
Sylviinæ	20	15
Parinæ	15	11
Corvinæ	10	12
Laniinæ	5	7
Sturninæ	3	4
Motacillinæ	12	7
Alaudinæ	6	2
Fringillinæ	32	32
Hirundininæ	4	5
	134	123
SCANSORES	4	11
UPUPÆ	1	1
COLUMBÆ	4	9
CUCULI	3	4
HALCYONES	2	3
CORACLÆ	7	4
STRIGES	10	11
ACCIPITRES	25	21
PLATALEÆ	2	4
HERODIONES	12	15
STEGANOPODES	3	8
TUBINARES	8	11
ANSERES	44	37
GAVIÆ	37	35
LIMICOLÆ	48	50
GRALLÆ	2	6
FULICARIÆ	10	9
PYGOPODES	9	7
GALLINÆ	9	7
	374	381

The birds of Japan do not differ very widely from the birds of the British Islands. It would be very remarkable if they did. The

Japanese Islands bear almost exactly the same relation to the east coast of the Palæarctic Region as the British Islands do to its west coast. The Palæarctic Region, as defined by Sclater and Wallace, is a very clearly defined one so far as the majority of birds are concerned. The range of many species of birds extends uninterruptedly from the British Islands across Europe and Siberia to Japan. Of course there is no species of bird which is found both in Britain and Japan but not in the intervening district. Cases of interrupted areas of distribution are almost unknown, though, as will hereafter appear, there are many cases in which West-European birds resemble more closely East-Asiatic ones, than the Siberian races which intervene. This is unquestionably the most remarkable fact connected with the birds of Japan, and it is one which has not been insisted upon as much as it ought to have been.

It is an undoubted fact that in most species where climatic variations of colour occur, the extreme of whiteness is not found in the examples from Central Siberia, but in those from Kamtschatka. The mean annual temperature of the former locality is nearly twenty degrees lower than that of the latter, and the mean winter temperature shows a much greater difference. Nearly all the species which appear to exhibit these climatic variations of colour are resident birds, which moult only once a year, in July and August; and the mean temperature of July, when the new feathers are forming, appears to coincide with the variation of colour so closely that it is difficult to resist the conclusion that they are cause and effect.

The Common Nuthatch (*Sitta cæsia*) ranges completely across Europe and Siberia from the British Islands to Japan. Throughout this extensive range very little variation occurs in the colour of its upper parts, which is a bright slate-grey. On the other hand, the variation in the colour of the underparts is very remarkable. In the West the range of this species extends as far south as Algeria, where the colour of the underparts is dark buff, paler on the throat. Proceeding in a north-easterly direction, little change is observable until the Baltic is reached, when the white on the throat gradually increases, until at Dantzig it has covered the breast, and at St. Petersburg it has spread over the belly. In Central Siberia the underparts, except the extreme flanks and the under tail-coverts, are snow-white, but in the valley of the Amoor and in Southern Japan the buff has reappeared on the belly, and the Dantzig bird is reproduced. The southern limit of the eastern range of this species appears to be South

China, where the colour of the underparts resembles exactly that of birds of Western Europe and Algeria. The young in first plumage of the European race closely resemble their parents, but those of the Central-Siberian race closely resemble the adults of the Baltic and Amoor races, leading to the supposition that the Central-Siberian race is the one which has changed most recently.

The Kamtschatkan race of the Nuthatch resembles the Central-Siberian race in the whiteness of its underparts, but is paler on the upper parts, especially on the forehead.

These climatic variations correspond to a remarkable degree with the July Isothermal Lines. The palest race (from Kamtschatka) moults in a mean temperature of $54°$ to $58°$; the Central-Siberian race enjoys a mean temperature at that season of $58°$ to $62°$. The Western race in the Baltic Provinces, which is scarcely distinguishable from the Eastern race in the valley of the Amoor, moults in a mean temperature of $65°$ to $70°$; whilst the dark race in Southern Europe, and its prototype in China, enjoys a mean temperature of $75°$ to $80°$ during the moulting-season.

Not only is the Kamtschatkan race of *Sitta cæsia* whiter than any other climatic race of that species, but the Kamtschatkan races of *Pyrrhula vulgaris*, *Pica caudata*, and *Parus palustris* exhibit the same peculiarity.

It has been stated (Stejneger, 'Orn. Expl. Commander Islands and Kamtschatka,' pp. 230, 231) that the Kamtschatkan races of *Picus major* and *Picus minor* are whiter than the Central-Siberian races of those species, but this does not appear to be the case.

In both those species, however, the Japanese races are darker than the Siberian races, and more nearly resemble those of Western Europe. The Japanese race of *Picus major* closely resembles the South-European race, whilst the Japanese race of *Picus minor* scarcely differs from the South-Scandinavian race of that species. The Japanese race of *Gecinus canus* is not known to differ from the European race of that species, but there is good reason to believe that Siberian examples are on an average greyer.

The Japanese race of *Falco tinnunculus* scarcely differs in colour from the race which breeds on the islands off the coast of West Africa. These races are darker and more richly coloured than those in the intervening country.

Siberian examples of *Certhia familiaris* are whiter than those from Japan or Europe, and the same remark applies to Siberian examples of *Picus leuconotus*, *Strix uralensis*, *Strix brachyotus*, and *Nucifraga caryocatactes*.

There are several instances in which Japanese species resemble European species more closely than they resemble their nearest Asiatic allies: for example, *Accentor rubidus* and *Accentor modularis*, *Garrulus japonicus* and *Garrulus glandarius*, *Acredula trivirgata* and *Acredula rosea*, &c.

The Common Jay (*Garrulus glandarius*) ranges across Europe, north of the Mediterranean, as far east as the valley of the Volga. In the valley of the Kama it is said to intergrade with the Siberian Jay (*Garrulus brandti*), which ranges eastwards from the Ural Mountains across Southern Siberia to Yezzo, the north island of Japan. In Southern Japan it is replaced by the Japanese Jay (*Garrulus japonicus*), a species so nearly allied to the European form that Schlegel only admitted it to be subspecifically distinct. The young in first plumage of the European Jay differ very slightly from their parents; but those of the Siberian Jay are less streaked on the crown, resembling in this respect the adults of the Chinese Jay. There can, however, be little doubt that the Japanese Jay is more nearly related to the Siberian than to the Chinese species. The three semitropical forms of the Common Jay are, *Garrulus bispecularis* from the Himalayas, *Garrulus sinensis* from China, and *Garrulus taivanus* from Formosa. These three species differ from the semi-arctic Jays in having no white on the outer webs of the secondaries. These facts can only be explained by the assumptions that Formosa received its Jay from China, and that Hondo received its Jay from Siberia. These assumptions also account for the absence (so far as is known) of a Jay on the Loo-Choo Islands. To explain the distribution of the two species on the Japanese islands, we can only assume that when the Jay which formerly ranged across the Palæarctic Region was driven southwards, the island of Yezzo was temporarily incapacitated from serving as a residence for Jays, and that it remained without a Jay until the Siberian Jays in their changed climate had differentiated into *Garrulus brandti*, which eventually emigrated to Yezzo. The Japanese Jay is not known to intergrade with the Siberian Jay, and can always be distinguished from the Common Jay by its black lores. In spite of its superficial resemblance to the Common Jay, the Japanese Jay is probably more nearly allied to the Siberian Jay, inasmuch as the colour of the crown varies with age much more than the colour of the lores.

The similarity between the British and Japanese Long-tailed Tits has often been remarked. They are so nearly allied that some examples are very difficult to determine; nevertheless, between their respective ranges an apparently distinct species occurs. The Continental Long-tailed Tit (*Acredula caudata*) is found in Northern and Central Europe, and across Siberia to the island of Yezzo. It can scarcely be called the arctic race of the Long-tailed Tit, firstly, because it occurs in Central Europe, and secondly, because in Central Siberia a real arctic race of *Acredula caudata* occurs. There is, however, much evidence to prove that it intergrades with the British Long-tailed Tit, and it may possibly do so with the Japanese Long-tailed Tit. One fact is absolutely certain, that the immature birds of all these races differ widely from the Siberian race, and approach very near the other two races, from which it may reasonably be assumed that it is the Siberian race which has become whiter, and not the British and Japanese races which have become darker.

The existence of a pale Siberian race between a darker Eastern and a darker Western race is found almost exclusively amongst resident birds. Migratory species either range with little or no local variation across the Palæarctic Region from the British Islands to Japan, or are represented by an Eastern and a Western race. Very many Japanese birds belong to this category. The following British birds are represented in Japan by Eastern races which are regarded as only subspecifically distinct because they are connected by intermediate forms :—*Buteo vulgaris, Regulus cristatus, Lanius excubitor, Motacilla boarula, Anthus spinoletta, Alcedo ispida, Ardea alba, Anser segetum, Tringa alpina, Tringa minuta, Charadrius cantianus, Numenius arquatus, Numenius phæopus, Limosa rufa, Limosa melanura, Rallus aquaticus, Podiceps rubricollis,* and some others.

CLASSIFICATION AND IDENTIFICATION OF JAPANESE BIRDS.

Subclass PASSERIFORMES.

The Passeriformes are the most numerous and the most highly developed of birds, though they contain some archaic families. So far as is known, they are the only birds which combine the following characters:—

Young born with a few scattered tufts of down, but never possessing a continuous downy covering before acquiring feathers : *flexor longus hallucis* (and not *flexor perforans digitorum*) leading to hallux, or in default of that digit to fourth digit reversed to take its place.

To these characters others may be added to strengthen the diagnosis :—The young are born helpless, and require to be fed in the nest by their parents for many days. The spinal feather-tract on the neck is well defined by lateral bare tracts, and is not split by a spinal bare tract. The number of the cervical (including the cervico-dorsal) vertebræ does not exceed 15.

The Subclass Passeriformes contains three Orders.

Order PICO-PASSERES.

The Pico-Passeres possess, of course, the five characters which have already been described as found in all the Passeriformes ; but in order to diagnose them it is only necessary to add to the two characters which are diagnostic of the larger group the following :—

Ambiens and accessory femoro-caudal muscles absent.

The Order Pico-Passeres contains six Suborders.

Suborder 1. *PASSERES*.

Palate ægithognathous; deep plantar tendons not united by a vinculum.

The Passeres comprise nearly half the known species of birds, and are represented in every part of the world capable of producing food upon which a land-bird can exist. They may be divided into several families, but all those found in Japan belong to the Passeridæ (or Acromyodi, if the group be regarded as of more than family rank).

The subfamilies of the Passeridæ are very difficult to define, and the following attempts at definitions of such as are represented in the Japanese Empire can only be regarded as provisional.

TURDINÆ.

Sexes generally different; young in first plumage (which is moulted in the first autumn) spotted, streaked, and barred on the underparts, and generally also on the upper parts; first primary very variable, always present, but never as long as the second. The feathering of the nostril, the development of the rictal bristles, and the width of the bill vary considerably, but it is impossible to draw any line between the Muscicapine and Turdine genera.

The Turdinæ are almost cosmopolitan, and are well represented in Japan.

1. GEOCICHLA VARIA.
(WHITE'S GROUND-THRUSH.)

Turdus varius, Pallas, Zoogr. Rosso-Asiat. i. p. 449 (1826).

White's Ground-Thrush differs from every other Japanese Thrush in having black concentric markings on both the upper and the under parts. It is the largest Japanese Thrush, and has fourteen tail-feathers.

Figures: Gould, Birds of Great Britain, ii. pl. 39; Dresser, Birds of Europe, ii. pl. 10.

White's Ground-Thrush was known to inhabit the mountains of Japan at least as long ago as 1840 (Temminck, Man. d'Orn. iv. p. 604); and was obtained in some numbers by the Siebold Expedition, presumably near Nagasaki (Temminck and Schlegel, Fauna Japonica, Aves, p. 67).

It has only once occurred on the island of Yezzo (Blakiston and Pryer, Ibis, 1878, p. 241), but it must be a very common bird on the more southerly Japanese islands, as great numbers are exposed for sale in the Yokohama market during winter (Swinhoe, Ibis, 1877,

p. 144). There are a score or more examples in the Pryer collection, and Mr. Ringer has procured it near Nagasaki.

White's Thrush breeds in East Siberia and North China, and is an accidental visitor to Europe and the British Islands.

2. GEOCICHLA SIBIRICA.
(SIBERIAN GROUND-THRUSH.)

Turdus sibiricus, Pallas, Reise Russ. Reichs, iii. p. 694 (1776).

The male of the Siberian Ground-Thrush is greyish black, with a white eye-stripe; the female is olive-brown, with white spots on the underparts; both sexes are typically Geocichline in the white pattern on the under surface of the wings.

Figures: Temminck and Schlegel, Fauna Japonica, Aves, pl. 31 (female); Dresser, Birds of Europe, ii. pl. 12 (male and female).

There is no authentic record of the occurrence of the Siberian Ground-Thrush on the island of Yezzo (Blakiston, Amended List of the Birds of Japan, p. 58); but it breeds in some numbers on the mountains of the main island. I have three examples collected by Mr. Jouy on Fuji-yama, and there are twelve examples in the Pryer collection from the same locality (Blakiston and Pryer, Trans. As. Soc. Japan, 1882, p. 161).

The Siberian Ground-Thrush breeds in Eastern Siberia, and is only an accidental visitor to Europe. It is said to have occurred once in the British Islands.

Eggs said to be of this species in the Pryer collection resemble those of the Ring-Ouzel, but are smaller.

Mr. Jouy found this bird quite as shy and retiring on Fuji-yama as I found it in the valley of the Yenesay. It frequents the deep woods, and in Japan is found as high as 5000 feet above the sea-level (Jouy, Proc. United States Nat. Mus. 1883, p. 278). It is a fine songster.

3. GEOCICHLA TERRESTRIS.
(KITTLITZ'S GROUND-THRUSH.)

Turdus terrestris, Kittlitz, Mém. présentés à l'Acad. Imp. des Sciences de St. Pétersb. par divers savans, 1830, p. 244.

Kittlitz's Ground-Thrush agrees with every other species of *Geo-*

cichla in the white pattern on the under surface of its wing, but differs from them all in having uniform brown axillaries. It is a brown bird, conspicuously streaked with black on the mantle, and obscurely spotted with dark brown on the breast.

Figures: Kittlitz, Mém. présentés à l'Acad. Imp. des Sciences de St. Pétersb. par divers savans, 1830, pl. 17.

Kittlitz's Ground-Thrush is supposed to be peculiar to the Bonin Islands, where it was discovered in 1828. Besides the type specimen in the Museum of the Imperial Academy of Sciences in St. Petersburg, there is a second example in the Leyden Museum, and a third in the Vienna Museum.

It appears to be allied to *Geocichla sibirica*, and quite as closely to *Geocichla pinicola* and *Geocichla nævia*.

4. MERULA CARDIS.
(GREY JAPANESE OUZEL.)

Turdus cardis, Temminck, Planches Coloriées, no. 518 (1813).

The male Grey Japanese Ouzel somewhat resembles the male of *Geocichla sibirica*, but it has no white eye-stripe, or white pattern on the underside of the wings. The female is olive-grey above, and white spotted with dark brown below. Adult males have slate-grey axillaries; female and immature males have these feathers orange-chestnut, but at no age is there any chestnut on the tail.

Figures: Temminck and Schlegel, Fauna Japonica, Aves, pl. 29 (male adult and first winter), pl. 30 (female adult and first winter).

The Grey Japanese Ouzel appears to be confined to the Japanese Islands during the breeding-season. It is a common summer visitor to Yezzo, whence I have three adult males, one adult female, and two immature males, collected by Mr. Henson between the 27th of May and the 20th of September. I have also a young bird in first plumage, collected by Captain Blakiston at Hakodadi in August (Seebohm, Ibis, 1884, p. 41), and two others collected by Mr. Jouy on Fuji-yama in July, so that there can be no doubt that it breeds on both islands. In the Pryer collection there are three adult males and two adult females from Fuji-yama, in addition to two immature males and one immature female. I have other examples, both of adults and immature birds, collected in the same locality by Mr. Heywood Jones and Mr. Jouy; and in the British Museum there is an example collected by Mr. Whitely at Nagasaki.

The Grey Japanese Ouzel is a lowland bird, and breeds abundantly at the base of Fuji-yama. The nest is generally placed in the fork of a small tree overhanging a stream, and is composed of moss, roots, and dry leaves, with a foundation of mud. It is lined with grass, fine roots, and horsehair (Jouy, Proc. United States Nat. Mus. 1883, p. 277). Eggs in the Pryer collection resemble those of the Missel-Thrush, but are slightly smaller. This bird is a fine songster, and is much valued by the Japanese as a cage-bird (Blakiston and Pryer, Trans. As. Soc. Japan, 1882, p. 165).

It leaves Japan in autumn to winter in South China and Hainan. I have been unable to find any evidence in favour of the statement (David and Oustalet, Ois. Chine,[1] p. 150) that it migrates to the valley of the Amoor in spring.

5. MERULA FUSCATA.
(DUSKY OUZEL.)

Turdus fuscatus, Pallas, Zoogr. Rosso-Asiat. i. p. 451 (1826).

The Dusky Ouzel may be recognized by the chocolate-chestnut colour of its axillaries, by the chestnut on its tertials and greater wing-coverts, and by the brown of its upper parts, which is russet rather than olive.

Figures: Gould, Birds of Asia, iii. pl. 1 (male and female); Dresser, Birds of Europe, ii. pl. 7 (male and female).

The Dusky Ouzel is a winter visitor to Japan, arriving from the north in great numbers. A few remain to winter in the northern island, but most of them pass onwards, and winter in the more southerly islands. They are very common in winter near Yokohama, whence there are nine examples in the Pryer collection. It also occurs near Nagasaki (Blakiston and Pryer, Trans. As. Soc. Japan, 1882, p. 167), whence examples have been sent by Mr. Ringer to the Norwich Museum, and whence those erroneously recorded as *Turdus naumanni* in the Report of the Siebold Expedition were probably obtained (Temminck and Schlegel, Fauna Japonica, Aves, p. 61). One of these examples was figured in 1831 under the name of *Turdus cinomus* (Temminck, Planches Coloriées, no. 514).

The Dusky Ouzel breeds in Eastern Siberia, above the limit of forest-growth, and winters in South China as well as in Japan. It arrives in Yezzo in great numbers soon after the middle of October (Blakiston, Ibis, 1862, p. 319), but a few stray birds occasionally migrate westwards, and occur during winter in various parts of Europe.

6. MERULA NAUMANNI.
(RED-TAILED OUZEL.)

Turdus naumanni, Temminck, Man. d'Orn. i. p. 170 (1820).

The Red-tailed Ouzel may be recognized by the pale chestnut on the inner webs of its tail-feathers, on the centres of the feathers of its breast and flanks, and on its axillaries and under tail-coverts. The upper parts are nearly uniform olive-brown.

Figures: Dresser, Birds of Europe, ii. pl. 6 (male and female); Blakiston, Ibis, 1862, pl. 10.

The Red-tailed Ouzel is a rare winter visitor to Japan. Dr. Henderson procured it at Hakodadi in October, 1857 (Cassin, Proc. Acad. Nat. Sc. Philad. 1858, p. 194); and in the Pryer collection there is one example obtained by Captain Blakiston at Hakodadi in March, and two examples from Yokohama. It has also occurred on the Loo-Choo Islands (Stejneger, Proc. United States Nat. Mus. 1886, p. 646).

It breeds in Eastern Siberia, and winters in China as well as in Japan. To Europe it is only an accidental visitor.

This species is almost as variable in the colour of the upper parts as *Merula fuscata*, and the distinguished naturalist who presided over the Museum at Warsaw, than whom no ornithologist had more opportunities of judging, was of opinion that they intergrade (Taczanowski, Journ. Orn. 1872, p. 437). The amount of rusty red on the upper parts and on the tail varies much; but the predominant colour of the upper parts is always olive and that of the breast rusty red in *Merula naumanni*, whilst the upper parts are always more or less russet-brown and the centres of the breast-feathers very dark brown in *Merula fuscata*. I have seen large series from China and Japan, but have never found them to intergrade, though they often approach each other.

7. MERULA PALLIDA.
(PALE OUZEL.)

Turdus pallidus, Gmelin, Syst. Nat. i. p. 815 (1788).

The Pale Ouzel has a large patch of white on the tip of the outer tail-feathers, and pale grey axillaries, a combination found in no other

Japanese Thrush. It has no eye-stripe. The upper parts are very russet, and the breast and flanks are almost grey.

Figures: Temminck, Planches Coloriées, no. 515 (male); Temminck and Schlegel, Fauna Japonica, Aves, pl. 26.

The Pale Ouzel is principally known as a winter visitor to Japan, and is not uncommon in the bamboo-thickets near Yokohama (Blakiston and Pryer, Trans. As. Soc. Japan, 1882, p. 164). It is a rare bird in Yezzo (Whitely, Ibis, 1867, p. 199), and there is no record of its having been found breeding in any of the Japanese islands. There are seven examples in the Pryer collection from Yokohama; and Mr. Ringer has sent an example to the Norwich Museum obtained at Nagasaki, whence those figured in the 'Fauna Japonica' as *Turdus daulias* were probably procured. There is an example in the Pryer collection obtained in the central group of the Loo-Choo Islands during January (Seebohm, Ibis, 1887, p. 174); and it has also been obtained in the southern group (Stejneger, Proc. United States Nat. Mus. 1887, p. 405).

The Pale Ouzel breeds in the valley of the Lower Amoor, and winters in South China and Formosa as well as in Japan.

8. MERULA CHRYSOLAUS.
(BROWN JAPANESE OUZEL.)

Turdus chrysolaus, Temminck, Planches Coloriées, no. 537 (1831).

The Brown Japanese Ouzel has pale grey axillaries, rusty-red breast and flanks, no eye-stripe, very little white on the outer tail-feathers, and almost uniform brown upper parts, slightly suffused with russet.

Figures: Temminck and Schlegel, Fauna Japonica, Aves, pl. 28 (male and female).

The Brown Japanese Ouzel is a resident in all the Japanese islands. It is common in Yezzo, congregating in large flocks in winter (Whitely, Ibis, 1867, p. 199). On the main island it breeds on Fuji-yama, and winters in the plains near Yokohama; but many of the young birds migrate in autumn to Formosa and South China, and it has once occurred in the Lower Amoor (Schrenck, Reis. und Forsch. im Amur-Lande, i. p. 352). I have two examples of the young in first plumage obtained by Mr. Jouy on Fuji-yama, and in the Pryer collection there are fourteen adult birds from the Yokohama market (Blakiston and Pryer, Ibis, 1878, p. 241). I have an example collected by Mr. Ringer at Nagasaki; and Mr. Pryer has recorded it from the central group of the Loo-Choo Islands (Seebohm, Ibis, 1887, p. 174).

Eggs of this species in the Pryer collection resemble finely streaked examples of those of the Blackbird.

The nest is made of much coarser materials than that of *Merula cardis*, and is composed of twigs bound together with long fibres of grass. It is placed in bushes. This bird is said to be a sweet songster (Blakiston and Pryer, Trans. As. Soc. Japan, 1882, p. 166).

Dr. Stejneger has described what he supposes to be a new species of Thrush from the mountains north of Yokohama, under the name of *Turdus jouyi* (Stejneger, Proc. United States Nat. Mus. 1887, p. 4). It is said to have a smaller bill, and to be more or less suffused with chestnut on the axillaries and under wing-coverts. The examples obtained were a breeding pair, but both appear to have been in female plumage. They were probably birds of the previous year of *Merula chrysolaus*, possibly of a late brood, and more immature than usual. The colour of the axillaries is more liable to variation than that of some other parts. In *Merula fuscata* it varies from pale grey to deep chestnut, and in *Merula cardis* from slate-grey to orange-chestnut. I have five examples of *Merula obscura* in which the axillaries are suffused with buff; and there is an example of *Merula chrysolaus* itself in the Paris Museum, collected by l'Abbé Faurie near Hakodadi, in which the axillaries and under wing-coverts are considerably suffused with buff.

9. MERULA OBSCURA.
(DUSKY OUZEL.)

Turdus obscurus, Gmelin, Syst. Nat. i. p. 816 (1788).

The Dusky Ouzel has pale grey axillaries and a white eye-stripe, a combination found in no other Japanese Thrush. The white patch at the tip of the outer tail-feathers is small, and the upper parts are olive-brown.

Figures: Temminck and Schlegel, Fauna Japonica, Aves, pl. 27 (male adult and bird of the year); Dresser, Birds of Europe, ii. pl. 9 (male adult, and young in first plumage).

The Dusky Ouzel is a very rare winter visitor to Japan. There are two examples in the Pryer collection from Yokohama (Blakiston and Pryer, Trans. As. Soc. Japan, 1882, p. 165); Mr. Jouy obtained examples (one of which I have in my collection) at Tate-yama, north of Yokohama, on the autumn migration (Jouy, Proc. United States Nat. Mus. 1883, p. 277); and there is an example in the British Museum, collected by Capt. St. John at Nagasaki. It has not been recorded from Yezzo.

It breeds in Eastern Siberia, and winters in the Burma Peninsula and in the islands of the Malay Archipelago. Stragglers occasionally wander to Europe.

10. MERULA CELÆNOPS.
(SEVEN-ISLAND OUZEL.)

Turdus celænops, Stejneger, Proc. United States Nat. Mus. 1887, p. 484.

The male of the Seven-Island Ouzel has a black head and neck: the female resembles that of *Merula chrysolaus*, but the chestnut of the breast and flanks is much deeper in colour.

The Seven-Island Ouzel was originally described from the island of Miaco-shima, one of the Seven Islands, about 50 miles from the mainland, and about 100 miles south of Yokohama. The types are in the museum of the Smithsonian Institution at Washington, and there is a skin of a male in the Pryer collection. It was afterwards procured by Mr. Holst on Fatsizio, an island about seventy miles further south, and a skin of a female as well as of a male from that locality are in my collection (Seebohm, Ibis, 1890, p. 98).

11. ERITHACUS AKAHIGE.
(JAPANESE ROBIN.)

Sylvia akahige, Temminck, Planches Coloriées, no. 571 (1835).

The Japanese Robin has an orange-chestnut throat and tail. The lower breast and flanks are grey in the male, and brown in the female.

Figures: Temminck and Schlegel, Fauna Japonica, Aves, pl. 21 n (male and female).

The Japanese Robin is not known to have occurred in Yezzo in a wild state *, but in Southern Japan it breeds on the mountains and winters in the plains. There is an example in the Pryer collection from Yokohama, and I have four examples from Nagasaki, for which I am indebted to the kindness of Mr. Ringer. It breeds on the Seven Islands (Stejneger, Proc. United States Nat. Mus. 1887, p. 486). L'Abbé David found it at Pekin in April, and at Fokien in November, so that it is probably a resident in North China as well as in Japan.

12. ERITHACUS NAMIYEI.
(STEJNEGER'S ROBIN.)

Icoturus namiyei, Stejneger, Proc. United States Nat. Mus. 1886, p. 645.

The adult male of Stejneger's Robin has the black chin and throat of Temminck's Robin, but its under wing-coverts, flanks, and axillaries are grey as in the female of that species. The female has a brownish-grey breast.

Stejneger's Robin was described by Dr. Stejneger from an adult male obtained by Mr. Namiye on the mountain of Nagogatake in Okinawa Shima. There is an example (a female) in the Pryer collection from the same island, which differs from the female of Temminck's Robin in various characters which are pointed out on the next page. No other examples are known. *Erithacus sibilans* may be distinguished from both the Japanese species by its resemblance to *Erithacus akahige* in the colour of its upper parts. The measurements of the two examples of Stejneger's Robin are as follows:—Wing from carpal joint, ♂ 2·85 inches, ♀ 2·75 ; tail, ♂ 2·05, ♀ 1·8 ; exposed culmen, ♂ ·55, ♀ ·5 ; tarsus, ♂ 1·15, ♀ 1·1 ; middle toe with claw, ♂ ·2, ♀ ·83 ; gradation of tail ♂ ·2, ♀ ·15.

Erithacus komadori, *Erithacus akahige*, *Erithacus sibilans*, *Erithacus namiyei*, and *Erithacus rubecula*, all belong to the same subgeneric group of the genus *Erithacus*. In the concavity of the wing, in the comparative length of the first and second primaries, in the feathering of the nostrils, and in the development of the rictal bristles they are almost identical. I cannot therefore admit the validity of *Icoturus* (Stejneger, Proc. United States Nat. Mus. 1886, p. 613) even as a subgenus.

* I have an example collected by Mr. Henson at Hakodadi on the 28th of June. As its wings are very much abraded and its tail is in moult, I assume it to be an escaped cage-bird.

13. ERITHACUS KOMADORI.

(TEMMINCK'S ROBIN.)

Sylvia komadori, Temminck, Planches Coloriées, no. 570 (1835).

Temminck's Robin has a black chin and throat and black flanks in the adult male, and in both sexes the upper parts are orange-chestnut. The female has the feathers which are black in the male creamy white with grey margins.

Figures: Temminck and Schlegel, Fauna Japonica, Aves, pl. 21 c.

Temminck's Robin was originally described from Japan, from examples procured by the Siebold Expedition; but later ornithologists asserted that it was only known as a cage-bird in that country, and that the Japanese imported it from the Corea (Blakiston and Pryer, Ibis, 1878, p. 239); nevertheless, no collectors on that peninsula have been able to discover it. Its home remained a mystery until it was brought from Yaye-yama Island in the southern group of the Loo-Choo chain (Stejneger, Proc. United States Nat. Mus. 1887, p. 404). I have two fine males from Japan, but, like most cage-birds, the tips of the quills and tail-feathers are imperfect.

It appears to be quite distinct from *Erithacus namiyei*. In the adult male the flanks are black instead of grey; the under wing-coverts are black margined with white, instead of grey margined with rufous, and the axillaries are white with dark centres instead of uniform grey. In the female the feathers of the throat and breast are creamy white margined with grey, instead of being uniform greyish brown, and the under tail-coverts are white instead of grey. Very little reliance can be placed upon the alleged structural differences between the two species, unless a much larger series of each could be obtained, to correct the amount of individual variation that usually occurs. It is, however, probable that *Erithacus komadori* has a slightly shorter tail, a slightly shorter tarsus, a slightly shorter bastard-primary, and a somewhat flatter and more pointed wing than *Erithacus namiyei*.

14. ERITHACUS CALLIOPE.

(SIBERIAN RUBY-THROATED ROBIN.)

Motacilla calliope, Pallas, Reise Russ. Reichs, iii. p. 697 (1776).

The male of the Siberian Ruby-throated Robin has a gorgeous

metallic ruby-coloured throat. The female is a plain brown bird like a Nightingale, with an olive-brown tail.

Figures: Gould, Birds of Asia, iv. pl. 38.

The Siberian Ruby-throated Robin is a summer visitor to the Kurile Islands (whence I have an example collected by Mr. Snow in June) and to Yezzo (Seebohm, Ibis, 1884, p. 182). There is a single example in the Pryer collection from Yokohama, and Mr. Ringer obtained it near Nagasaki, where the examples procured by the Siebold Expedition were probably obtained (Temminck and Schlegel, Fauna Japonica, Aves, p. 57). It has been recorded from the southern group of the Loo-Choo Islands (Stejneger, Proc. United States Nat. Mus. 1887, p. 406).

This fine songster breeds in Siberia from the Ural Mountains to Kamtschatka, and winters in South China, the Philippine Islands, Burma, and India.

15. ERITHACUS CYANEUS.
(SIBERIAN BLUE ROBIN.)

Motacilla cyane, Pallas, Reise Russ. Reichs, iii. p. 697 (1776).

The male Siberian Blue Robin is blue above and white below. The female is olive-brown above, suffused with blue on the upper tail-coverts, and rufous-brown below.

Figures: Radde, Reisen Süd. v. Ost-Sibir. ii. pl. 10.

The Siberian Blue Robin is a summer visitor to Yezzo (Seebohm, Ibis, 1884, p. 182), but probably does not winter there. In the Pryer collection there are four males and a female from Fuji-yama, where it breeds (Jouy, Proc. United States Nat. Mus. 1883, p. 281).

This species is a summer visitor to East Siberia from Lake Baikal to the mouth of the Amoor. It winters in China, Burma, North India, the Malay Peninsula, and Borneo.

16. MONTICOLA CYANUS.
(BLUE ROCK-THRUSH.)

Turdus cyanus, Linnæus, Syst. Nat. i. p. 296 (1766).

The Eastern Blue Rock-Thrush with chestnut belly intergrades with the typical form, with the belly blue like the rest of the plumage, in

North-east China, and can only be regarded as subspecifically distinct from its western representative, though it has been described as *Turdus solitarius* (Müller, Natursyst. Suppl. p. 142) as long ago as 1776. The length of wing varies from 4½ to 5 inches.

Figures: Daubenton, Planches Enluminées, no. 636 (male), no. 564, fig. 2 (female).

The eastern race of the Blue Rock-Thrush is a common summer visitor to all the Japanese Islands, and is occasionally seen in winter in Southern Japan (Blakiston and Pryer, Trans. As. Soc. Japan, 1882, p. 163). There are two examples in the Swinhoe collection from Hakodadi (Swinhoe, Ibis, 1874, p. 157), and eight in the Pryer collection from Yokohama. Mr. Ringer has obtained it at Nagasaki, where the examples procured by the Siebold Expedition, and recorded as *Turdus manillensis*, were probably obtained (Temminck and Schlegel, Fauna Japonica, Aves, p. 67). It is very common on the Bonin Islands, whence I have a series in various stages of plumage collected by Mr. Holst (Seebohm, Ibis, 1890, p. 98). Capt. Rodgers procured it from the Loo-Choo Islands (Cassin, Proc. Acad. Philad. 1862, p. 314); and there are three examples in the Pryer collection from the same locality (Seebohm, Ibis, 1887, p. 174).

The range of the Blue Rock-Thrush extends from Spain across Southern Europe and Central Asia to China. The eastern form breeds in Japan, in the valley of the Ussuri in Eastern Siberia, and in Formosa, wintering in South-east China and the islands of the Malay Archipelago.

Intermediate forms between the Eastern and Western races are very common in China, and an example in the Norwich Museum sent by Mr. Ringer from Nagasaki, as well as one in the British Museum, probably from the same locality, show traces of blue on many feathers of the belly apparently derived from a strain of Western blood.

17. CINCLUS PALLASI.

(SIBERIAN BLACK-BELLIED DIPPER.)

Cinclus pallasii, Temminck, Man. d'Orn. i. p. 177 (1820).

The Siberian Black-bellied Dipper, like its Himalayan ally, is chocolate-brown above and below, but is darker and less rufous than that species.

Figures: Temminck and Schlegel, Fauna Japonica, Aves, pl. 31 c (adult and young).

The Siberian Black-bellied Dipper is a common resident on the mountain-streams of all the Japanese islands (Blakiston and Pryer, Ibis, 1878, p. 239). There are two examples in the Swinhoe collection from Hakodadi (Swinhoe, Ibis, 1875, p. 419), and there are five examples in the Pryer collection from Yokohama. I have also an example from Nagasaki.

The range of the Siberian Black-bellied Dipper extends northwards to Kamtschatka and the Aleutian Islands, westwards to Lake Baikal, and southwards to Central China.

18. ACCENTOR ALPINUS.
(ALPINE ACCENTOR.)

Motacilla alpina, Gmelin, Syst. Nat. i. p. 804 (1788).

The Japanese race of the Alpine Accentor has the throat white spotted with black; and the upper tail-coverts are chestnut with dark centres.

Figures: Swinhoe, Proc. Zool. Soc. 1870, pl. 9; Gould, Birds of Asia, iv. pl. 43.

The Japanese Alpine Accentor is not known to occur in Yezzo, but there are two examples in the Pryer collection from Fuji-yama.

Dybowski procured it on a mountain near the southern shore of Lake Baikal; Maack obtained it in the valley of the Amoor; Middendorff found it on the southern shore of the Sea of Okhotsk; and Swinhoe described it from North China. I have examples collected by Prjevalski in Kansu, which are slightly chestnut on the upper tail-coverts and much streaked on the flanks, but in this respect they are intermediate between *A. alpinus* and *A. erythropygius*. The Japanese Alpine Accentor has been described as a distinct species under the latter name (Swinhoe, Proc. Zool. Soc. 1870, p. 124), but can scarcely be regarded as more than subspecifically distinct, in which case it may be known as *Accentor alpinus erythropygius*.

The Alpine Accentors appear completely to intergrade. Typical forms differ as follows:—

	alpinus . . .	} Upper tail-coverts grey.
Flanks uniform chestnut. {	*rufilatus* . . .	
	nipalensis . .	} Upper tail-coverts chestnut.
	erythropygius .	

The most interesting fact concerning them is that the Japanese form resembles the European one in having the chestnut flank-feathers edged with grey, and appears to be connected with it by intermediate forms in South-east Mongolia.

The habits of the Japanese Alpine Accentor resemble those of its European ally. It is described as flitting around on the rocks, uttering a low soft chuckling note, and as being very tame. It has been found both on Fuji-yama and on Tate-yama (Jouy, Proc. United States Nat. Mus. 1883, p. 300).

19. ACCENTOR RUBIDUS.
(JAPANESE HEDGE-SPARROW.)

Accentor modularis rubidus, Temminck and Schlegel, Fauna Japonica, Aves, p. 69 (1847).

The Japanese Hedge-Sparrow has an unstreaked brown throat and breast. It is much more rufous than its British representative.

Figures: Temminck and Schlegel, Fauna Japonica, Aves, pl. 32 (in very abraded plumage); Gould, Birds of Asia, iv. pl. 42 (in newly moulted plumage).

The Japanese Hedge-Sparrow is peculiar to Japan. There are nine examples in the Pryer collection from the neighbourhood of Yokohama; and there are two examples in the British Museum collected by Mr. Whitely near Hakodadi in winter. It is therefore probable that this species is a resident in all the Japanese islands. The examples figured in the 'Fauna Japonica' without the broad chestnut stripes on the flanks are probably birds in abraded plumage.

This bird ascends Fuji-yama in summer as high as 8000 feet, where it frequents the scrub willows, and has a sparrow-like chirping note (Jouy, Proc. United States Nat. Mus. 1883, p. 300). Eggs in the Pryer collection do not differ from those of its European ally.

The nearest ally of the European Hedge-Sparrow (*Accentor modularis*) and the Japanese Hedge-Sparrow (*Accentor rubidus*) is the Maroone-backed Hedge-Sparrow (*Accentor immaculatus*), a species which ranges from Nepal, through the Eastern Himalayas to Eastern Thibet and Setchuen in Western China. As is the case with several other species of European and Japanese or Chinese birds, the Central form appears to have changed more than the extreme Western and Eastern forms, probably in consequence of a greater change of climate.

20. PRATINCOLA MAURA.
(SIBERIAN STONECHAT.)

Motacilla maura, Pallas, Reise Russ. Reichs, ii. p. 708 (1773).

The Siberian Stonechat differs from its close ally in Western Europe in having the upper tail-coverts white without any dark streaks, and the axillaries black without any white tips.

Figures: Gould, Birds of Asia, iv. pl. 34.

The Siberian Stonechat is a common summer visitor to Yezzo (Whitely, Ibis, 1867, p. 197), and I have an example obtained by Mr. Snow on the Kurile Islands. In the Swinhoe collection there is a pair obtained by Captain Blakiston at Hakodadi in April; and in the Pryer collection there are five examples from Fuji-yama. I have also three examples collected in the latter locality by Mr. Heywood Jones; and it has been obtained by Mr. Ringer at Nagasaki, where the examples procured by the Siebold Expedition, and recorded as *Saxicola rubicola*, were probably obtained (Temminck and Schlegel, Fauna Japonica, Aves, p. 58).

This species breeds in Eastern Europe in the valley of the Petchora and eastwards across Siberia to Kamtschatka. It winters in India, Burma, and South China.

The habits of the Siberian Stonechat are precisely the same as those of our Common Stonechat; but eggs in the Pryer collection from Japan, said to be those of this species, are not nearly so blue as British or Siberian examples.

21. RUTICILLA AUROREA.
(DAURIAN REDSTART.)

Motacilla aurorea, Gmelin, Syst. Nat. i. p. 976 (1788).

The Daurian Redstart has a white patch on the wing, caused by white bases to the secondaries and tertials. The male has a black back and throat, a chestnut breast and rump, and a grey crown and nape.

Figures: Temminck and Schlegel, Fauna Japonica, Aves, pl. 21 D (male and female); David and Oustalet, Ois. Chine, pl. 26 (male).

The Daurian Redstart is a resident on all the Japanese islands, breeding in the mountains and wintering in the plains, many doubtless migrating southwards in autumn (Blakiston and Pryer, Trans. As. Soc. Japan, 1882, p. 162). There are nine examples in the Pryer collection from the neighbourhood of Yokohama. In the British Museum is an example from Hakodadi, and Mr. Ringer has sent examples from Nagasaki to the Norwich Museum.

The Daurian Redstart also breeds in South-east Siberia, East Mongolia, and North China. It winters in Formosa, South China, Hainan, and occasionally in Assam, the Malay Peninsula, Java, and Timor.

It is generally found in low bushes or tangled thickets, and has a loud piping note (Jouy, Proc. United States Nat. Mus. 1883, p. 282).

22. TARSIGER CYANURUS.
(SIBERIAN BLUE-TAIL.)

Motacilla cyanurus, Pallas, Reise Russ. Reichs, ii. p. 709 (1776).

The male Siberian Blue-tail is blue above, with a white eye-stripe; and white below with orange-chestnut flanks. The female is olive-brown above with no eye-stripe, and in addition to the orange flanks there is an obscure broad brown band across the breast.

Figures: Temminck and Schlegel, Fauna Japonica, Aves, pl. 21 (male and female); David and Oustalet, Ois. Chine, pl. 28 (male).

The Siberian Blue-tail is a summer visitor to Yezzo (Whitely, Ibis, 1867, p. 197); but in the more southerly islands of Japan it is a resident, breeding on the mountains and wintering in the plains. There are twelve examples in the Pryer collection from Yokohama, and Mr. Ringer has sent skins to the Norwich Museum obtained at Nagasaki (Blakiston and Pryer, Trans. As. Soc. Japan, 1882, p. 161). There is an example in the Pryer collection from the central group of the Loo-Choo Islands (Seebohm, Ibis, 1887, p. 174), and another in the Museum of the Smithsonian Institution at Washington from the same locality (Stejneger, Proc. United States Nat. Mus. 1886, p. 646).

The range of the Siberian Blue-tail extends from the Ural Mountains, whence I have seen examples in the Moscow Museum, to Kamtschatka. It is a winter visitor to China and Formosa.

I found this bird in the valley of the Yenesay as far north as the Arctic Circle, and Mr. Jouy describes it as one of the commonest birds in the mountains of Japan during summer, often the only one seen on some of the higher passes. It is very familiar in its ways and easily approached. Seated on a low branch of a tree or shrub, with its head on one side, it utters a low guttural chuckling note (Jouy, Proc. United States Nat. Mus. 1883, p. 281). Mr. Jouy procured examples of this species on Fuji-yama in June and on Tate-yama in December.

23. NILTAVA CYANOMELÆNA.
(JAPANESE BLUE FLYCATCHER.)

Muscicapa cyanomelana, Temminck, Planches Coloriées, no. 470 (1829).

The male Japanese Blue Flycatcher is blue on all the upper parts, black on the throat and breast, white on the rest of the underparts and at the base of the tail. The female is a brown bird, with white belly and under tail-coverts, and a large pale patch on the throat.

Figures: Temminck and Schlegel, Fauna Japonica, Aves, pl. 17 D (male), pl. 16 (female); David and Oustalet, Ois. Chine, pl. 81.

The Japanese Blue Flycatcher is a summer visitor to all the Japanese islands. I have ten examples collected by Mr. Henson at Hakodadi in May, and three collected by Mr. Heywood Jones on Fuji-yama in summer (Seebohm, Ibis, 1884, p. 180). There are eight examples in the Pryer collection from the latter locality, including a young male in first plumage collected by Mr. Jouy in July. In the British Museum there is a male collected by Mr. Whitely at Nagasaki, whence the examples figured in the 'Fauna Japonica,' the male as *Muscicapa melanoleuca* and the female as *M. gularis*, were probably also procured.

This handsome bird also breeds in Manchuria near the mouth of the Ussuri (Taczanowski, Journ. Orn. 1875, p. 251). It passes along the coast of China on migration to winter in Borneo.

The Japanese Blue Flycatcher appears to be nearly allied to *Niltava vivida* from Formosa, which may be regarded as an island form of *Niltava sundara*. Neither of these species has any white on the tail, but both have the curious pale patch on the throat.

It is common in the deep woods on Fuji-yama, breeding early in June, and being easily attracted by imitating its mellow whistling note (Jouy, Proc. United States Nat. Mus. 1883, p. 306).

The female may always be distinguished from the other Japanese

Flycatchers by its large size (wing 3½ inches or more), and by its large pale patch on the throat.

Young in first plumage are, like young Thrushes, spotted with buff and barred with black on both the upper and under parts.

24. SIPHIA LUTEOLA.
(MUGIMAKI FLYCATCHER.)

Motacilla luteola, Pallas, Zoogr. Rosso-Asiat. i. p. 470 (1827).

The male Mugimaki Flycatcher is slate-grey above, with a white eye-stripe, a white patch on the shoulder, and white at the base of most of the tail-feathers. The throat and breast are orange-chestnut, shading into white on the belly and under tail-coverts. In the female the slate-grey of the male is replaced by olive, but the white on the wings and tail remains.

Figures: Temminck and Schlegel, Fauna Japonica, Aves, pl. 17 B (male).

The Mugimaki Flycatcher appears to be an accidental visitor to Japan on migration. The Siebold Expedition only obtained a solitary example, probably at Nagasaki; a single example is in the museum at Sapporo, in Yezzo (Blakiston and Pryer, Trans. As. Soc. Japan, 1882, p. 148); and a young male was obtained at Tate-yama, in the centre of the main island, in autumn (Jouy, Proc. United States Nat. Mus. 1883, p. 305).

This species breeds in Eastern Siberia from Lake Baikal to the mouth of the Amoor, passes through China and Formosa on migration, and winters in Borneo.

The Mugimaki Flycatcher belongs to the genus *Siphia*, in which, although the sexes differ in colour, they agree in having the base of the tail more or less white and the upper tail-coverts nearly black. The genus was established in 1837 (Hodgson, Indian Review, i. p. 651), and *Siphia strophiata* is the type.

It is the only Japanese Flycatcher which has white at the base of the tail in both sexes. The male of the Japanese Blue Flycatcher has white at the base of the tail, but neither sex has dark upper tail-coverts.

25. XANTHOPYGIA NARCISSINA.
(NARCISSUS FLYCATCHER.)

Muscicapa narcissina, Temminck, Planches Coloriées, no. 577, fig. 1 (1835).

The male Narcissus Flycatcher is orange on the rump and throat, shading into yellow on the centre of the breast, yellow on the supercilium, white on the greater wing-coverts and under tail-coverts, and nearly black on the rest of the plumage. The female is olive above, shading into russet on the upper tail-coverts and tail, and greyish white below, suffused with yellow and brown in immature examples.

Figures: Temminck and Schlegel, Fauna Japonica, Aves, pl. 17 c (male), pl. 17 (female under the name of *Muscicapa hylocharis*).

The Narcissus Flycatcher is a common summer visitor to Yezzo, but in the more southerly Japanese islands it breeds on the mountains and has been known to winter in the plains (Blakiston and Pryer, Ibis, 1878, p. 234). In the Pryer collection are sixteen examples from Fuji-yama and Yokohama; in the Swinhoe collection is an example from Hakodadi obtained by Captain Blakiston; and I have an example from Nagasaki, for which I am indebted to the kindness of Mr. Ringer.

It is abundant on Fuji-yama in June and July, but is very shy in its habits, frequenting the deep woods (Jouy, Proc. United States Nat. Mus. 1883, p. 306). Its song is described as very sweet.

The Narcissus Flycatcher breeds in South China as well as in Japan, and has occurred in the Philippine Islands in winter.

The Narcissus and the Tricoloured Flycatchers appear to have no relations. The genus *Xanthopygia* was established for their reception in 1847 (Blyth, Journ. As. Soc. Bengal, xvi. p. 123), to which some other species have been referred on what appear to be insufficient grounds.

Xanthopygia tricolor has no claim to be regarded as a Japanese bird. It is not included in the 'Fauna Japonica,' a fact which condemns the stuffed specimen (a male) in the British Museum, which is labelled "Japan, Leyden Museum;" and there can be no doubt that the female figured in the 'Fauna Japonica' as *Muscicapa hylocharis* is an immature female of the Narcissus Flycatcher, and not, as has been suggested (Sharpe, Cat. Birds Brit. Mus. iv. p. 250), of the Tricoloured Flycatcher.

26. MUSCICAPA SIBIRICA.
(SIBERIAN FLYCATCHER.)

Muscicapa sibirica, Gmelin, Syst. Nat. i. p. 936 (1788).

The Siberian Flycatcher is a little brown bird, very closely allied to the Brown Flycatcher, but differing from it in being rather darker in colour, especially on the breast. It differs from the females of the two species of *Xanthopygia* in having no trace of green on the upper parts.

Figures: Hume and Henderson, Lahore to Yarkand, pl. 4.

The Siberian Flycatcher appears to be a common bird in Japan. Captain Blakiston sent me an example from Sapporo, in Yezzo, dated May 26, 1877; and in the Pryer collection are five adult birds and three young in first plumage from Fuji-yama, proving that it breeds in the main island.

This species breeds in Dauria and the valley of the Amoor and also in the Himalayas. It passes through China on migration to winter in India, Burma, and the Malay Peninsula.

The Siberian Flycatcher so closely resembles the Brown Flycatcher that they are often confounded together. It is a slightly larger bird, the upper parts are brown instead of ashy brown, and the sides of the neck and breast are brown instead of pale brown.

Young in first plumage are spotted and barred, both on the upper and under parts, like young Thrushes.

27. MUSCICAPA LATIROSTRIS.
(BROWN FLYCATCHER.)

Muscicapa latirostris, Raffles, Trans. Linn. Soc. xiii. p. 312 (1821).

The Brown Flycatcher is a little grey bird, very closely allied to the Siberian Flycatcher, but differing from it in being rather paler in colour, especially on the breast. It differs from the females of the two species of *Xanthopygia* in having no trace of green on the upper parts.

Figures: Temminck and Schlegel, Fauna Japonica, Aves, pl. 15; Hume and Henderson, Lahore to Yarkand, pl. 5.

The Brown Flycatcher is a common summer visitor to Yezzo and the Kurile Islands, and in the more southerly Japanese islands it is common during the breeding-season on the mountains. Dr. Henderson procured it at Hakodadi in October 1857 (Cassin, Proc. Acad. Nat. Sc. Philad. 1858, p. 194); and there is an example in the Swinhoe collection collected by Mr. Whitely in the same locality on the 24th of September, and another collected by Captain Blakiston in May (Swinhoe, Ibis, 1874, p. 159). There are nine examples in the Pryer collection from Fuji-yama and Yokohama. The example figured in the 'Fauna Japonica' as *Muscicapa cinereo-alba* was probably obtained at Nagasaki.

The Brown Flycatcher breeds in the valley of the Yenesay and the valley of the Amoor, and probably in the Himalayas and the mountains of China. In winter it is found in India, Ceylon, Sumatra, Java, Borneo, and Malacca.

28. TERPSIPHONE PRINCEPS.
(JAPANESE PARADISE FLYCATCHER.)

Muscipeta princeps, Temminck, Planches Coloriées, no. 584 (1836).

The adult male Paradise Flycatcher may be recognized by its long central tail-feathers (10 to 11 inches). The female looks like a Red-tailed Shrike with the head of a Flycatcher.

Figures: Temminck and Schlegel, Fauna Japonica, Aves, pl. 17 E (male and female).

The Japanese Paradise Flycatcher is a common summer visitor to the southern islands of Japan, but is not known to migrate as far north as Yezzo. I have five examples procured on Fuji-yama by Mr. Heywood Jones, and there are five examples in the Pryer collection from the same locality and one from the central group of the Loo-Choo Islands. Messrs. Blakiston and Pryer record its occurrence near Nagasaki, where the examples figured in the 'Fauna Japonica' as *Muscipeta principalis* were probably obtained.

It passes along the coasts of South China on migration to winter in the Malay peninsula.

It is very abundant around Fuji-yama in summer, and builds in the deep fork of a small tree, sometimes supported by the swaying branches of a *Wisteria*, eight or ten feet from the ground, and generally near running water. The nest is made of dry grass, strips of bark, and fresh moss, lined with fine moss roots, and sometimes garnished with lichen or spiders' webs (Jouy, Proc. United States

Nat. Mus. 1883, p. 304). Eggs in the Pryer collection resemble the rufous variety of the eggs of the Red-backed Shrike, but are not so round.

The Paradise Flycatchers are an African genus of birds, no less than ten species being found in the Ethiopian Region. Two others are found in India, and the remaining two in China and Japan. They are very conspicuous objects, as they fly from bush to bush with their long tails streaming behind them.

CRATEROPODINÆ.

Sexes alike; young in first plumage only differing from that of the adult in being slightly paler; first primary generally rather more than half the length of the second; nostrils exposed.

The range and number of species of the Crateropodinæ are very difficult to determine, but they are represented in most of the tropical and subtropical parts of the Old World, including the Pacific Islands. Four species are found in Japan.

29. HYPSIPETES AMAUROTIS.

(BROWN-EARED BULBUL.)

Turdus amaurotis, Temminck, Planches Coloriées, no. 497 (1830).

The Brown-eared Bulbul is smaller than its close ally on the Loo-Choo Islands (wing from carpal joint 4·8 to 5·3), but scarcely differs in size from the Bonin-Island form. It differs from both in having no chestnut-brown on the throat, breast, or belly.

Figures: Temminck and Schlegel, Fauna Japonica, Aves, pl. 31 a.

The Brown-eared Bulbul may possibly be peculiar to Japan during the breeding-season. In Yezzo it is principally known as a summer visitor, but a few remain during winter. In Southern Japan it breeds on the mountains and winters in the plains (Blakiston and Pryer, Ibis, 1878, p. 240). There are examples in the Swinhoe collection from Hakodadi and Nagasaki (Swinhoe, Ibis, 1874, p. 158); and there are five examples in the Pryer collection from Yokohama, and two from the central group of the Loo-Choo Islands. It is probable that some of the Yezzo birds migrate to the Loo-Choo Islands in autumn, returning northwards in spring (Stejneger, Zeitschr. ges.

Orn. 1887, p. 173). Others wander as far as the Corea, where they have been obtained in December, January, and February (Taczanowski, Proc. Zool. Soc. 1887, p. 603).

The nest of the Brown-eared Bulbul is built in a bush, and made of twigs, moss, and coarse roots, lined with fine roots (Blakiston and Pryer, Trans. As. Soc. Japan, 1882, p. 163). Eggs in the Pryer collection are pinkish white spotted with reddish brown, and with lilac underlying markings; they resemble eggs of the European Blackbird in size, but in colour they scarcely differ from eggs of the Chinese and Indian Bulbuls belonging to the genus *Pycnonotus*.

30. HYPSIPETES SQUAMICEPS.
(BONIN-ISLAND BULBUL.)

Oriolus squamiceps, Kittlitz, Mém. prés. à l'Acad. Imp. des Sciences de St. Pétersbourg, par divers savans, 1830, p. 241.

The Bonin-Island Bulbul is larger than its Japanese ally (wing from carpal joint 5·5 to 4·8), and is suffused with chestnut-brown, not only on the ear-coverts and flanks, but also on the throat and belly.

Figures: Kittlitz, Kupfertafeln zur Naturgeschichte der Vögel, pl. 12. fig. 1, under the name of *Galgulus amaurotis*.

This Bulbul is only known from the Bonin and the Loo-Choo Islands; but inasmuch as examples from the latter locality are on an average smaller than the typical form from the former, they may be regarded as subspecifically distinct. The comparative measurements of the two races are as follows:—

	Bonin Islands.	Loo-Choo Islands.
Wing	5·45 to 4·85	4·9 to 4·46
Tail	5·05 to 4·15	4·6 to 4·0
Bill	1·05 to ·85	·9 to ·76
Tarsus	1·0 to ·9	·9 to ·8

The measurements are in English inches; the wing is measured from the carpal joint, and the bill from the frontal feathers. The two races do not differ in colour.

The typical form was discovered by Kittlitz in 1828, and was described by him in 1830 as an *Oriolus*. When he figured it in 1832, he identified it with the Japanese species, but doubting Temminck's assertion that it was a *Turdus*, he decided that it must be a Roller, and called it *Galgulus amaurotis*. Since that date the two species remained confused together until 1881, when the Bonin-Island Bulbul reappeared in ornithological literature under the name of *Hypsipetes squamiceps* (Meyer, Zeitschrift ges. Orn. i. p. 211).

The Loo-Choo form appears completely to intergrade with the typical form, from which it may be distinguished as *Hypsipetes squamiceps pryeri*. It was originally described from an example collected by Mr. Namiye on Okinawa-Shima, the largest of the middle group of the Loo-Choo Islands, under the name of *Hypsipetes pryeri* (Stejneger, Proc. United States Nat. Mus. 1886, p. 642). Pryer described its attempts at song as an almost melodious connected whistle, whilst those of its Japanese ally are said to be most discordant (Stejneger, Zeitschrift ges. Orn. 1887, p. 173).

There are two examples of the typical form from the Bonin Islands in the Pryer collection, and I have lately received twelve more from the same locality collected by Mr. Holst (Seebohm, Ibis, 1890, p. 98). There are seven examples in the Pryer collection of the race which inhabits the Loo-Choo Islands.

31. HAPALOPTERON FAMILIARE.
(BONIN WHITE-EYED WARBLER.)

Ixos familiaris, Kittlitz, Mém. prés. à l'Acad. Imp. des Sciences de St. Pétersbourg, par divers savans, 1830, p. 235.

The Bonin White-eyed Warbler has a round wing with large first primary. Upper parts olive, underparts yellow, a ring of white feathers round the eye; lores yellow; forehead and superciliary stripe black; ear-coverts black on anterior half, yellow on posterior half.

Figures: Kittlitz, Mém. prés. à l'Acad. Imp. des Sciences de St. Pétersb. par divers savans, 1830, pl. 13.

The Bonin White-eyed Warbler was discovered in 1828 by Kittlitz, and remained unknown until it was rediscovered in 1889 by Mr. Holst (Seebohm, Ibis, 1890, p. 100). The only record that I can find of any example having been seen between these dates is that of two live birds in the National Museum at Tokio (Blakiston and Pryer, Trans. As. Soc. Japan, 1882, p. 138, no. 180⅔). I have twelve examples collected by Mr. Holst on the Parry Islands and on one of the Baily Islands.

It is a Timeline Warbler, probably allied to *Stachyris*. In 1818 it was doubtfully referred to the genus *Iora* (Gray, Genera of Birds, i. p. 199); but in 1854 the genus *Apalopteron* was invented for its reception (Bonaparte, Compt. Rend. xxxix. p. 59).

32. ZOSTEROPS PALPEBROSA.
(INDIAN WHITE-EYE.)

Zosterops palpebrosa, Temminck, Planches Coloriées, no. 293, fig. 3 (1824).

The Chinese form of the Indian White-eye is a little bird, not much larger than a Golden-crested Wren, with a white ring round its eye, olive above and white below, shading into pale grey on the flanks and breast, and into yellow on the throat and under tail-coverts.

Figures: Gould, Birds of Asia, ii. pl. 34 (Chinese form).

The Chinese White-eye is said to share with the Tree-Sparrow the honour of being the commonest bird in the Loo-Choo Islands (Seebohm, Ibis, 1888, p. 231).

Its range extends to Formosa, South China, and Hainan. It is not nearly so yellow a green on the upper parts as the typical form, which inhabits India and Burma, but intermediate forms occasionally occur.

The examples from the Loo-Choo Islands are rather large (wing from carpal joint 2·15 to 2·25 inches), and the bills are large (·4 inches from frontal feathers); they closely resemble examples from the Eastern Himalayas, Andaman Islands, and the Nicobars, the *Zosterops nicobarica* of Blyth.

The Chinese form has been named *Zosterops simplex* (Swinhoe, Ibis, 1861, p. 331), and is fairly entitled to be regarded as sub-specifically distinct under the name of *Zosterops palpebrosa simplex*. Possibly the examples from the Loo-Choo Islands, which have been named *Zosterops loochooensis* (Tristram, Ibis, 1889, p. 229), ought to be recognized as *Zosterops palpebrosa nicobarica*.

33. ZOSTEROPS JAPONICA.
(JAPANESE WHITE-EYE.)

Zosterops japonicus, Temminck and Schlegel, Fauna Japonica, Aves, p. 57 (1847).

The Japanese White-eye is easily distinguished from its Chinese ally by the colour of its breast and flanks, which are pale chestnut-brown instead of pale grey.

Figures: Temminck and Schlegel, Fauna Japonica, Aves, pl. 22.

The Japanese White-eye is a resident in all the Japanese Islands, and is peculiar to Japan. It is not very common in Yezzo, but was obtained at Hakodadi as long ago as 1853 by the Perry Expedition (Cassin, Exp. Am. Squad. China Seas and Japan, ii. p. 221). There are eight examples in the Pryer collection from Yokohama, and I have two examples collected by Mr. Heywood Jones on Fuji-yama. I have also three examples obtained by Mr. Ringer at Nagasaki.

The Japanese White-eye is so absolutely intermediate between the species which inhabits South China and that found in North China, that it is impossible to say to which it is most nearly allied. The latter species, *Zosterops erythropleura*, has once occurred in the valley of the Amoor; the brown on its underparts is deepened into chestnut and restricted to the flanks.

The nest of the Japanese White-eye is a beautiful structure composed entirely of moss, patched outside with large pieces of lichen, and lined inside with horse-hair. It is rather flat in shape, and is evidently a ground nest (Jouy, Proc. United States Nat. Mus. 1883, p. 288). Eggs in the Pryer collection are unspotted bluish white, of the dimensions of full-sized Willow-Warbler's eggs.

SYLVIINÆ.

Sexes generally alike; young in first plumage (which is retained during the first winter) the same but brighter; first primary very variable, always present, but never as long as the second; feathering of nostrils very variable.

There are probably from 300 to 400 species that may be referred to this subfamily, which is nearly cosmopolitan. Fifteen species have occurred in Japan.

34. PHYLLOSCOPUS CORONATUS.
(TEMMINCK'S CROWNED WILLOW-WARBLER.)

Ficedula coronata, Temminck and Schlegel, Fauna Japonica, Aves, p. 48 (1847).

Temminck's Crowned Willow-Warbler differs from the other Japanese Willow-Warblers in having a pale mesial line on the crown, and in having the under tail-coverts bright yellow, in strong contrast to the rest of the underparts, which are nearly white.

Figures: Temminck and Schlegel, Fauna Japonica, Aves, pl. 18.

Temminck's Crowned Willow-Warbler is a very common summer visitor to all the Japanese Islands. Dr. Henderson obtained it at Hakodadi in October 1857 (Cassin, Proc. Acad. Nat. Sc. Philad. 1858, p. 193), and I have several examples from the same locality (Whitely, Ibis, 1867, p. 197). There are thirteen examples in the Pryer collection from Yokohama, and Mr. Jouy found it on Fujiyama in July (Jouy, Proc. United States Nat. Mus. 1883, p. 282).

Temminck's Crowned Willow-Warbler breeds in Eastern Siberia as well as in Japan, and passes along the coasts of Formosa and China on migration, to winter in the islands of the Malay Archipelago.

35. PHYLLOSCOPUS BOREALIS.
(ARCTIC WILLOW-WARBLER.)

Phyllopneuste borealis, Blasius, Naumannia, 1858, p. 313.

The Arctic Willow-Warbler differs from its Japanese allies in having a very small and pointed bastard primary, and in having the underparts nearly white, very slightly tinged with yellow on the breast and under tail-coverts.

Figures: Dresser, Birds of Europe, i. pl. 79.

The Arctic Willow-Warbler passes the Japanese coasts in spring and summer on its migration from its breeding-grounds in Kamtschatka to its winter-quarters. I have an example collected by Wossnesensky on the Kurile Islands, and it has been obtained in Yezzo, but appears to be rare (Blakiston, Am. List Birds of Japan, p. 56). There is an example in the Pryer collection from Yokohama; and there is one in the Leyden Museum from Nagasaki (Blakiston and Pryer, Trans. As. Soc. Japan, 1882, p. 159).

The Arctic Willow-Warbler breeds in the Arctic Regions from

Finmark across Siberia to Alaska, and passes in great numbers on migration along the coasts of China and Formosa, to winter in the islands of the Malay Archipelago, the Burma peninsula, and the South Andaman Islands.

36. PHYLLOSCOPUS XANTHODRYAS.

(SWINHOE'S WILLOW-WARBLER.)

Phylloscopus xanthodryas, Swinhoe, Proc. Zool. Soc. 1863, p. 296.

Swinhoe's Willow-Warbler differs from its Japanese allies in having all the underparts much suffused with yellow.

The Japanese representative of the Arctic Willow-Warbler, better known as Swinhoe's Willow-Warbler, breeds in the Kurile Islands, in Yezzo, and in the mountains of Southern Japan, migrating southwards in autumn. I have an example collected by Wossnesensky on the Kurile Islands; there is an example in the British Museum obtained by Capt. St. John at Hakodadi (Seebohm, Cat. Birds Brit. Mus. v. p. 43); and there are eleven examples in the Pryer collection from Fuji-yama.

Swinhoe's Willow-Warbler is only known to breed in Japan, where it is common. It passes the coast of China on migration and winters in Borneo.

37. PHYLLOSCOPUS TENELLIPES.

(PALE-LEGGED WILLOW-WARBLER.)

Phylloscopus tenellipes, Swinhoe, Ibis, 1860, p. 53.

The Pale-legged Willow-Warbler has very pale legs and feet. It has two pale bars across the wing, and the 2nd primary is equal to or slightly longer than the 7th. Like most of its allies it is olive-brown above, but it differs from them in having the rump and upper tail-coverts russet-brown.

There is an undoubted example of this species in the British Museum, which was formerly in the Tweeddale collection. It is sexed a female, and was procured by Mr. Henry Whitely at Hakodadi on the 5th of May, 1865; and there is a second example in the Paris Museum, procured by l'Abbé Fauré in the same locality.

The Pale-legged Willow-Warbler probably breeds in Japan and

China. The type was procured at Amoy during the autumn migration and is in the Swinhoe collection. It has recently been obtained in North Fokien in May and October, and there are several skins in the British Museum obtained by Mr. Oates in its winter-quarters in the Burma peninsula.

38. ACROCEPHALUS ORIENTALIS.
(CHINESE GREAT REED-WARBLER.)

Salicaria turdina orientalis, Temminck and Schlegel, Fauna Japonica, Aves, p. 50 (1847).

The Chinese Great Reed-Warbler is a large bird, the length of wing varying from 3 to $3\frac{1}{2}$ inches.

Figures: Temminck and Schlegel, Fauna Japonica, Aves, pl. 20 B.

The Chinese Great Reed-Warbler is a common summer visitor to all the Japanese Islands wherever reed-beds are found. There is an example in the Swinhoe collection from Hakodadi (Swinhoe, Ibis, 1874, p. 153); whence an example had been procured by the Perry Expedition twenty years previously (Cassin, Exp. Am. Squad. China Seas and Japan, ii. p. 221), and whence examples have been recently sent by Mr. Henson. There are five examples in the Pryer collection from Yokohama, and both Mr. Heywood Jones and Mr. Jouy obtained it on Fuji-yama.

The Chinese Great Reed-Warbler breeds in Eastern Siberia and North China as well as in Japan, and passes through South China on migration, to winter in the islands of the Malay Archipelago, the Burma peninsula, and the South Andaman Islands.

39. ACROCEPHALUS BISTRIGICEPS.
(SCHRENCK'S REED-WARBLER.)

Acrocephalus bistrigiceps, Swinhoe, Ibis, 1860, p. 51.

Schrenck's Reed-Warbler has a broad dark-brown band on each side of the crown, abruptly defined over the pale eye-stripe, but gradually fading into the plain brown of the top of the head.

Figures: Schrenck, Reisen und Forsch. im Amur-Lande, i. pl. 12. fig. 4.

Schrenck's Reed-Warbler is a common summer visitor to all the Japanese Islands. There are four examples in the Swinhoe collection obtained by Captain Blakiston in May and June at Hakodadi (Swinhoe, Ibis, 1874, p. 154); and there are two examples in the Pryer collection from Yokohama. I have examples collected during the breeding-season on Fuji-yama by Mr. Heywood Jones in 1878 and by Mr. Jouy in 1882. Mr. Ringer has obtained it from Nagasaki (Blakiston and Pryer, Trans. As. Soc. Japan, 1882, p. 156) and kindly presented me with an example.

Schrenck's Reed-Warbler breeds in Eastern Siberia as well as in Japan, and passes along the coast of China on migration, to winter in the Burma peninsula.

It is a rather shy bird, but is very common in the meadows round Fuji-yama in summer, the males mounting the tops of the long grass and disappearing on the other side (Jouy, Proc. United States Nat. Mus. 1883, p. 288).

40. LOCUSTELLA FASCIOLATA.
(GRAY'S GRASSHOPPER-WARBLER.)

Acrocephalus fasciolatus, Gray, Proc. Zool. Soc. 1860, p. 349.

Gray's Grasshopper-Warbler has the upper parts nearly uniform in colour and is a large bird (wing from the carpal joint 2·9 to 3·2 inches).

Figures: Seebohm, Cat. Birds Brit. Mus. v. pl. 5 (adult and young).

Gray's Grasshopper-Warbler is a rare visitor on migration to the Japanese Islands. There is an example in the Swinhoe collection from Hakodadi (Swinhoe, Ibis, 1876, p. 332); but it has not yet been recorded from Southern Japan. There can, however, be no reasonable doubt that it passes Hondo as well as Yezzo on migration.

This species breeds near Lake Baikal and in the valley of the Amoor. It passes along the coasts of China and Japan on migration, to winter in the islands of the Malay Archipelago.

41. LOCUSTELLA OCHOTENSIS.
(MIDDENDORFF'S GRASSHOPPER-WARBLER.)

Sylvia (Locustella) ochotensis, Middendorff, Sibirische Reise, ii. p. 185 (1853).

Middendorff's Grasshopper-Warbler has uniform upper parts; and the tail-feathers on the under surface become gradually nearly black towards the apex, and are finally tipped with greyish white.

Figures: Middendorff, Sibirische Reise, ii. pl. 16. fig. 7 (bird of the year); Swinhoe, Ibis, 1876, pl. 8. fig. 1 (young in first plumage).

Middendorff's Grasshopper-Warbler probably breeds in the Kurile Islands. I have an example collected by Wossnesensky on Urup Island in 1844, and another sent me by Captain Blakiston from Ishurup (an island between Urup and Yezzo), shot on the 28th of June. There are four examples in the Pryer collection obtained by Mr. Snow on the Kurile Islands; and one (the type of *Arundinax blakistoni*) in the Swinhoe collection (Swinhoe, Ibis, 1876, p. 332), a bird in first plumage obtained at Hakodadi in October. The type of *Locustella subcerthiola* (Swinhoe, Ibis, 1874, p. 154) was also procured at Hakodadi, but it is not in the Swinhoe collection; it appears to have been an adult bird of this species. A third example from Hakodadi is in the Philadelphia Museum (Seebohm, Ibis, 1880, p. 275), and is the type of *Lusciniopsis japonica* (Cassin, Proc. Acad. Nat. Sc. Philad. 1858, p. 193).

Middendorff's Grasshopper-Warbler breeds in Eastern Siberia as well as on the Kurile Islands, and passes along the coasts of China and Japan, to winter in the islands of the Malay Archipelago.

42. LOCUSTELLA LANCEOLATA.
(TEMMINCK'S GRASSHOPPER-WARBLER.)

Sylvia lanceolata, Temminck, Man. d'Orn. iv. p. 614 (1840).

Temminck's Grasshopper-Warbler has clearly defined streaks on the upper parts, but the tail-feathers are plain russet-brown, with no markings on either the upper or the under surface.

Figures: Dresser, Birds of Europe, ii. pl. 92. fig. 2.

Temminck's Grasshopper-Warbler is probably a rare visitor on migration to all the Japanese Islands. It was originally discovered in Japan, during the cruise of the 'Portsmouth,' by Dr. Henderson at

Hakodadi in October 1857, and described as a new species under the name of *Lusciniopsis hendersonii* (Cassin, Proc. Acad. Nat. Sc. Philad. 1858, p. 194). I have examined the type in the Philadelphia Museum; it is streaked on the breast and lower throat, and slightly so on the under tail-coverts. There was an example in the Swinhoe collection from Hakodadi (Swinhoe, Ibis, 1875, p. 449), but it cannot now be found.

This species breeds in Siberia and in North Russia as far west as St. Petersburg. It passes through China on migration, and winters in the Burma peninsula and the Andaman Islands.

43. CETTIA SQUAMICEPS.
(SWINHOE'S BUSH-WARBLER.)

Tribura squamiceps, Swinhoe, Proc. Zool. Soc. 1863, p. 292.

Swinhoe's Bush-Warbler is a small bird, with the tail only about half as long as the wing. It has a very conspicuous pale stripe above the eye, and a dark stripe through the eye.

Figures: Swinhoe, Ibis, 1877, pl. 4.

Swinhoe's Bush-Warbler is a summer visitor to Japan. There is an example in the Swinhoe collection from Hakodadi (Swinhoe, Ibis, 1874, p. 155); and I have two examples collected by Mr. Henson from the same locality in May. There are four examples in the Pryer collection from Fuji-yama, where it is said to be rather rare (Jouy, Proc. United States Nat. Mus. 1883, p. 284).

Swinhoe's Bush-Warbler is probably confined to Japan during the breeding-season, and winters in Formosa and South China. It is represented in Eastern Siberia by a very nearly allied species, *Cettia ussurianus*, which only differs from its Japanese ally in having the upper parts olive-brown instead of chocolate-brown.

44. CETTIA CANTANS.
(LARGE JAPANESE BUSH-WARBLER.)

Salicaria cantans, Temminck and Schlegel, Fauna Japonica, Aves, p. 51 (1847).

The Large Japanese Bush-Warbler is dull olive-brown on the

upper parts, and greyish white on the underparts. It varies in length of wing from 2·8 to 2·5 inches.

Figures: Temminck and Schlegel, Fauna Japonica, Aves, pl. 19.

The Large Japanese Bush-Warbler is a summer visitor to Yezzo, but in Southern Japan it is a common resident (Blakiston and Pryer, Trans. As. Soc. Japan, 1882, p. 156). I have a female (wing from carpal joint 2·55 inches), collected at Hakodadi on the 19th of April, 1865 (Whitely, Ibis, 1867, p. 197); and there are fourteen unsexed examples in the Pryer collection from Yokohama (wing varying from 2·75 to 2·5 inches). I have a female collected by Mr. Ringer at Nagasaki (wing 2·5 inches) which is as russet as examples of *Cettia minuta* from Formosa, but the tail is longer than the wing. There are two examples in the Pryer collection from the central group of the Loo-Choo Islands (wing 2·7 inches), which are both typically olive in colour.

The Large Japanese Bush-Warbler is only known from Japan and the Loo-Choo Islands, and is everywhere found in company with the Small Japanese Bush-Warbler, which I thought to be its female when I wrote the fifth volume of the 'Catalogue of Birds in the British Museum.' Since then further evidence has been collected, which, as far as it goes, leads to the conclusion that the two forms are distinct species.

Mr. Jouy collected a series of these birds on Fuji-yama and on Tate-yama, and came to the conclusion that the large form is distinct from the smaller one, but unfortunately his evidence is rather meagre.

Of the large form he enumerates 5 adult males and 1 adult female, whilst of the small form he only mentions 1 adult male and no females. He further states that the young in first plumage of the larger form have darker legs than those of the smaller form, but he is unable to detect any other difference in colour either in adult or young birds.

The Japanese Bush-Warbler is a favourite cage-bird with the Japanese, who value it for its song, which is not extensive, though the few notes are sweet (Blakiston and Pryer, Ibis, 1878, p. 237). I am informed that the Japanese do not recognize the existence of two species. In its habits it evidently resembles its European representative, Cetti's Warbler, being found along the banks of streams and in brush heaps. It utters a harsh scolding note when disturbed, and has a Wren-like habit of cocking its tail over its back (Jouy, Proc. United States Nat. Mus. 1883, p. 283).

Eggs of *Cettia cantans* from Yokohama in the Pryer collection are uniform brick-red in colour, and very closely resemble eggs of *Cettia cetti* from South Europe, of *Cettia canturians* from Lake Kinkiang in Central China, and of *Cettia fortipes* from India.

45. CETTIA CANTILLANS.
(SMALL JAPANESE BUSH-WARBLER.)

Salicaria cantillans, Temminck and Schlegel, Fauna Japonica, Aves, p. 62 (1847).

The Small Japanese Bush-Warbler differs from its larger ally only in size (wing from carpal joint 2·3 to 2·1 inches).

Figures: Temminck and Schlegel, Fauna Japonica, Aves, pl. 20.

The Small Japanese Bush-Warbler is a summer visitor to Yezzo, but is a resident in Southern Japan. It is common in the plantations at Hakodadi (Whitely, Ibis, 1867, p. 197); there are six examples in the Pryer collection from Yokohama, and four from the central group of the Loo-Choo Islands.

It does not differ in colour from the larger species; and in both the tail, when in perfect condition, is slightly longer than the wing.

It is somewhat remarkable that two species so nearly allied should have precisely the same geographical distribution, but this apparently anomalous fact is capable of explanation. The large species (wing 2·8 to 2·5 inches) is probably the result of an emigration to Yezzo of a party of *Cettia canturians*, which breeds in the valley of the Ussuri, the island of Askold, and North China, and winters in South China and Formosa. The Chinese Bush-Warbler is slightly larger than its Japanese ally (wing 3·1 to 2·8 inches); its tail is proportionately shorter, and its colour is more russet, especially on the crown. The small species (wing 2·3 to 2·1 inches) is probably the result of an emigration of a party of *Cettia minuta*, which reached Southern Japan *viâ* Formosa and the Loo-Choo Islands. The Hainan Bush-Warbler is a resident in South China and Hainan, and is represented on the island of Formosa by intermediate forms (wing 2·56 to 2·35 inches) which intergrade in colour with both forms. To explain the present condition of these closely allied species, it is necessary to assume, first, that the two emigrating colonies increased and spread, the one northwards and the other southwards, until they both ranged over the whole Japanese group; and secondly, that the effect of the changed climatic and other conditions was the same on each species, reducing the size, lengthening the tail, and altering the colour from russet to olive.

46. CETTIA DIPHONE.
(BONIN BUSH-WARBLER.)

Sylvia diphone, Kittlitz, Mém. prés. à l'Acad. Imp. des Sciences St. Pétersb. par divers savans, 1830, p. 237.

The Bonin Bush-Warbler resembles the Small Japanese Bush-Warbler in colour; but it differs from it in having a longer tail (2·56 to 2·46 instead of 2·3 to 2·1), a longer tarsus (·98 to ·93 instead of ·9 to ·89), and a longer bill (·7 to ·69 instead of ·6 to ·5 inches).

Figures: Kittlitz, Mém. prés. à l'Acad. Imp. des Sci. St. Pétersb. par divers savans, 1830, pl. 14.

The Bonin Bush-Warbler was discovered by Kittlitz in 1828, and remained almost unknown until it was rediscovered in 1889 by Mr. Holst (Seebohm, Ibis, 1890, p. 99). A mutilated example was procured by Mr. N. Ota in February 1883 (Blakiston, Amended List of the Birds of Japan, p. 56, no. 231½), and there is an example in the Pryer collection probably from the same source.

I have three examples from Peel Island, and two from one of the Parry Islands, collected by Mr. Holst.

47. CISTICOLA CISTICOLA.
(FAN-TAILED WARBLER.)

Sylvia cisticola, Temminck, Man. d'Orn. i. p. 228 (1820).

The Fan-tailed Warbler has a large first primary, conspicuous streaks on the back, and pale tips to the tail-feathers.

Figures: Temminck and Schlegel, Fauna Japonica, Aves, pl. 20 c (summer plumage of Japanese race); Dresser, Birds of Europe, iii. pl. 99 (winter plumage of typical race).

The Fan-tailed Warbler is a resident in Southern Japan, but is not known to have occurred in Yezzo. There are a dozen examples in the Pryer collection from Yokohama, and three from the central group of the Loo-Choo Islands. Mr. Ringer obtained it at Nagasaki. It is very remarkable that of these fifteen skins, one only (from the Loo-Choo Islands) is a male in summer plumage, with unstriped crown.

All the examples are large, varying in length of wing (from carpal joint) from 2 to 2·23 inches ; and the example in summer plumage has a broad buff band across the tail (Seebohm, Ibis, 1887, p. 175).

There is a male in summer plumage from the neighbourhood of Yokohama in the Smithsonian Institution at Washington ; and it has been obtained in the southern group of the Loo-Choo Islands (Stejneger, Proc. United States Nat. Mus. 1887, p. 408).

This extreme form of the Eastern or tropical race of the Fantail Warbler has been called *Salicaria* (*Cisticola*) *brunneiceps* (Temminck and Schlegel, Fauna Japonica, Aves, p. 134), and may possibly have a right to the name of *Cisticola cisticola brunneiceps* on account of its large size. The buff band across the tail appears to be characteristic of the summer plumage of the tropical form, which ranges through Formosa, South China, Burma, India, and Ceylon, to tropical Africa. Examples in the Swinhoe collection from Formosa vary in length of wing from 2·15 to 1·85 inches. The Eastern race appears to be entitled to the name of *Cisticola cisticola cursitans*, the latter name having been bestowed upon examples from the neighbourhood of Calcutta (Franklin, Proc. Zool. Soc. 1831, p. 118).

Winter examples of the two forms are not easy to distinguish, but the Eastern form has on an average a shorter first primary and a longer second primary than its Western representative. In an example from Smyrna and one from Yokohama the wing is of the same length, 2 inches. In the Japanese example the first primary measures ·45, the second is ·9 longer, only ·15 shorter than the longest, which is ·3 inch longer than the tenth. In the Asia-Minor example the first primary measures ·67, the second is ·62 longer, ·2 shorter than the longest, which is only ·25 inch longer than the tenth. It must be admitted, however, that there is considerable individual variation in these structural characters, but on an average they appear to be sufficiently reliable to serve as a foundation for a subspecies.

48. LUSCINIOLA PRYERI.
(PRYER'S GRASS-WARBLER.)

Megalurus pryeri, Seebohm, Ibis, 1884, p. 40.

Pryer's Grass-Warbler has a plain and much graduated tail, a concave wing with a large first primary as in *Cisticola*, and streaked upper parts as in that genus, or as in a typical *Locustella*.

Pryer's Grass-Warbler does not appear to me to differ in any generic character from the other Grass-Warblers. It cannot be far removed from *Lusciniola melanopogon* or from *Lusciniola luteiventris*. Its tail consists of twelve feathers and is much graduated; its wings are much concaved, and the first primary is very large; its bill is small, and the rictal bristles are very small; its under tail-coverts are very long, but its tail is shorter than the wing.

I only know of the existence of three skins of this species: the type and a second skin from the Pryer collection are in my possession; the third skin is in the British Museum. All three were obtained by Mr. Pryer near Yokohama.

The statement (Blyth, Ibis, 1867, p. 25) that *Phylloscopus fuscatus* is common in China, Formosa, and Japan is not confirmed by recent collectors. I have never seen a Japanese example of this species, but if it winters in Formosa (Swinhoe, Ibis, 1863, p. 306) it probably passes Japan on migration.

PARINÆ.

Sexes alike; young in first plumage the same, but paler; first primary not more (generally much less) than half the length of the second; nostrils more or less concealed by feathers or hairs, but varying much in this respect. Scarcely worthy of separation from the Corvinæ.

The Parinæ include the Tits, the Nuthatches, the Creepers, and the Goldcrests, and number about 125 species, of which eleven are represented in Japan. They are almost cosmopolitan, but are absent from South America, Madagascar, and the Pacific Islands.

49. REGULUS CRISTATUS.
(GOLDCREST.)

Regulus cristatus, Koch, Syst. baier. Zool. p. 199 (1816).

The Goldcrest is easily recognized by the yellow (*female*) or orange (*male*) mesial line on the crown. The Japanese race differs from its European ally in having the nape and upper back more or less suffused with slaty brown.

Figures: Gould, Birds of Asia, vi. pl. 60 (very bad).

The Goldcrest is a resident on all the Japanese Islands (Blakiston and Pryer, Ibis, 1878, p. 238). There are no examples in the Swinhoe collection from Hakodadi, but there are eight in the Pryer collection from Yokohama. It has been recorded from Kiu-siu (Soller, Arch. Miss. Scientifiques, 3rd series, xv. p. 277), where the examples obtained by the Siebold Expedition were probably procured (Temminck and Schlegel, Fauna Japonica, Aves, p. 70). On Fuji-yama it breeds at an elevation of 7000 feet (Jouy, Proc. United States Nat. Mus. 1883, p. 284).

The breeding-range of the Goldcrest extends from the British Islands across Europe and Southern Siberia to the Himalayas, China, and Japan. Asiatic examples are greyer on the nape and on the upper back than European ones, and may fairly be regarded as subspecifically distinct. The species has been split into three; but the supposed three forms appear to be merely three points in a series which completely intergrade. The typical form was described by Linneus from Europe. In 1856 the Japanese race was separated under the name of *Regulus japonicus* (Bonaparte, Compt. Rend. xliii. p. 767), and in 1863 the Himalayan race was separated under the name of *Regulus himalayensis* (Jerdon, Birds of India, ii. p. 206); but it is impossible to recognize three races. Examples from Asia Minor, Samarcand, the Himalayas, and Japan are scarcely distinguishable. The alleged difference in size and in the colour of the crown is a myth. Examples from St. Petersburg agree precisely with others from Western Europe. Possibly the wisest course is to coin a new trinomial for the eastern race of the Goldcrest, and call it *Regulus cristatus orientalis*.

50. PARUS PALUSTRIS.
(MARSH-TIT.)

Parus palustris, Linneus, Syst. Nat. i. p. 341 (1766).

In the Marsh-Tits the black on the crown extends to the bill and covers the nape, and the black on the throat is very restricted.

Figures: Dresser, Birds of Europe, iii. pls. 108, 109.

The Marsh-Tit is a resident on all the Japanese Islands (Blakiston and Pryer, Trans. As. Soc. Japan, 1882, p. 150). It was first described as a Japanese bird from examples obtained by Dr. Henderson, during the cruise of the 'Portsmouth,' at Hakodadi in October 1857 (Cassin, Proc. Acad. Nat. Sc. Philad. 1858, p. 193). There are three examples in the Swinhoe collection from Hakodadi procured by Captain Blakiston in winter (Swinhoe, Ibis, 1874, p. 156); and Mr. Snow obtained it on the Kurile Islands. There are six examples in the Pryer collection from Yokohama, and it is common in Central Hondo both in summer and winter (Jouy, Proc. United States Nat. Mus. 1883, p. 286).

The range of the Marsh-Tit extends across the Palæarctic Region from the British Islands to Japan, embracing a variety of climates, each of which possesses a more or less distinct race of Marsh-Tit. The two extremes appear to have become specifically distinct, as it is not known that either of them completely intergrades with the typical race. The Marsh-Tits of Kamtschatka have the upper parts sandy white, and the flanks pure white, and may be regarded as distinct under the name of *Parus kamtschatkensis* of Bonaparte. In Turkestan and Mongolia the other extreme, *Parus songarus* of Severtzow, occurs, with very brown upper parts and flanks. The other races of Marsh-Tit appear completely to intergrade and to be climatic rather than local races. *Parus palustris baikalensis* is the Arctic form with the widest range, extending from Archangel across Siberia to Vladivostok. The eastern examples are on an average slightly larger than the western, but they seem to have smaller bills. They are all very grey, and the black on the head is prolonged to the upper back. The two semi-arctic forms, *Parus palustris borealis* in Scandinavia, and *Parus palustris japonicus* (Seebohm, Ibis, 1879, p. 32) in Southern Japan, are almost identical in colour, but the latter are slightly more *sandy* brown on the upper parts and flanks.

Examples from Yezzo may be on an average slightly more sandy than those from Southern Japan; and examples from St. Petersburg may be on an average slightly greyer than those from Southern Sweden. Examples from the Kurile Islands may be referred to *Parus palustris japonicus* or to *Parus palustris baikalensis*, according to the caprice of the collector, or according to the individual variation of the skins.

Examples from North China are indistinguishable from those obtained in Greece. They are browner than examples from Japan and Scandinavia, but they are more sandy and not quite so brown as those from the Pyrenees. British examples are on an average a shade browner still, but some examples from Denmark are quite as brown.

51. PARUS ATER.
(COLE TIT.)

Parus ater, Linneus, Syst. Nat. i. p. 341 (1766).

In the Cole Tits the black on the crown extends to the bill, but there is a white patch on the nape; and the black on the throat extends downwards to the breast and sideways to the shoulders.

Figures: Dresser, Birds of Europe, iii. pl. 107. fig. 3.

The Cole Tit is a resident on all the Japanese Islands (Blakiston and Pryer, Trans. As. Soc. Japan, 1882, p. 149). There are three examples in the Swinhoe collection from Hakodadi (Swinhoe, Ibis, 1874, p. 155), and there are nine examples in the Pryer collection from Yokohama. It is exceedingly abundant in winter in Central Hondo (Jouy, Proc. United States Nat. Mus. 1883, p. 285), and Mr. Pryer has recorded it from the central group of the Loo-Choo Islands (Seebohm, Ibis, 1887, p. 176).

The breeding-range of the Cole Tit extends from the British Islands across Europe and Siberia to Japan. The typical form appears to range from the British Channel across Europe, and across Asia from the Arctic Circle to the southern slopes of the Himalayas, and through North China to Japan. It varies in three directions: in the blueness of the grey of the upper parts; in the pureness of the white on the breast; and in the elongation of the feathers of the crown into a crest.

Parus ater æmodius has a decided crest, and is also darkest on the breast, which is sandy buff, but its back is not quite so brown as that of British examples. It inhabits the southern slopes of the Himalayas. It appears to lessen its crest and to become paler on the breast in China; and in Japan the crest is almost obsolete, the breast has become sandy white, and the grey on the back very blue. The same change takes place as it ranges westwards. Examples from the Thian-Shan mountains (the *Parus piceæ* of Severtzow) have small crests, the breast is very slightly buffer than in Japanese birds, and the grey on the back is almost as blue. In Russian Turkestan (*Parus rufipectus* of Severtzow) the crest is all but obsolete, the breast is a shade paler, but the colour of the back remains the same.

Parus ater britannicus, from the British Islands, represents the extreme of brownness on the back, the entire absence of a crest, and the extreme of whiteness on the breast. The two latter characters are, however, common to examples from Europe and Western Siberia.

Parus ater in its typical form ranges across continental Europe and Siberia, but in the eastern half of its range a tendency to develop a crest is more or less observable, and the breast is slightly sandy in colour. If Chinese examples be distinguished as *Parus ater pekinensis*, those from Japan must be described as intermediate between the Chinese and European forms.

52. PARUS ATRICEPS.
(INDIAN GREAT TIT.)

Parus atriceps, Horsfield, Trans. Linn. Soc. xiii. p. 160 (1820).

The Manchurian race of the Indian Great Tit, like its British representative, has a black band down the underparts, and a green mantle, but its flanks are nearly white.

Figures: Temminck and Schlegel, Fauna Japonica, Aves, pl. 33, under the name of *Parus minor*; Gould, Birds of Asia, ii. pl. 56.

The Manchurian race of the Indian Great Tit is a resident in Japan, whence it was originally described by Temminck and Schlegel from examples obtained by Dr. Siebold. It was first procured in Yezzo by Dr. Henderson, who found it abundant near Hakodadi in October (Cassin, Proc. Acad. Nat. Sc. Philad. 1858, p. 192); and there is an example in the Swinhoe collection obtained by Captain

Blakiston at Hakodadi in February (Swinhoe, Ibis, 1874, p. 156). There are examples in the Paris Museum procured at Aomori, in the north of Hondo, by l'Abbé Fauire; and there are five examples in the Pryer collection from Yokohama. There is an example in the Norwich Museum, collected by Mr. Ringer at Nagasaki (Blakiston and Pryer, Trans. As. Soc. Japan, 1882, p. 151); and there are three examples in the Pryer collection from the central group of the Loo-Choo Islands (Seebohm, Ibis, 1887, p. 176). The latter are intermediate in colour between the Manchurian race of this species and the typical form.

The Indian Great Tit, *Parus atriceps*, has a wide range. It is generally distributed throughout India from the Himalayas to Ceylon. It is also found in Burma, Sumatra, Java, Lombock, Flores, and Hainan. In South China as far north as Foo-chow it appears completely to intergrade with the Manchurian Great Tit, *Parus atriceps minor*, which only differs from it in having the mantle suffused with yellowish green, instead of being pure slate-grey. These intermediate forms were called *Parus commixtus* by Swinhoe (Ibis, 1868, p. 63), and it is to this form that the examples in the Pryer collection from the Loo-Choo Islands belong.

Parus atriceps also intergrades with a northern race which ranges from Afghanistan and Gilgit to Turkestan, South-western Siberia, and Western Mongolia, whence I have several examples collected by General Prjevalski in the oasis of the Urungu River. *Parus atriceps boccharensis* is a desert form: it is rather larger in size, a little paler in colour, and has a much longer tail than the typical form. All three forms differ from *Parus major* in having no trace of yellow on the underparts when adult; but examples of young in first plumage in the Swinhoe collection from South-west Fokien are suffused with yellow on the underparts.

The breeding-range of the Manchurian race of the Indian Great Tit extends from Japan across China as far south as the valley of the Yangtse-kiang, as far west as East Mongolia (whence I have an example collected by General Prjevalski in Kansu), and as far north as the valley of the Ussuri (whence I have an example collected by Monsieur Jankoff). The Manchurian birds are probably migratory, as there are several examples in the Swinhoe collection obtained in winter at Amoy.

The climatic variations of the Great Tit and its allies are very anomalous. *Parus major*, so common in the British Islands, appears to range across Europe and Southern Siberia as far as the Stanovoi Mountains on the shores of the Sea of Okhotsk, with little or no variation in colour. South of the Amoor the yellow suddenly disappears from the underparts, and in South China it gradually disappears from the mantle, leaving the Indian or tropical form white, black, and slate-grey (the usual characteristics of an Arctic race), to be suddenly represented in Persia by the species found in the British Islands. Neither in the east nor in the west does the Common Great Tit intergrade with the Indian Great Tit; and although the Japanese birds are intermediate in the colour of the upper parts, they are not in the least so as regards the colour of the underparts. The Loo-Choo Islands appear to have received their Great Tits from South China. The Japanese Great Tits may have come from the Corean Peninsula, since so far as is known there are no Great Tits in Sakhalien or in the valley of the Amoor north of its junction with the Ussuri. I can see no difference between examples from Yezzo and those from Yokohama.

The probable explanation of this anomalous variation is that the Japanese birds are the modified descendants of *Parus atriceps boccharensis*, which was differentiated as a desert form in Mongolia, and that the true tropical representative of *Parus major* is *Parus monticola*, which ranges from the Himalayas across Southern China to Formosa.

The Manchurian Great Tit is described as the commonest Tit in Japan, abundant everywhere on the mountains in summer and very common in the plains in winter (Jouy, Proc. United States Nat. Mus. 1883, p. 286).

53. PARUS VARIUS*.
(JAPANESE TIT.)

Parus varius, Temminck and Schlegel, Fauna Japonica, Aves, p. 71 (1847).

---

* The name of *Parus varius* having been applied to the *Parus americanus* of Linneus (now known as *Parula americana*) as long ago as 1791 (Bartram, Trav. Florida, p. 292), will probably be rejected by the devotees of the Stricklandian code, who may, if they like, substitute for it the name of *Parus sieboldi*; but I can see no reason whatever for abandoning the name already in use.

The Japanese Tit may always be recognized by its chestnut flanks and buff forehead.

Figures : Temminck and Schlegel, Fauna Japonica, Aves, pl. 35.

The Japanese Tit is supposed to be only a summer visitor to Yezzo, whence there is an example in the Swinhoe collection obtained by Captain Blakiston at Hakodadi in April (Swinhoe, Ibis, 1874, p. 155); but it is a resident in Hondo, whence there are four examples from Yokohama in the Pryer collection.

It has occurred in February in the Corean Peninsula, but it is not known whether it breeds there or not (Taczanowski, Proc. Zool. Soc. 1887, p. 604).

It is a favourite cage-bird with the Japanese. Its note is described as resembling that of the Little Woodpecker. Like the other Tits it frequents the pines, but it is much less sociable and is generally seen alone or in pairs (Jouy, Proc. United States Nat. Mus. 1883, p. 287).

The Japanese Tit is represented in Formosa by a smaller race, which is figured in Gould's 'Birds of Asia,' ii. pl. 49, and was originally described as *Parus castaneiventris* (Gould, Proc. Zool. Soc. 1862, p. 280).

An example of the Formosan race of the Japanese Tit was collected by Mr. Namiye at Nagogatake, in the central group of the Loo-Choo Islands, on the 16th of March (Stejneger, Proc. United States Nat. Mus. 1886, p. 650).

The Formosan Tit is only known from three or four examples which vary slightly in size (wing from carpal joint 2·35 to 2·4 inches). The example obtained on the Loo-Choo Islands is rather larger (wing 2·64 inches), and it is probable that a larger series would bridge over the distance between it and the Japanese species, which is larger still (wing 2·8 to 3·1 inches). The example from the Loo-Choo Islands is described as agreeing with the Formosan race in having less chestnut on the upper mantle, as being intermediate between the two in having indications of a creamy patch on the upper breast, and as agreeing with pale examples of the Japanese race in the colour of its flanks.

The Formosan Tit and its close ally the Japanese Tit appear to have no near relations; but it is possible that the latter is the Japanese representative of the Blue Tit (*Parus caeruleus*), of which the Azure Tit (*Parus cyanus*) is the Siberian representative, and *Parus ultramarinus* the North-African representative. All these species, which appear to be quite distinct from each other, agree with the Japanese Tit in having white foreheads. It is possible, however, that the white forehead is not an important character, and that the affinities of the Japanese Tit may be with *Parus rufonuchalis*, *Parus melanolophus*, and *Parus beavani*.

54. ACREDULA CAUDATA.
(CONTINENTAL LONG-TAILED TIT.)

Parus caudatus, Linnæus, Syst. Nat. i. p. 342 (1766).

The Continental Long-tailed Tit may be recognized by its long tail and white head and neck.

Figures: Dresser, Birds of Europe, iii. pl. 104.

The Continental Long-tailed Tit is a common resident in the island of Yezzo, but is not known to cross the Strait of Tsugaru to the more southerly islands. I have an example collected by Captain Blakiston at Hakodadi in February, and a second collected by Mr. Henson in the same locality on the 25th of October (Swinhoe, Ibis, 1874, p. 156).

The Continental Long-tailed Tit is only an occasional visitor to the British Islands, but its breeding-range extends across Europe and Southern Siberia to Japan.

55. ACREDULA TRIVIRGATA.
(JAPANESE LONG-TAILED TIT.)

Parus trivirgatus, Temminck and Schlegel, Fauna Japonica, Aves, p. 71 (1847).

The Japanese Long-tailed Tit has a black stripe on each side of the crown and on the lores.

Figures: Temminck and Schlegel, Fauna Japonica, Aves, pl. 34; Dresser, Birds of Europe, iii. pl. 103.

The Japanese Long-tailed Tit is a resident in Southern Japan, breeding on the mountains and wintering in the plains (Blakiston and Pryer, Ibis, 1878, p. 235). It is not known to have occurred in Yezzo, where its place is taken by the Continental Long-tailed Tit (*Acredula caudata*). There is a large series in the Pryer collection from Yokohama, and Mr. Ringer has obtained it at Nagasaki, whence he has sent examples to the Norwich Museum.

A single example has been recorded from the Corea (Taczanowski, Proc. Zool. Soc. 1887, p. 601), where it may possibly be only an occasional winter visitor.

The supposed intergradation of *Acredula caudata* and *Acredula rosea* has frequently been recorded in Hesse (Berlepsch, Journ. Orn. 1880, p. 218) and other parts of West Germany (Tauber, Journ. Orn. 1880, p. 421); but it has not been suggested that *Acredula caudata* intergrades with *Acredula trivirgata*. On the other hand, the dark markings on the lores are often found in British examples, and are often very obscure in Japanese examples, so that *Acredula rosea* and *Acredula trivirgata* may possibly intergrade. In Central Siberia there is so much white on the tertials and on the outer webs of the secondaries, and the length of the tail is so great, that it is difficult to avoid recognizing an *Acredula caudata sibirica*.

56. ÆGITHALUS CONSOBRINUS.
(SWINHOE'S PENDULINE TIT.)

Ægithalus consobrinus, Swinhoe, Proc. Zool. Soc. 1870, p. 133.

Swinhoe's Penduline Tit has a grey crown, with a rufous band round the nape and a black band across the forehead, which extends through the eye and across the ear-coverts.

Figures: Gould, Birds of Asia, ii. pl. 70.

Swinhoe's Penduline Tit is a resident at Nagasaki, whence I have seen two males and two females collected by Mr. Ringer. I am indebted to Captain Blakiston for having one of the former in my collection (Seebohm, Ibis, 1884, p. 37), and to Mr. Ringer for one of the latter. The type from Central China is also in the Swinhoe collection. Having seen five examples of this rare bird, it appears to me impossible not to recognize its apparent distinctness, a conclusion to which Dr. Stejneger has also arrived (Proc. United States Nat. Mus. 1886, p. 389). It is, however, very probable that it may eventually be proved to be the female of the western species, or be degraded to subspecific rank.

The great variability of this species and the close resemblance of the alleged males from China and Japan with the females of the European species, and the fact that the alleged females of the Japanese birds are apparently immature, are all arguments against the validity of the species. The characters relied upon are the smallness of the ear-patch (which agrees with that of the typical female), the much narrower black frontal band (which is not narrower than usual in my skins from China), the absence of the chestnut on the forehead (which is scarcely perceptible in an example from Asia Minor), the white eye-stripe (which may be a good character), the buff throat (which is white in the typical form), and the absence of the concealed chestnut bases of the breast-feathers (which has every appearance of being a really good character).

57. TROGLODYTES FUMIGATUS.
(JAPANESE WREN.)

Troglodytes fumigatus, Temminck, Man. d'Orn. iii. p. 161 (1835).

The Japanese Wren principally differs from the Common Wren in the colour of the underparts, which is darker and more rufous than that of the underparts of the western species.

Figures: Sharpe, Cat. Birds Brit. Mus. vi. pl. 16. fig. 2.

The Japanese Wren is a resident in all the Japanese Islands. Captain Blakiston has sent examples from Yezzo (Swinhoe, Ibis, 1874, p. 152); there are twelve in the Pryer collection from Yokohama; I have one collected by Mr. Heywood Jones on Fuji-yama in summer, and three collected by Mr. Ringer at Nagasaki in winter, where it was also procured by the Siebold Expedition (Temminck and Schlegel, Fauna Japonica, Aves, p. 69).

It is common in Central Hondo, near the peaks of the high mountains in summer, and frequents bushes near streams in the lowlands in winter. Its song is described as low, delicious, and warbling, exactly like that of the American Winter Wren (Jouy, Proc. United States Nat. Mus. 1883, p. 287).

The Japanese representative of the Common Wren is on an average a paler and more rufous form than the Himalayan race, but the darkest example from Nagasaki is scarcely distinguishable from the palest from Sikkim, both dated January.

In the colour of the upper parts the various species of Wrens completely intergrade. It is impossible to draw a line anywhere between the palest desert forms from Algeria or Turkestan, and the darkest tropical forms from Cashmere and Sikkim. The barring is on an average most conspicuous in the tropical form, but examples showing the extreme amount of barring occur in France, Norway, Mongolia, and other localities.

In the colour of the underparts it seems possible to draw a line, which may be a natural one. *Troglodytes parvulus* and its subspecific allies form a pale group, which range across Europe to Algeria in the south and to Russian Turkestan in the east; whilst *Troglodytes fumigatus* and its subspecific allies form a dark group, which range from Japan across Asia to the Himalayas and the Altai Mountains. On Bering Island a pale form occurs, *Troglodytes fumigatus pallescens*, which probably came from Alaska. The underparts are generally much more barred in the dark species than in the pale one, but the amount of individual variation in this respect is very great. The variations in the size of the bill and feet are considerable, but no genetic value can be attached to them. The large bill and feet characteristic of the races of St. Kilda, the Faroe Islands, Bering Island, and the Kuriles have probably each been independently acquired.

The Kurile Island race of the Common Wren is remarkable for its long bill, the exposed culmen measuring ·55 inch (14 millimetres), a length exceeding that of the Faroese Wren and equalling that of the Commander Island Wren. The length of the hind toe, ·4 inch (10 millimetres), or with the claw ·6 inch (15½ millimetres), agrees with that of the Commander Island Wren and that of the St. Kilda Wren, but exceeds that of the European Wren and that of the Japanese Wren. In colour it agrees with the least rufous of the Japanese Wrens, but is much less rufous than the ordinary type of that race, and much more rufous than the Commander Island Wren. It is more rufous than typical examples from Europe, but scarcely differs in colour from an example collected by General Prjevalski in the Chuan-Che range of mountains in Mongolia. The bars on the upper parts are not quite obsolete on the mantle, and on the breast are well marked, but this is probably only a sign of summer plumage.

I have two examples collected by Mr. Snow in June on Uschisir, one of the small central islands of the Kurile range, and Dr. Stejneger has described a third, to which he has given the name of *Troglodytes fumigatus kurilensis* (Stejneger, Proc. United States Nat. Mus. 1888, p. 548), a name which I had already given it in manuscript.

58. CERTHIA FAMILIARIS.
(COMMON CREEPER.)

TYPICAL FORM.
Certhia familiaris, Linneus, Syst. Nat. i. p. 184 (1766).

ARCTIC FORM.
Certhia scandulaca, Pallas, Zoogr. Rosso-Asiat. i. p. 432 (1826).

The Creeper has a curved bill like a Hoopoe, and stiff pointed tail-feathers like a Woodpecker.

Figures: Dresser, Birds of Europe, iii. pl. 122 (typical form).

The Creeper is common in the woods and plantations near Hakodadi (Whitely, Ibis, 1862, p. 196), and appears to be a resident on the Kuriles, as well as in the other Japanese Islands. I have an example collected by Mr. Snow on the Kurile Islands, and there is an example in the Swinhoe collection obtained by Captain Blakiston at Hakodadi in February (Swinhoe, Ibis, 1874, p. 152). I have also two examples collected by Mr. Henson at Hakodadi on the 13th of April, and there is an example in the Pryer collection from Yokohama.

The Common Creeper is one of those unsatisfactory birds that seem to vary with every variation of climate; so that the ornithologist is obliged either to become a "lumper" of the old school of binomialists, and confuse all the races together under the name of *Certhia familiaris*, or to become a "splitter" of the new school of trinomialists, and give to each geographical race a third name, until he has made so many that he is frightened at the ever-increasing number.

Examples from Central Siberia are so much whiter than the typical European form that it is impossible not to allow *Certhia familiaris scandulaca* subspecific rank.

Examples from Yokohama agree very closely with those from North China, Kansu, the Chuan-Che mountains, and Asia Minor. They are too white to be regarded as belonging to the typical form, but they are not white enough to belong to the Arctic race. Examples from the Kurile Islands and from Yezzo are quite as difficult to place. They agree very closely with examples from the Amoor, and are distinctly whiter than those from Southern Japan; but they are not so white as examples from Central Siberia.

The Common Creeper is a circumpolar species, and its range extends across North America, where it seems to be subject to the same climatic variation. Tropical forms occur in the Himalayas and in Mexico.

59. SITTA CÆSIA*.
(NUTHATCH.)

TYPICAL FORM.
Sitta cæsia, Wolf, Taschenbuch, i. p. 128 (1810).

ARCTIC FORM.
Sitta uralensis, Lichtenstein, Gloger's Handb. Vög. Deutschl. pp. 377, 358 (1834).

EASTERN SEMI-ARCTIC FORM.
Sitta amurensis, Swinhoe, Proc. Zool. Soc. 1871, p. 350.

KAMTSCHATKAN FORM.
Sitta albifrons, Taczanowski, Bull. Soc. Zool. France, 1882, p. 385.

The Nuthatch has the bill of a Woodpecker with the tail of a Tit.

Figures: Dresser, Birds of Europe, iii. pl. 119 (typical form), pl. 118 (Arctic form, but feet coloured wrong).

The Nuthatch is a resident in all the Japanese Islands (Blakiston and Pryer, Ibis, 1878, p. 236). I have an example collected by Mr. Snow on the Kurile Islands. Dr. Henderson obtained it at Hakodadi in October 1857 (Cassin, Proc. Acad. Nat. Sci. Philad. 1858, p. 195); and there are two examples in the Swinhoe collection from Yezzo (Swinhoe, Ibis, 1874, p. 152). There are seven examples in the Pryer collection from Yokohama.

The range of the Nuthatch extends from the British Islands across Europe and Siberia to Japan and China.

The local races of the Common Nuthatch are the despair of the ornithological nomenclator.

Sitta cæsia, a smallish bird (wing 3·15 to 3·3 in.), with chestnut breast, in the British Islands and Western Europe, intergrades with *Sitta cæsia homeyeri* in Pomerania, the Baltic Provinces of Russia,

---

* According to the law of priority *Sitta europæa* ought to be accepted as the typical form, but to avoid the absurdity of calling a Japanese bird *Sitta europæa uralensis*, it is necessary to make *Sitta cæsia* the typical form. To do otherwise would be misleading.

Poland, and the Crimea, which in turn intergrades on the one hand with

Sitta cæsia europæa in Scandinavia, a slightly larger bird (wing 3·35 to 3·5 in.) with nearly white breast; and on the other with

Sitta cæsia uralensis in Northern Siberia, a small bird (wing 2·9 to 3·2 in.) with pure white underparts, which intergrades on the one hand with

Sitta cæsia albifrons in Kamtschatka; a small bird, with the head and nape slightly paler, the forehead white, and the greater wing-coverts tipped with white.

On the other hand it intergrades with

Sitta cæsia amurensis, which only differs from *Sitta cæsia homeyeri* in having the smaller dimensions of *Sitta cæsia uralensis*. This race inhabits the valley of the Amoor, Manchuria, and Southern Japan, and doubtless intergrades with

Sitta cæsia sinensis in China and Eastern Thibet, a race which only differs from the typical form in being slightly smaller (wing 2·9 to 3·0 in.).

The Nuthatches from Southern Japan have the throat and upper breast white, and the lower breast and belly pale chestnut, and may be regarded as *Sitta cæsia amurensis*.

Those from Yezzo are intermediate between *Sitta cæsia uralensis* and *Sitta cæsia albifrons*, but are so near to the former that they may be reasonably included in that race, though Dr. Stejneger has called them *Sitta amurensis clara* (Stejneger, Proc. United States Nat. Mus. 1886, p. 393). Those from the Kurile Islands are also intermediate between *Sitta cæsia uraleneis* and *Sitta cæsia albifrons*, but are so near the latter that even Dr. Stejneger has not thought it necessary to create a new subspecies for their reception.

CORVINÆ.

Sexes alike; young in first plumage the same, but paler; first primary more than half the length of the second; nostrils concealed by feathers. Very doubtfully distinct from the Parinæ.

The Corvinæ scarcely number 200 species, of which 12 occur in Japan. The subfamily may be regarded as cosmopolitan.

60. CORVUS CORAX.
(RAVEN.)

Corvus corax, Linnæus, Syst. Nat. i. p. 155 (1766).

The Raven is a large bird (wing from carpal joint 19½ to 16 inches) with a very thick bill. The feathers of the mantle resemble those of the Carrion-Crow in being glossed with greenish purple and in having nearly white bases. The feathers of the upper breast as well as those of the throat are lanceolate.

Figures: Dresser, Birds of Europe, iv. pl. 265. fig. 1.

The Raven breeds in the Kurile Islands, where its presence was long ago recorded (Pallas, Zoogr. Rosso-Asiat. i. p. 380), but it is not known to have occurred in Japan proper. I have an example which Captain Blakiston sent me (Seebohm, Ibis, 1879, p. 31), and there is a second in the Pryer collection, both obtained by Mr. Snow in the Kurile Islands.

The Raven is a circumpolar species, inhabiting the whole of the Palæarctic and Nearctic Regions, except Japan, where its place appears to be taken by *Corvus macrorhynchus*, the Raven of the Oriental Region.

The Raven from the Commander Islands has been described by Dr. Dybowski and by Dr. Stejneger as a distinct species, on the ground of a difference in the wing-formula. An example from the Kurile Islands and four examples from the valley of the Yenesay agree in having the 4th primary the longest, the 3rd a little shorter, and the 5th a trifle shorter still; whilst the 1st primary is between the 7th and 8th. This agrees with European examples.

61. CORVUS MACRORHYNCHUS.
(ORIENTAL RAVEN.)

Corvus macrorhynchus, Wagler, Systema Avium, p. 311 (1827).

The Japanese race of the Oriental Raven is intermediate in size, and in some other characters, between the Common Raven and the Common Crow (wing from carpal joint 14½ to 13½ inches); but its bill is as thick as that of the Common Raven (upper mandible at nostrils ·8 inch high). The feathers of the mantle differ from both these species in having a greener gloss and much darker bases.

The feathers of the throat are lanceolate, but not those of the upper breast.

Figures: Temminck and Schlegel, Fauna Japonica, Aves, pl. 39 B.

The Japanese Raven is a resident on all the Japanese Islands, and is the common Crow of Japan. I have examples collected by Mr. Snow on the Kurile Islands. In the Swinhoe collection is an example collected by Mr. Whitely at Hakodadi (Whitely, Ibis, 1867, p. 200); and in the Pryer collection there are several examples from Yokohama. In the British Museum there is an example from Nagasaki; and in the Pryer collection there is one from the central group of the Loo-Choo Islands. Mr. Holst procured it on the Bonin Islands (Seebohm, Ibis, 1890, p. 97).

It is said to be more of a maritime species than the Carrion-Crow, and to have much harsher and more varied notes (Jouy, Proc. United States Nat. Mus. 1883, p. 302).

The Oriental Raven represents the Common Raven in India, Ceylon, the Burma peninsula, the islands of the Malay Archipelago, South China, and Japan; but in North China, the Kurile Islands, and in Eastern Siberia both species occur, the Japanese race of the Oriental Raven being found in company with the Common Raven. The typical form of the Oriental Raven (often called the Indian Jungle-Crow) is on an average smaller than its Siberian and Japanese ally (wing $13\frac{1}{2}$ to $11\frac{1}{2}$ inches), and becomes smaller still in Ceylon (wing $12\frac{1}{4}$ to $10\frac{3}{4}$ inches). Examples from Siberia, Japan, China, and Ceylon are supposed always to have dark grey bases to the feathers of the back; those from the islands of the Malay Archipelago are supposed to have nearly white bases to these feathers; but Mr. Hume has conclusively shown (Stray Feathers, 1877, p. 461) that both forms are found in India.

Corvus japonensis (Bonaparte, Consp. Generum Avium, i. p. 386) was described in 1850 from Japan, but it appears to be only a subspecific form of the Indian Jungle-Crow, and may therefore be called *Corvus macrorhynchus japonensis*. The example from the Loo-Choo Islands is smaller (wing $12\frac{1}{2}$ inches) and has a more slender bill (height of upper mandible at centre of nostrils ·5 inch), and agrees exactly with the type in the Swinhoe collection of *Corvus colonorum* from Formosa. Intermediate forms (wing $13\frac{1}{2}$; upper mandible ·55 inch) occur near Yokohama, so that the Loo-Choo race, if it be regarded as separable, must be known as *Corvus macrorhynchus levaillanti* (Lesson, Traité d'Orn. p. 328), a name dating from 1831, whereas Swinhoe's name only dates from 1864.

62. CORVUS CORONE.
(CARRION-CROW.)

Corvus corone, Linnæus, Syst. Nat. i. p. 155 (1766).

The Carrion-Crow is almost as large as the Japanese Raven (wing from carpal joint 11 to 12½ inches), but it has a very much slenderer bill. The feathers of the mantle are glossed with greenish purple and have pale grey bases.

Figures: Dresser, Birds of Europe, iv. pl. 263. fig. 1.

The Carrion-Crow is a resident in all the Japanese Islands, but is not so abundant as its thicker-billed ally. I have three examples collected by Mr. Snow on the Kurile Islands. In the Swinhoe collection there are two examples from Hakodadi, one collected by Captain Blakiston (Swinhoe, Ibis, 1874, p. 159) and the other by Mr. Whitely; and in the Pryer collection there are three examples from Yokohama. Mr. Ringer has procured it at Nagasaki, where the examples obtained by the Siebold Expedition were doubtless procured (Temminck and Schlegel, Fauna Japonica, Aves, p. 79).

The range of the Carrion-Crow extends westward from Japan across Siberia as far as the valley of the Yenesay; thence it continues along the mountains of Southern Siberia across Turkestan, the Caucasus, and the valley of the Danube into Europe, west of the Elbe, as far south as Spain, and as far north as the British Islands.

The Carrion-Crow of East Asia has been separated from that of Europe by Eversmann, Dybowski, and others, on the ground of its larger size, more rounded tail, and more brilliant plumage. Examples from Japan vary in length of wing from carpal joint from 12½ to 14 inches; the outer tail-feathers are from ½ to 1 inch shorter than the longest, and the gloss of the back looks green when contrasted with that of the Rook, and purple when contrasted with that of the Japanese Raven. It can scarcely be regarded even as subspecifically distinct.

The attempt to place the Raven, Crow, Jackdaw, and Rook each in a separate genus has been almost universally abandoned, much to the credit of ornithologists, and greatly to the advantage of ornithology, which has been exposed to much well-deserved derision from the invention of so many pseudo-genera.

63. CORVUS DAURICUS.
(PALLAS'S JACKDAW.)

Corvus dauuricus, Pallas, Reise Russ. Reichs, iii. p. 694 (1776).

Pallas's Jackdaw is about the size of its British representative (wing $9\frac{1}{4}$ to 9 inches), but it has a white collar round the neck; the breast, belly, and flanks are white, and the bases of the feathers of the mantle are dark grey.

Figures: Temminck and Schlegel, Fauna Japonica, Aves, pl. 41.

Pallas's Jackdaw appears to be a rare bird in Japan, and is confined (so far as is known) to the extreme south. I have never seen a Japanese example except those obtained by Dr. Siebold, probably near Nagasaki, and now in the Leyden Museum. It is said to have occurred near Yokohama (Blakiston and Pryer, Ibis, 1878, p. 232).

The range of Pallas's Jackdaw extends from Eastern Siberia to North China and Eastern Mongolia.

64. CORVUS NEGLECTUS.
(SWINHOE'S JACKDAW.)

Corvus neglectus, Schlegel, Bijdr. Dierk. Amsterd., fol., art. *Corvus*, p. 16.

Swinhoe's Jackdaw very closely resembles its British representative, but its belly is dark grey instead of greyish black, and the first eight secondaries are nearly equal in length.

Figures: Temminck and Schlegel, Fauna Japonica, Aves, pl. 40.

Swinhoe's Jackdaw is a rare bird in collections. The only examples which I have seen from Japan are the types obtained by Dr. Siebold, presumably near Nagasaki, and now in the Leyden Museum, and an example from Osaka in the extreme south of Hondo, and now in the Hakodadi Museum (Blakiston and Pryer, Trans. As. Soc. Japan, 1882, p. 143). The latter specimen is paler on the underparts than typical examples in the Swinhoe collection from China, and resembles an example in the same collection from Shanghai, which Swinhoe regarded as a hybrid between *Corvus dauricus* and *Corvus neglectus* (Swinhoe, Proc. Zool. Soc. 1871, p. 383).

The range of Swinhoe's Jackdaw extends from Eastern Siberia to North China.

65. CORVUS PASTINATOR.

(EASTERN ROOK.)

Corvus pastinator, Gould, Proc. Zool. Soc. 1845, p. 1.

The Eastern Rook is slightly smaller than the Common Crow (wing from carpal joint $12\frac{3}{4}$ to $11\frac{1}{4}$ inches), and its bill is quite as slender. The feathers of the mantle are glossed with greenish purple, and have dark-grey bases. In adult examples the forehead and lores are bare of feathers.

The Eastern Rook is a resident in Southern Japan, but has not been known to have occurred in Yezzo. Captain Blakiston sent me an example from Yokohama for examination (Seebohm, Ibis, 1879, p. 31); and there is an example in the British Museum collected by Captain St. John at Nagasaki, where those procured by Dr. Siebold, and erroneously recorded as *Corvus frugilegus*, were probably obtained (Temminck and Schlegel, Fauna Japonica, Aves, p. 79).

The range of the Eastern Rook extends from Irkutsk, across South-eastern Siberia, to North China and Japan. It is not known that either species occurs in the valley of the Western Yenesay or in that of the Obb, but the range of the European Rook extends eastwards to the valley of the Irtisch.

The Rooks appear to be much less hardy than the Crows, as their range does not extend nearly so far north; but if we may judge from the bareness of their nostrils, their food is much more exclusively obtained in the ground, and they are consequently soon starved out by a frost. As the mean temperature of January in Hakodadi is seven degrees below freezing-point, whilst in Yokohama it is seven degrees above it, there is no difficulty in explaining why the Eastern Rook is not a resident in Yezzo.

The Western Rook, *Corvus frugilegus*, agrees with the Eastern Rook, and differs from the Common Crow, in having dark bases to the feathers of the mantle. The Western Rook when adult has the throat, as well as the forehead and lores, bare of feathers, which is never the case with its Eastern ally. All three species differ in the colour of the head—in *Corvus frugilegus* the purple of the crown is glossed with blue, in *Corvus pastinator* with red, and in *Corvus corone* with green.

66. NUCIFRAGA CARYOCATACTES.
(NUTCRACKER.)

Corvus caryocatactes, Linneus, Syst. Nat. i. p. 157 (1766).

The Nutcracker is a well-known bird, about the size of a Jay, dark brown, spotted with white.

Figures: Dresser, Birds of Europe, iv. pl. 252.

The Nutcracker is a resident in Japan, both in Yezzo (Blakiston, Ibis, 1862, p. 326) and on the mountains of Southern Japan. There are six examples in the Pryer collection from Fuji-yama, where they occur at an elevation of about 5000 feet, descending nearer the plains in winter (Jouy, Proc. United States Nat. Mus. 1883, p. 302).

The Nutcracker is an occasional visitor to the British Islands, and its breeding-range extends across Europe and Siberia to Japan. In Siberia the white spots, both on the upper and under parts and on the ends of the tail-feathers, are much more developed than they are in Europe and somewhat more so than they are in Japan. In Siberian examples the bill is much slenderer than in European and Japanese examples, and consequently they have been called *Nucifraga caryocatactes leptorhynchus* (Seebohm, Ibis, 1888, p. 236).

67. CYANOPOLIUS CYANUS.
(EASTERN BLUE MAGPIE.)

Corvus cyanus, Pallas, Reise Russ. Reichs, iii. p. 694 (1776).

The Eastern Blue Magpie is much smaller than the Common Magpie (wing from carpal joint 5¼ to 5¾ inches). The head is black above, the body grey above and nearly white below, and the tail and most of the wing are azure blue.

Figures: Temminck and Schlegel, Fauna Japonica, Aves, pl. 42.

The Eastern Blue Magpie is a resident in Southern Japan. There is an example in the Swinhoe collection from Yokohama (Swinhoe, Ibis, 1877, p. 145); and there are four examples in the Pryer collection from the same locality. It is rather remarkable that it has not been recorded from Yezzo, as it is common in the valley of the Amoor and in North China.

The Spanish Blue Magpie, *Cyanopolius cooki*, is so closely allied to the Eastern Blue Magpie, that it is doubtful whether they are more than subspecifically distinct. The Spanish bird is browner, and the white tips to the central tail-feathers only occur accidentally. It has also been recorded from Morocco. It has no other nearer ally than the species belonging to the genera *Pica* and *Urocissa*. The Spanish Blue Magpie was unknown to Temminck in 1820, but is included in the third volume of his 'Manuel d'Ornithologie' published in 1835, having been mentioned in 1827 in Wagler's 'Systema Avium' from an example in the Paris Museum. That the area of distribution of these two species was once continuous is a self-evident proposition. That the range of the Blue Magpie once extended from Spain to Japan, but that the species has been exterminated in the rest of Southern Europe and in Western Siberia, is a possible but highly improbable hypothesis. That once upon a time there was an emigration of Blue Magpies from Eastern Siberia to Western Europe, as there has been twice within the present century of Sand-Grouse, is a much more probable theory; but the most probable explanation of this anomalous fact of geographical distribution is the obvious one that the Chinese Blue Magpie was brought from China to Spain, precisely in the same manner as the Chinese Ringed Pheasant was introduced into England. It has probably become browner since its introduction in consequence of the greater rainfall of Spain, and it may have lost the white tips to the centre tail-feathers by protective selection. The young in first plumage of the Japanese Blue Magpie have not only *all* the tail-feathers but also the *tertials* tipped with white.

68. GARRULUS BRANDTI.

(BRANDT'S JAY.)

Garrulus brandtii, Eversmann, Add. Pallas. Zoogr. iii. p. 8 (1843).

Brandt's Jay differs from the Common and Japanese Jays in having the ground-colour of the forehead, crown, nape, and mantle chestnut-buff, and the outer webs of the primaries pale grey.

Figures: Dresser, Birds of Europe, iv. pl. 255.

Brandt's Jay is a resident in Yezzo, but has not been known to cross the Straits of Tsugaru. There are several examples in the Swinhoe collection from Hakodadi (Swinhoe, Ibis, 1875, p. 450), and one in the Pryer collection from the same locality.

69. GARRULUS JAPONICUS.
(JAPANESE JAY.)

Garrulus glandarius japonicus, Temminck and Schlegel, Fauna Japonica, Aves, p. 83 (1847).

The Japanese Jay differs from Brandt's Jay in having black lores, in having the ground-colour of the forehead and crown nearly white, and in having the outer webs of the primaries nearly white towards the apex and nearly black towards the base.

Figures: Temminck and Schlegel, Fauna Japonica, Aves, pl. 43.

The Japanese Jay is a common resident in Southern Japan, but has not been known to occur in Yezzo. There are several examples in the Swinhoe collection from Yokohama (Swinhoe, Ibis, 1877, p. 144), and a large series in the Pryer collection from the same locality. Mr. Ringer has obtained it at Nagasaki, whence he has sent examples to the Norwich Museum. It has not been recorded from any part of the Asiatic continent.

It is quite a mistake to suppose that the European Jay is more nearly allied to the Japanese than to the Siberian or Chinese Jays. It certainly resembles it most in the colour and markings of the crown; but this is a very variable character in adults, and still more so in young in first plumage. The Japanese Jay is unique in having black lores, the other three species having them coloured like the crown. The Japanese and Chinese species agree in having the terminal portions of the outer webs of the primaries white and the basal portions black, whilst the European and Siberian species agree in having the whole outer webs of the primaries for the most part grey. The Chinese species possesses two unique characters: it has no black streaks on the crown, and the central portion of the outer webs of its first four secondaries are barred with white, black, and blue. The European Jay is nearest allied to the Siberian Jay, and, according to Bogdanow, intergrades with it.

70. GARRULUS SINENSIS.
(CHINESE JAY.)

Garrulus sinensis, Gould, *fide* Swinhoe, Proc. Zoo Soc. 1863 p. 304.

The Chinese Jay has all the body-feathers uniform vinaceous, shading into white on the upper and under tail-coverts, except a broad black malar stripe. The central portions of the outer webs of the first five secondaries are barred with white, black, and blue, and the basal portions of the outer webs of the primaries are black.

The Chinese Jay is very closely allied to the Himalayan Jay, *Garrulus bispecularis*, and to the Formosan Jay, *Garrulus taivanus*. In the Himalayan Jay the nasal bristles are of the same colour as the crown; in the Chinese Jay they are tipped with black; and in the Formosan Jay they are entirely black. In this character the Chinese Jay agrees with the Japanese, Siberian, and European Jays.

There is an example of the Chinese Jay in the British Museum which is labelled as having been bought at Stevens's Sale-rooms in 1865 as part of a collection of birds procured by Captain St. John at Nagasaki (Sharpe, Cat. Birds Brit. Mus. iii. p. 101). This is the only record of the occurrence of this species in Japan.

The Chinese Jay has a very restricted range, and, with the above-named exception, has only been recorded from South China. The Himalayan Jay is found as far east as Eastern Thibet, and on the island of Formosa a third allied species occurs.

71. PICA CAUDATA.
(COMMON MAGPIE.)

Pica caudata, Gerini, Orn. Meth. Dig. ii. p. 40 (1769).

The Common Magpie is black and white; the black more or less bronzed with green and purple, and the white confined to the scapulars, the belly, the rump, and the centres of the quills.

Figures: Dresser, Birds of Europe, iv. pl. 260.

The Common Magpie was long ago recorded both from the Kurile Islands and from Japan (Pallas, Zoogr. Rosso-Asiat. i. p. 390), but it has not been obtained from either locality of recent years. It is not known to have occurred in Yezzo or in the main island of Japan, but there can be little doubt that it breeds on Kiusiu (Blakiston, Amended List of the Birds of Japan, p. 48). Dr. Rein obtained the nests and eggs and found the bird to be common near the Shimbara Gulf, east of Nagasaki; and skins were brought to Leyden by the Siebold Expedition (Temminck and Schlegel, Fauna Japonica, Aves, p. 81).

It is probably the typical form of the Magpie which is found in Southern Japan. I am indebted to the kindness of Mr. Dresser for several examples of the Magpie said to have been procured by Mr. Snow on the Kurile Islands. Of these one might be called *Pica caudata leucoptera*, if the recognition of such an intermediate form be allowable; the others are *Pica caudata kamtschatkensis*, and probably came from Kamtschatka. It is, however, possible that this local race may migrate to the Kurile Islands in autumn.

LANIINÆ.

Sexes alike or nearly so; first primary about half the length of the second; young in first plumage transversely barred on the underparts, and in some genera on the upper parts also. Rictal bristles well developed.

If the Laniinæ be regarded as consisting of the Shrikes, the Cuckoo-Shrikes, and the Swallow-Shrikes, to the exclusion of the Drongo Shrikes (which may possibly be a natural arrangement), the subfamily will contain about 300 species, of which 7 have been recorded from the Japanese Empire. With the exception of the Arctic Region, they are distributed throughout the Old World. The true Shrikes, being the most arctic, have found their way across Bering Straits and have spread over North America.

72. LANIUS MAJOR.
(PALLAS'S GREY SHRIKE.)

Lanius major, Pallas, Zoogr. Rosso-Asiat. i. p. 401 (1826).

Pallas's Grey Shrike has the crown and back grey, shading into white on the forehead and upper tail-coverts.

The claim of Pallas's Grey Shrike to be regarded as a Japanese bird rests upon a single example procured by Captain Blakiston near Hakodadi about the year 1873 (Seebohm, Ibis, 1884, p. 37).

Pallas's Grey Shrike breeds in Southern Siberia from the Ural Mountains to Kamtschatka, and is an occasional winter visitor to the British Islands.

73. LANIUS MAGNIROSTRIS.
(THICK-BILLED SHRIKE.)

Lanius magnirostris, Lesson, Zool. Voy. Indes-Or. p. 251 (1831).

The Thick-billed Shrike has the crown and nape grey, the back, rump, and upper tail-coverts chestnut, barred with black, and the tail plain russet.

Figures: Walden, Ibis, 1867, pl. 6.

The Thick-billed Shrike is a very rare bird in Japan. There is an example in the Pryer collection from Yokohama (Seebohm, Ibis, 1884, p. 37); and a second example was obtained by Mr. Jouy on Fuji-yama during July (Blakiston, Am. List Birds of Japan, p. 49).

The Thick-billed Shrike breeds on the shores of the Japanese Sea, north of Vladivostok (Taczanowski, Journ. Orn. 1876, p. 197), and in Central China; it winters in the Malay Peninsula.

74. LANIUS SUPERCILIOSUS.
(JAPANESE RED-TAILED SHRIKE.)

Lanius superciliosus, Latham, Index Orn. Suppl. p. xx (1801).

The Japanese Red-tailed Shrike has the crown, nape, back, rump, upper tail-coverts, and tail rich chestnut.

Figures: Walden, Ibis, 1867, pl. 5. fig. 2 (erroneously named *Lanius phœnicurus*).

The Japanese Red-tailed Shrike is probably only a summer visitor to Yezzo and Southern Japan. There is an example from Hakodadi in the Swinhoe collection (Swinhoe, Ibis, 1875, p. 450); and there are eight examples in the Pryer collection from Yokohama.

It winters in some numbers in Malacca. So far is certain, but whether it also breeds in Malacca, or whether any of the Japanese birds remain in Southern Japan during the winter, is unknown.

The Japanese Red-tailed Shrike appears to be a rufous island form of the Indian Red-tailed Shrike, *Lanius cristatus*, with which it almost seems to intergrade; that is to say, that the brightest examples from Eastern Siberia scarcely differ from the dullest examples from Japan, though a series of the one are very different on an average from a series of the other. There can be no doubt that the Siberian birds (the *Lanius phœnicurus* of Pallas), of which I have a large series from the Yenesay, the Amoor, and the Ussuri, are identical with the Indian birds (in their winter-quarters) and not with the Japanese birds. The western form is so pale that it may be called a desert form—*Lanius isabellinus*, breeding in Turkestan and Mongolia, and wintering in Scinde, Arabia, and Abyssinia. This form appears to be specifically distinct, inasmuch as the males have a white bar across the wing formed by the white bases of the primaries. This species is also subject to much climatic variation, and may be separated into two or more subspecies.

The Japanese Red-tailed Shrike makes a large nest in the fork of a small tree or bush, composed of roots, the stems of plants, and dry grass, lined with finer grass and rootlets (Jouy, Proc. United States Nat. Mus. 1883, p. 292). Eggs in the Pryer collection resemble rufous eggs of the Woodchat.

75. LANIUS LUCIONENSIS.
(CHINESE RED-TAILED SHRIKE.)

Lanius lucionensis, Linnæus, Syst. Nat. i. p. 135 (1766).

The Chinese Red-tailed Shrike has the crown and nape grey in the adult and brown in the young, shading into greyish white on the forehead and into chestnut on the upper tail-coverts and tail.

Figures: Walden, Trans. Zool. Soc. ix. pl. 29. fig. 1.

The Chinese race of the Red-tailed Shrike appears to winter in some of the Loo-Choo Islands, and may possibly be a resident there. A nearly adult example in the Pryer collection was procured near Naha, the capital of the largest island (Okinawa) of the central group, in January.

This race breeds in North China and winters in the Philippine Islands and on some of the islands of the Malay Archipelago.

Lanius cristatus is intermediate in colour between *Lanius superciliosus* and *Lanius lucionensis*, but it does not appear quite to intergrade with either. Its breeding-grounds are by no means intermediate between those of its allies, either geographically or climatically. The Rufous Shrikes, like the Swallows, appear to moult in our winter, and it is very probable that the winter-quarters of *Lanius cristatus* may be climatically intermediate between those of its allies.

76. LANIUS BUCEPHALUS.
(BULL-HEADED SHRIKE.)

Lanius bucephalus, Temminck and Schlegel, Fauna Japonica, Aves, p. 39 (1847).

The Bull-headed Shrike has the head and nape rufous, and the tail for the most part grey.

Figures : Temminck and Schlegel, Fauna Japonica, Aves, pl. 14.

The Bull-headed Shrike is a very common resident in Southern Japan, but to Yezzo it is only a summer visitor. There is an example in the Swinhoe collection from Hakodadi (Swinhoe, Ibis, 1875, p. 450) ; and there are eight examples in the Pryer collection from Yokohama. I have three examples collected by Mr. Heywood Jones on Fuji-yama ; and Mr. Ringer has procured it at Nagasaki, and has presented examples to the Norwich Museum from that locality.

The Bull-headed Shrike breeds in the valley of the Ussuri (Taczanowski, Journ. Orn. 1876, p. 197), as well as in Japan and North China, and winters in South China.

Eggs in the Pryer collection resemble blue varieties of eggs of the Woodchat. The nest is described as made of twigs and dead grass, lined with finest grass (Blakiston and Pryer, Trans. As. Soc. Japan, 1882, p. 146).

77. PERICROCOTUS CINEREUS.
(SIBERIAN MINIVET.)

Pericrocotus cinereus, Lafresnaye, Rev. Zool. 1845, p. 94.

The Siberian Minivet is grey above and white below. The forehead is white, the crown and nape black, and the wings and tail are partly black and partly white.

Figures : Gould, Birds of Asia, ii. pl. 11.

It is rather remarkable that the Siberian Minivet has not been recorded from Yezzo. I have an example collected by Mr. Heywood Jones on Fuji-yama, and in the Pryer collection are five examples from the same locality.

This species also breeds in Manchuria, whence examples have been sent by Dybowski from the mouth of the Ussuri River ; and passes through China on migration to winter in the Philippine Islands, Borneo, Sumatra, and the Malay Peninsula.

78. PERICROCOTUS TEGIMÆ.
(LOO-CHOO MINIVET.)

Pericrocotus tegimæ, Stejneger, Proc. United States Nat. Mus. 1886, p. 618.

The Loo-Choo Minivet differs from its Siberian and Japanese ally in having a grey instead of a white breast, and in having the white on the forehead restricted to a narrow line at the base of the bill and over each eye, instead of occupying the whole of the forehead and the front part of the crown as far back as the eyes.

Figures: Stejneger, Zeitschr. ges. Orn. 1887, pl. 2.

The Loo-Choo Minivet was described by Dr. Stejneger from examples collected in March by Mr. Namiye on Okinawa Shima, one of the central group of the Loo-Choo Islands, where it is supposed to be a resident. There are three examples in the Pryer collection from the same locality.

STURNINÆ.

Sexes alike, or nearly so; first primary very small; no rictal bristles; young in first plumage (which is moulted in the first autumn) sometimes very different from the adult, sometimes the same. Autumn plumage of adult changed in spring, not by a moult, but by casting the tips of the feathers.

The Sturninæ probably number about 150 species, if the Waxwings be included in the group. The Starlings are distributed over most parts of the Old World except in the Arctic Region, where their place is taken by the Waxwings. The range of the latter extends across the Arctic Regions of both continents. Two species of each group occur in Japan.

79. STURNUS CINERACEUS.
(GREY STARLING.)

Sturnus cineraceus, Temminck, Planches Coloriées, no. 556 (1835).

The Grey Starling is a large bird (wing from carpal joint about 5 inches), with yellow bill and feet.

Figures: Temminck and Schlegel, Fauna Japonica, Aves, pl. 45.

The Grey Starling is a common summer visitor to Yezzo; but in Southern Japan it is a resident. There are several examples in the Swinhoe collection from Hakodadi (Swinhoe, Ibis, 1874, p. 159), where a single specimen was obtained twenty years previously by the Perry Expedition (Cassin, Exp. Am. Squad. China Seas and Japan, ii. p. 220). There are four examples in the Pryer collection from Yokohama, and it has been obtained by Mr. Ringer at Nagasaki (Blakiston and Pryer, Trans. As. Soc. Japan, 1882, p. 186).

It breeds in holes in fir-trees (Blakiston and Pryer, Ibis, 1878, p. 233). Eggs in the Pryer collection resemble rather small and rather dark eggs of the European Starling.

The range of the European Starling extends from the British Islands across Europe and Siberia as far east as Western Dauria. In Eastern Dauria, the lower valley of the Amoor, and southwards into North China it is replaced by the Grey Starling, which winters in South China, Formosa, and Hainan. There is a smaller resident species in South China, *Sturnus sericeus*, which is said to have been once procured in Japan, but it is very doubtful that it has occurred there in a wild state (Blakiston and Pryer, Trans. As. Soc. Japan, 1882, p. 116).

80. STURNIA PYRRHOGENYS.
(RED-CHEEKED STARLING.)

Lamprotornis pyrrhogenys, Temminck and Schlegel, Fauna Japonica, Aves, p. 86 (1847).

The Red-cheeked Starling is a small bird (wing from carpal joint about 4¼ inches) with dark-blue bill and feet.

Figures: Temminck and Schlegel, Fauna Japonica, Aves, pl. 46, as *Lamprotornis pyrrhopogon*.

The Red-cheeked Starling is one of a small section of Japanese birds belonging to different families, and some of them to different orders, but agreeing in the remarkable peculiarity, that whilst they are, so far as is known, absolutely confined to the Japanese Islands during the breeding-season, they migrate southwards in autumn, some to one country and others to another.

It has been recorded from Eturop, one of the Kurile Islands (Blakiston and Pryer, Trans. As. Soc. Japan, 1882, p. 116); and there are several examples in the Swinhoe collection from Hakodadi

(Swinhoe, Ibis, 1874, p. 159), where it had been found in abundance twenty years previously by the Perry Expedition (Cassin, Exp. Am. Squad. China Seas and Japan, ii. p. 220). There are half a dozen examples in the Pryer collection from Yokohama; and Mr. Ringer has procured it at Nagasaki. It has also been recorded from the southern group of the Loo-Choo Islands (Stejneger, Proc. United States Nat. Mus. 1887, p. 413).

In winter it has been recorded from the Philippine Islands, and from Celebes and Borneo. As it is not known to have occurred in China or Formosa; it appears to take a short cut from the Loo-Choo Islands to the Philippines.

It was described and figured as long ago as 1760 (Brisson, Orn. iii. p. 446), from an example obtained on the Philippine Islands; but Brisson mistook it for a large species of Stonechat, and named it *Ficedula rubetra philippensis major*! Ten or twelve years later Buffon and Montbeillard's great work appeared, accompanied by the 'Planches Enluminées,' in which d'Aubenton figured the adult (pl. 185. fig. 2) and the young (pl. 627. fig. 2); the former being described as a species of Stonechat (Buffon, Hist. Nat. Ois. v. p. 230), and the latter as a species of Blackbird! (Montb. Hist. Nat. Ois. iii. p. 396). In 1783 the name of *Motacilla violacea* was based upon the figure of the adult (Boddaert, Table Planches Enl. p. 11), and that of *Turdus dominicanus* upon the figure of the young (Boddaert, Table Planches Enl. p. 38). The nomenclature of the Red-cheeked Starling was further complicated in 1788, the adult being named *Motacilla philippensis* (Gmelin, Syst. Nat. i. p. 968); and again in 1829, though a step towards its correct systematic position was made when the bird was named *Pastor ruficollis* (Wagler, Syst. Av. p. 92). In 1847 Temminck and Schlegel, in the 'Fauna Japonica,' gave it a new generic name, and two new specific names, one in the text and one on the plate. In 1850 two more names—*Heterornis pyrrhogenys* (said to reside in Japan and Borneo) and *Heterornis ruficollis* (said to be a Philippine species)—were added to the synonymy of this bird (Bonaparte, Conspectus Avium, i. p. 418). Two more names were added in 1870, *Temenuchus pyrrhogenys* and *Temenuchus ruficollis* (Gray, Hand-list of Birds, ii. p. 21); one more in 1872, *Acridotheres pyrrhogenys* (Giebel, Thes. Orn. i. p. 268); one more in 1875, *Sturnia violaceus* (Walden, Trans. Zool. Soc. ix. p. 203); and two more in 1877, *Sturnia pyrrhogenys* and *Sturnia ruficollis* (Giebel, Thes. Orn. iii. p. 550).

The Red-cheeked Starling is a very well marked species, but it is apparently nearest allied to *Sturnia daurica*, a Starling which breeds in Eastern Siberia and Mongolia, and passes through China on migration to winter in the Burma peninsula and Java.

81. AMPELIS GARRULUS.
(BOHEMIAN WAXWING.)

Ampelis garrulus, Linneus, Syst. Nat. i. p. 297 (1766).

In the Bohemian Waxwing the tip of the tail is yellow.

Figures: Dresser, Birds of Europe, iii. pl. 155.

The Bohemian Waxwing is a common winter visitor to Yezzo, and occasionally wanders into Southern Japan. There are two examples in the Swinhoe collection from Hakodadi (Swinhoe, Ibis, 1874, p. 158), and four in the Pryer collection from Yokohama. The examples procured by the Siebold Expedition were probably obtained at Nagasaki (Temminck and Schlegel, Fauna Japonica, Aves, p. 84).

The Bohemian Waxwing breeds in the Arctic Regions of both continents, and sometimes visits the British Islands in great numbers.

82. AMPELIS JAPONICUS.
(JAPANESE WAXWING.)

Bombycivora japonica, Siebold, Hist. Nat. Jap. St. no. 2 (1824).

In the Japanese Waxwing the tip of the tail is red.

Figures: Temminck and Schlegel, Fauna Japonica, Aves, pl. 44, as *Bombycilla phœnicoptera*.

The Japanese Waxwing is a winter visitor to Japan, but is less abundant than the European species. It is very rare in Yezzo (Whitely, Ibis, 1867, p. 200), but there are seven examples in the Pryer collection from Yokohama. Mr. Ringer has presented examples to the Norwich Museum obtained at Nagasaki (Blakiston, Am. List Birds of Japan, p. 50); and it has also been recorded from the central group of the Loo-Choo Islands (Stejneger, Proc. United States Nat. Mus. 1886, p. 648).

The Japanese Waxwing breeds in South-eastern Siberia, and winters in Japan, China, and Formosa.

In winter it often feeds on the berries of the mistletoe (Jouy, Proc. United States Nat. Mus. 1883, p. 291).

MOTACILLINÆ.

First primary obsolete; bill narrow and notched; tertials reaching very nearly or quite to the end of the wing.

The Motacillinæ scarcely number 100 species, but they are nearly cosmopolitan, being absent only from the Pacific Islands. Seven species have been recorded from the Japanese Empire.

83. MOTACILLA LUGENS.
(KAMTSCHATKAN WAGTAIL.)

Motacilla lugens, Pallas, *fide* Kittlitz, Kupfertafeln zur Naturgeschichte der Vögel, p. 16 (1832).

The Kamtschatkan Wagtail always has the sides of the head white, with a black band through the eye; and there is always much white on the inner webs of the first and second as well as of the remaining primaries.

Figures: Seebohm, Ibis, 1878, pl. 9 (male in first summer plumage).

The Kamtschatkan Wagtail was originally described from examples obtained in Kamtschatka. Under the impression that white secondaries were the peculiar character which distinguished the Japanese Wagtail, I named the bird in its first summer plumage (in which the secondaries are grey) *Motacilla amurensis* (Seebohm, Ibis, 1878, p. 315). Soon afterwards I discovered that black cheeks were the peculiar character which distinguished the Japanese Wagtail both summer and winter; and finding a series of Wagtails from Japan with white secondaries and white cheeks, I named them *Motacilla blakistoni* (Seebohm, Ibis, 1883, p. 91). In 1884 Captain Blakiston discovered that my *Motacilla amurensis* was the same species in first summer plumage which in the following summer and for the rest of its life became my *Motacilla blakistoni*. In 1885 both these names were shown to be synonyms of the *Motacilla lugens* of Kittlitz (Sharpe, Cat. Birds Brit. Mus. x. p. 474).

I have a large series of this species from the Kurile Islands, collected by Mr. Snow; from Yezzo, collected by Captain Blakiston;

from Yokohama, collected by Mr. Pryer; and from Nagasaki, collected by Mr. Ringer.

The Kamtschatkan Wagtail breeds in Kamtschatka, the Kurile Islands, and very sparingly in Yezzo. It migrates southward in autumn, and is common in winter at Nagasaki and Yokohama.

On the continent it appears to have a very restricted range, breeding in the valley of the Lower Amoor and wintering in South China.

The changes of plumage of the Kamtschatkan Wagtail have given rise to much confusion and to many synonyms, for some of which I am responsible. First, as regards season: in summer the lower throat is black, in winter white. Second, as regards age: adult birds have the secondaries and the greater part of the primaries white; in immature birds they are brown. The brown quills are not moulted until the second autumn, so that adult birds are always more than a year old. Third, as regards sex: in their first winter plumage the males resemble the females in having grey backs, but differ from them in having some black on the crown. In their first spring plumage the back and rump of both male and female are mottled black and grey, but in the male the black greatly preponderates and in the female the grey. In adult winter plumage the back and rump of the male are more black than grey, but those of the female are entirely grey. In adult summer plumage the back and rump of the male are entirely black, but those of the female are grey slightly mottled with black.

84. MOTACILLA JAPONICA.
(JAPANESE WAGTAIL.)

Motacilla japonica, Swinhoe, Proc. Zool. Soc. 1863, p. 275.

The Japanese Wagtail never has white on the ear-coverts or between the ear-coverts and the eye, the cheeks being grey in young in first plumage and black ever afterwards.

Figures: Temminck and Schlegel, Fauna Japonica, Aves, pl. 25.

The Japanese Wagtail was supposed to be peculiar to Japan, where it is a resident, both on Yezzo and the more southerly islands; but it has been found to breed in the upper valley of the Ussuri (Taczanowski, Journ. Orn. 1876, p. 194). I have one of the examples from the latter locality, besides a series from various localities in Japan. In the Swinhoe collection there are five examples from Yezzo; in the Pryer collection there are six examples from Yokohama; and in the British Museum is an example, presented by Mr. Ringer, from Nagasaki, where the example figured in the 'Fauna Japonica' as *Motacilla lugens* was probably obtained.

The changes which the Japanese Wagtail undergoes differ completely from those of its Kamtschatkan ally. First, as regards season: the plumage of winter does not differ from that of summer; the head is always entirely black, except the white forehead, chin, and eye-stripe. Second, as regards age: the amount of white on the quills of adult birds resembles that of the allied species; but in immature birds there is also a great deal of white on the basal halves of these feathers. Young in first plumage have the entire head, breast, and back grey. Third, as regards sex: females differ only from males in having the back a very dark slate-grey instead of black.

The Pied Wagtails appear to be of all birds the most unfortunate as regards nomenclature. The Japanese Wagtail was originally described in 1835 as *Motacilla lugubris* (Temminck, Man. d'Orn. iii. p. 175), and twelve years later as *Motacilla lugens* (Temminck and Schlegel, Fauna Japonica, Aves, p. 60). In 1863 Swinhoe discovered that both these names were preoccupied (though he did not place either of them correctly), and renamed the Japanese Wagtail *Motacilla japonica*. He, however, fell into the same error that the authors of the 'Fauna Japonica' committed, correctly regarding the plumage there figured as that of adult summer, but incorrectly regarding the Kamtschatkan Wagtail as the winter plumage. It was not until twenty-one years later that the Japanese Wagtail was correctly diagnosed (Seebohm, Ibis, 1884, p. 38). If Swinhoe was a lumper I was a splitter. Swinhoe confounded two species under the name *M. japonica*. I restricted his name to the Japanese Wagtail, and split the Kamtschatkan Wagtail into two supposed species, *M. amurensis* and *M. blakistoni*. This ought to have been a final settlement of the nomenclature of the Japanese Wagtail; but in 1885 the synonymy of this unfortunate bird was once more confused by the addition of a fourth name, *M. grandis* (Sharpe, Cat. Birds Brit. Mus. x. p. 492). It is unnecessary to say that there can be no excuse for the creation of this useless synonym.

85. MOTACILLA BOARULA.
(GREY WAGTAIL.)

Motacilla boarula, Linneus, Mantissa Plantarum, p. 527 (1771).

The Grey Wagtail has yellow under tail-coverts, much more brilliant than the yellow breast and belly. Japanese examples have, on an average, shorter tails than those from Europe.

Figures: Dresser, Birds of Europe, iii. pl. 128.

The Grey Wagtail is found on all the Japanese Islands, but is probably only a summer visitor to the Kuriles, whence I have examples collected by Mr. Snow, and to Yezzo, whence there are examples in the Swinhoe collection obtained by Captain Blakiston (Swinhoe, Ibis, 1874, p. 157). A pair were found at Hakodadi by the Perry expedition twenty years earlier (Cassin, Exp. Am. Squad. China Seas and Japan, ii. p. 221). There are several examples in the Pryer collection from Yokohama, and Mr. Ringer gave me two examples from Nagasaki, where those obtained by the Siebold expedition were doubtless procured (Temminck and Schlegel, Fauna Japonica, Aves, p. 59). It has been obtained early in March on the central group of the Loo-Choo Islands (Stejneger, Proc. United States Nat. Mus. 1886, p. 642), and there is an example in the Pryer collection from the same locality (Seebohm, Ibis, 1887, p. 176).

The Grey Wagtail has a very extensive breeding-range, from the British Islands across Russia and Siberia to Japan; but there can be no doubt that eastern examples have on an average shorter tails than western examples. Fifty Asiatic examples vary in the length of the tail from 3·1 to 3·75 inches; whilst forty examples from England, Europe, and Asia Minor vary from 3·5 to 4·4 inches. The eastern form was described in 1776 as *Motacilla melanope* (Pallas, Reis. Russ. Reichs, iii. p. 696), and may fairly claim to be subspecifically distinct under the name of *Motacilla boarula melanope*.

86. MOTACILLA FLAVA.
(BLUE-HEADED WAGTAIL.)

Motacilla flava, Linneus, Syst. Nat. i. p. 331 (1766).

The Blue-headed Wagtail has the breast, belly, and under tail-coverts of the same shade of yellow. It always has an eye-stripe, white in the male, dull white in the female, and buff in the young.

Figures: Dresser, Birds of Europe, iii. pl. 129. figs. 1, 2.

It is not known that any species of Yellow Wagtail (subgenus *Budytes*) occurs on any of the main islands of Japan; but examples occur on the Kurile Islands (Seebohm, Ibis, 1884, p. 39) which have dark olive-green heads and buff eye-stripes. There is one in the Pryer collection. They are probably females and immature males of *Motacilla flava*, which breeds in the Commander Islands, and ranges across Southern Siberia and Central Europe as far as Holland, but is only known as an accidental visitor on migration to the British Islands.

87. ANTHUS MACULATUS.
(EASTERN TREE-PIPIT.)

Anthus maculatus, Hodgson, Gray's Zool. Miscell. 1844, p. 83.

In the Eastern Tree-Pipit the hind toe is longer than its claw; the belly is always white, and the tail short (less than $2\frac{1}{2}$ inches). It is greener than the two other Japanese species.

Figures: Temminck and Schlegel, Fauna Japonica, Aves, pl. 23.

The Eastern Tree-Pipit is only a summer visitor to Yezzo; but in Southern Japan it breeds on the mountains and winters in great numbers in the pine-plantations in the plains (Blakiston and Pryer, Trans. As. Soc. Japan, 1882, p. 153). There are ten examples in the Pryer collection from Yokohama; Mr. Ringer has presented examples to the Norwich Museum from Nagasaki; and there is an example in the Pryer collection from the central group of the Loo-Choo Islands.

The breeding-range of the Tree-Pipit extends from the British Islands across Europe and South Siberia to Japan. Eastern examples are more suffused with green on the upper parts, and the spots on the mantle are so much more obscure that typical examples cannot be confounded together except in abraded plumage. Some examples from the valley of the Yenesay and from the Himalayas are, however, slightly intermediate; and it is possible that the two races may ultimately be regarded as only subspecifically distinct.

The Eastern Tree-Pipit is found on Fuji-yama as high up as the snow-line. The nest is placed on the ground, and is built of moss and coarse grass, lined with fine grass and rootlets (Jouy, Proc.

United States Nat. Mus. 1883, p. 289). Eggs in the Pryer collection are pinky grey, with light and dark sepia streaks, spots, and blotches, and exactly resemble a common variety of the eggs of the western species.

88. ANTHUS SPINOLETTA.
(ALPINE PIPIT.)

Anthus spinoletta, Linneus, Syst. Nat. i. p. 288 (1766).

The Japanese race of the Alpine Pipit has brown upper parts, suffused with grey in summer plumage. The dark centres to the feathers are very obscure, almost obsolete on the crown.

Figures: Temminck and Schlegel, Fauna Japonica, Aves, pl. 24 (under the name of *Anthus pratensis japonicus*).

The Japanese Alpine Pipit breeds on the Kurile Islands, whence I have three examples collected on Urup by Wossnesenski (Seebohm, Ibis, 1879, p. 34) and three collected by Mr. Snow. Dr. Henderson obtained it at Hakodadi in October 1857 (Cassin, Proc. Acad. Nat. Sc. Philad. 1858, p. 193); and I have six examples collected by Captain Blakiston in Yezzo in September, October, and November (Swinhoe, Ibis, 1875, p. 449). It is common in winter in Southern Japan, and there are no fewer than twelve examples in the Pryer collection from Yokohama. Its breeding-range extends to Eastern Siberia, and there are examples in the Swinhoe collection from South China, where it is only known as a winter visitor.

There seem to be four races of Alpine Pipit. The typical, or Western Palæarctic, form differs from the other three in being on an average slightly larger; the Eastern Palæarctic form, *Anthus spinoletta blakistoni*, only differs from the typical form in size; but the Japanese form, *Anthus spinoletta japonicus*, differs from most examples of the other three forms in having pale legs and feet; and the Nearctic form, *Anthus spinoletta pennsylvanicus*, generally (though not always) differs from the other three forms in having the outer web of the penultimate tail-feather on each side white for some distance from the tip.

In Captain Blakiston's collection there is a remarkably handsome specimen of the fully adult Japanese Alpine Pipit in summer plumage, which scarcely differs from that of the American Alpine Pipit (Swainson and Richardson, Faun. Bor-Amer. ii. pl. 44).

89. ANTHUS CERVINUS.
(RED-THROATED PIPIT.)

Motacilla cervina, Pallas, Zoogr. Rosso-Asiat. i. p. 511 (1826).

The Red-throated Pipit differs from the Japanese Alpine Pipit in having conspicuous dark centres to all the feathers of the upper parts; and from the Eastern Tree-Pipit in having the belly always buff, and the hind toe shorter than its claw.

Figures: Dresser, Birds of Europe, iii. pl. 136.

The Red-throated Pipit is a rare visitor on migration in spring and autumn to the Japanese Islands. I have never seen a specimen from Japan, but there cannot be much doubt that it does occur there, as I have one example (No. 2056) obtained by Mr. Snow on Shumshu (the most northerly of the Kurile Islands) on the 7th of June, 1876 (Seebohm, Ibis, 1879, p. 34); a second example obtained by Mr. Snow on Eturup (the most southerly but one of the same group) in September; a third example (in the Pryer collection) obtained by Mr. Snow on the Kurile Islands on the 29th of July; and a fourth and fifth example (also in the Pryer collection) obtained on the central group of the Loo-Choo Islands in January.

The breeding-range of the Red-throated Pipit extends on the tundras above the limit of forest-growth from the Atlantic to the Pacific; but in the eastern and western extremities of its range it is a rare bird, which accounts for the difficulty of procuring examples in Japan and the British Islands, where it only occurs on migration to its winter-quarters in North-east Africa and Burma.

ALAUDINÆ.

Sexes nearly alike; young in first plumage (which is moulted in the first autumn) spotted above and below; first primary generally very small, sometimes obsolete; back of tarsus scutellated.

There are about 70 species of Larks, chiefly confined to the Palæarctic, Ethiopian, and Oriental Regions. One of the Palæarctic species is circumpolar, and each of the Australian, Neotropical, and Nearctic Regions contains a solitary species.

One species (possibly two) is represented in the Japanese Empire.

90. ALAUDA ARVENSIS.
(SKY-LARK.)

Alauda arvensis, Linneus, Syst. Nat. i. p. 287 (1766).

The Sky-Lark is too well known to need description.

Figures: Temminck and Schlegel, Fauna Japonica, Aves, pl. 47 (under the name of *Alauda japonica*).

The Sky-Lark is found on all the Japanese Islands and on the Kurile Islands. To the latter and possibly to Yezzo it is only a summer visitor, but in Southern Japan it breeds on the mountains and winters in the plains. I have four examples procured on the Kurile Islands by Mr. Snow. They are large birds (wing from carpal joint 4·7 to 4·25 inches), and, in newly moulted autumn plumage, they have very white bellies and bright sandy-buff margins to the feathers of the upper parts. They belong to a large north-eastern race of the Sky-Lark, which breeds in Kamtschatka and the Kurile Islands, and winters in Japan and North China. If it be regarded as subspecifically distinct it must bear the name of *Alauda arvensis pekinensis* (Swinhoe, Proc. Zool. Soc. 1863, p. 89). I have seven examples procured in Yezzo by Captain Blakiston during March, April, May, June, and September; they are rather smaller birds (wing 4·1 to 3·7 inches), but they do not differ from the Kurile examples in colour. There are eight examples in the Pryer collection from Yokohama, which are on an average slightly smaller still (wing 4·0 to 3·55 inches), but no difference of colour is discoverable. Examples from every locality get very dark in summer from the abrasion of the buff margins of the feathers, and lose the brilliance of the white on the belly from stains. If this small race be regarded as subspecifically distinct it must bear the name of *Alauda arvensis japonica*. This race appears to be a resident in Japan, North China, and Thibet.

These three races have their exact parallels in Europe. The typical *Alauda arvensis* is a large north-western race, breeding in Scandinavia and wintering in England and Central Europe. The intermediate forms, which are resident in the British Isles, represent the intermediate forms found in Yezzo; whilst a small resident race inhabits Southern Europe, *Alauda arvensis cantarella* (Bonap. Comp. List B. Eur. & N. Amer. p. 37). The western races are buffer on the underparts and greyer on the upper parts than the eastern races;

but between them, breeding in Siberia and wintering in North-west India, Asia Minor, and Palestine, occurs an Arctic race, *Alauda arvensis dulcivox* (Hodgson, Gray's Zool. Misc. i. p. 84), of intermediate size between the northern and southern races, and intermediate in the colour of the underparts between the eastern and western races, but much greyer on the upper parts than either of them.

When we remember that the north, south, east, west, and arctic races intergrade in every direction, it is easy to understand how complicated a problem the nomenclature of the races of the Sky-Lark becomes. It is, however, still more complicated by the existence of a small race or species (wing 3·7 to 3·2 inches) in South China and Ceylon, of the colour of the European Sky-Lark, but having on an average a larger bill and a longer hind claw. This species intergrades with a pale grey race, which appears to be a resident in Turkestan and India. The pale small species has been called *Alauda gulgula* (Franklin, Proc. Zool. Soc. 1831, p. 119); and the rufous race of it might be called *Alauda gulgula celivox* (Swinhoe, Zoologist, 1859, p. 6723), were it not that on the Island of Formosa it appears to intergrade with *Alauda arvensis pekinensis*. Probably all the seven forms are only climatic races of one widely spread and very variable species.

91. ALAUDA ALPESTRIS.
(SHORE-LARK.)

Alauda alpestris, Linnæus, Syst. Nat. i. p. 289 (1766).

The Shore-Lark may be recognized by the black on the fore part of the crown, the ear-coverts, and the upper breast.

Figures: Dresser, Birds of Europe, iv. pl. 243.

The Shore-Lark, *Alauda alpestris*, has very slender claims to be regarded as a Japanese bird. It was included doubtfully from a drawing amongst the discoveries of Dr. Siebold (Temminck and Schlegel, Fauna Japonica, Aves, p. 138), but it has not been obtained by recent collectors. There is some evidence that it occurs on the Kurile Islands, as its local name there is recorded (Pallas, Zoogr. Rosso-Asiat. ii. p. 520).

The Shore-Lark is a circumpolar bird, breeding on the tundras of both hemispheres above the limit of forest-growth.

FRINGILLINÆ.

Sexes generally different; first primary obsolete; bill thick, conical, and unnotched; tertials reaching beyond the middle of the wing.

The Fringillinæ number about 500 species, of which 32 have been recorded from the Japanese Empire. This subfamily is almost cosmopolitan, but in the Australian Region it is only known from the Sandwich Islands.

92. COCCOTHRAUSTES VULGARIS.
(COMMON HAWFINCH.)

Coccothraustes vulgaris, Pallas, Zoogr. Rosso-Asiat. ii. p. 12 (1826).

The Hawfinch can always be recognized by its very thick bill and the curious shape of some of its innermost primaries, which are notched at the end of the inner webs and expanded at the end of the outer webs.

Figures: Temminck and Schlegel, Fauna Japonica, Aves, pl. 51.

The Common Hawfinch is a resident in Japan. I have an example collected by Mr. Henson near Hakodadi in February, and there are two examples in the Swinhoe collection obtained by Captain Blakiston in the same locality (Swinhoe, Ibis, 1874, p. 160). There are examples in the Paris Museum obtained near Aomori, in the north of Hondo, by l'Abbé Fauire; and there are five examples in the Pryer collection from the neighbourhood of Yokohama, where it is probably only a winter visitor, as it is said to appear in Central Hondo in considerable numbers in autumn about every third year (Jouy, Proc. United States Nat. Mus. 1883, p. 295); Mr. Ringer has obtained it at Nagasaki, whence he has sent examples to the Norwich Museum.

The breeding-range of the Common Hawfinch extends from the British Islands across Europe and Southern Siberia to Japan.

Eastern examples have been described as distinct, under the name of *Coccothraustes japonicus* (Bonaparte, Consp. Gen. Av. i. p. 506), under the impression that the ends of the wing-coverts were paler in

European than in Japanese and Chinese examples. There can be
no doubt that the two alleged forms are not specifically distinct.
The utmost that can be said is that on an average Eastern examples
may be a shade darker at the ends of the wing-coverts than Western
ones, but the extreme range of variation is so small that it is very
doubtful whether it ought to be recognized as a subspecific difference.

93. COCCOTHRAUSTES PERSONATUS.
(JAPANESE HAWFINCH.)

Coccothraustes personatus, Temminck and Schlegel, Fauna Japonica, Aves, p. 94 (1847).

The Japanese Hawfinch has quite as thick a bill as the Common
Hawfinch, but is easily distinguished from it by its black crown and
black tail.

Figures: Temminck and Schlegel, Fauna Japonica, Aves, pl. 52.

The Japanese Hawfinch appears to be a resident in Japan. It is
found in Yezzo (Whitely, Ibis, 1867, p. 201), and in the Pryer collection there are three examples from the neighbourhood of Yokohama.
It has been obtained on Fuji-yama in June and July (Jouy, Proc.
United States Nat. Mus. 1883, p. 293); and in the British Museum
there are three examples obtained by Mr. Ringer near Nagasaki.

On the continent this species breeds near the mouth of the Ussuri
River in Eastern Siberia, and winters in South China.

94. LOXIA CURVIROSTRA.
(COMMON CROSSBILL.)

Loxia curvirostra, Linnæus, Syst. Nat. i. p. 299 (1766).

The Common Crossbill is easily recognized by its crossed mandibles. It is the type of a group in which the males differ from the
females in being suffused with crimson.

Figures: Dresser, Birds of Europe, iv. pl. 203.

The Common Crossbill is a resident in the Japanese Islands.
Mr. Snow obtained it on the Kurile Islands (Blakiston and Pryer,
Trans. As. Soc. Japan, 1882, p. 176). I have two examples collected
by Mr. Henson near Hakodadi in June; and in the Pryer collection

there are nine from the neighbourhood of Yokohama. The examples procured by the Siebold expedition were probably obtained at Nagasaki (Temminck and Schlegel, Fauna Japonica, Aves, p. 93).

The Common Crossbill breeds in the pine-forests of Arctic Europe and Asia from Ireland and Scotland to Kamtschatka and Japan, wandering more or less irregularly southwards in winter.

95. CHAUNOPROCTUS FERREIROSTRIS.
(BONIN GROSBEAK.)

Coccothraustes ferreirostris, Vigors, Zool. Journ. 1828, p. 354.

The Bonin Grosbeak has a thicker bill than any other Finch. The female is a brown bird above and below, but the male is suffused with crimson on the head and underparts.

Figures: Vigors, Beechey's Voyage of the 'Blossom,' pl. 8.

The Bonin Grosbeak was discovered on one of the Bonin group during the voyage of the 'Blossom,' and the types are now in the British Museum (Sharpe, Cat. Birds Brit. Mus. xii. p. 31). It was rediscovered some years later, and redescribed as *Fringilla papa* (Kittlitz, Mém. prés. à l'Acad. Imp. des Sciences de St. Pétersb. par divers savans, 1830, p. 239); but it is not known that it has been obtained by any recent traveller, though Mr. Holst heard that it was to be found on the Bailly Islands (Seebohm, Ibis, 1890, p. 102).

The Bonin Grosbeak is probably related to the Pine-Grosbeak, and may possibly have originated in a flock of those birds which emigrated from the Arctic Regions many thousands of years ago, and which have gradually adapted themselves to the changed conditions of life.

Its nearest relation appears to be *Telespiza cantans* from Medway Island, about 300 miles north-west of the Hawaiian or Sandwich Islands, where another not distantly allied species occurs, *Psittirostra psittacea*.

96. PINICOLA ENUCLEATOR.
(PINE-GROSBEAK.)

Loxia enucleator, Linneus, Syst. Nat. i. p. 299 (1766).

The Pine-Grosbeak is a large Rose-Finch (wing from carpal joint nearly 4½ inches), with a somewhat hooked beak. The colour varies with sex and age very similarly to that of the Crossbills. There are always two pale bars across the wings formed by the pale tips of the greater and median wing-coverts.

Figures: Dresser, Birds of Europe, iv. pl. 201.

Two examples of the Pine-Grosbeak have occurred on the Kurile Islands (Blakiston, Amended List of the Birds of Japan, p. 63), one of which I have had an opportunity of examining.

This species breeds in the pine-forests of Arctic Europe and Asia, from Lapland to Kamtschatka, and migrates irregularly southwards in autumn, occasionally, but very rarely, visiting the British Islands. It also breeds in Arctic America.

97. CARPODACUS ROSEUS.
(ROSE-FINCH.)

Fringilla rosea, Pallas, Reise Russ. Reichs, iii. p. 699 (1776).

The Rose-Finch is a very brilliant bird and is larger than its nearest allies in Japan (wing from carpal joint about $3\frac{1}{2}$ inches). It resembles the Japanese Rose-Finch in having pearly-white plumes on the forehead and throat, but differs from it in having the tail much shorter than the wing.

Figures: Gould, Birds of Asia, v. pl. 33.

The Rose-Finch is a rare winter visitor to the Japanese Islands. It has been obtained in Yezzo (Blakiston and Pryer, Ibis, 1878, p. 245); in the Swinhoe collection there is an example from Tokio (Swinhoe, Ibis, 1877, p. 145); and in the Pryer collection there is a second example from the same district. A third example has been recorded from Tate-yama (Jouy, Proc. United States Nat. Mus. 1883, p. 294); whilst in the British Museum there is a fourth Japanese example probably collected near Nagasaki.

This species breeds in Eastern Siberia and winters in China.

98. CARPODACUS ERYTHRINUS.
(SCARLET ROSE-FINCH.)

Pyrrhula erythrina, Pallas, N. Comm. Acad. Sc. Imp. Petrop. xiv. p. 587 (1770).

The Scarlet Rose-Finch is intermediate in size between its two nearest allies in Japan (wing from carpal joint about $3\frac{1}{4}$ inches).

It resembles the Common Rose-Finch in having the tail much shorter than the wing, but it differs both from that species and from the Japanese Rose-Finch in having neither pearly-white plumes on the head nor white bars across the wings.

Figures : Dresser, Birds of Europe, iv. pl. 195.

The sole claim of the Scarlet Rose-Finch to be regarded as a Japanese bird rests upon an example in the Pryer collection, which was bought alive in the Yokohama market (Blakiston and Pryer, Trans. As. Soc. Japan, 1882, p. 175). The clean condition of its feet and the brilliancy of the deep scarlet breast and throat look like a wild bird, and the fact that its wings are much abraded also point to its not having become used to a cage; but it may nevertheless have been imported from China.

This species breeds in the subarctic regions, wherever forests are to be found, from Finland to Kamtschatka and further south at high elevations. It winters in India, Burma, and China, and has twice been known to wander as far as the British Islands.

99. CARPODACUS SANGUINOLENTUS.
(JAPANESE ROSE-FINCH.)

Pyrrhula sanguinolentus, Temminck and Schlegel, Fauna Japonica, Aves, p. 92 (1847).

The Japanese Rose-Finch is the smallest of the three Rose-Finches found in Japan, but it has a relatively longer tail (wing from carpal joint about 2¾ inches, tail slightly longer). It has two very conspicuous white bars across the wing, and pearly-white plumes on the forehead and throat.

Figures : Temminck and Schlegel, Fauna Japonica, Aves, pl. 54 (male), pl. 54 B (female).

The Japanese Rose-Finch is a common resident of the group of islands to which it was formerly supposed to be peculiar. Mr. Snow obtained it from the Kurile Islands (Blakiston and Pryer, Trans. As. Soc. Japan, p. 174); and in the Swinhoe collection there are two examples from Hakodadi (Swinhoe, Ibis, 1874, p. 160). In the Paris Museum there are examples procured by l'Abbé Faurie near Aomori in Northern Hondo; and in the Pryer collection there are nine from the neighbourhood of Yokohama.

Westwards the range of this species extends to Manchuria, Eastern Siberia, and the Corean Peninsula.

100. FRINGILLA SPINUS.
(SISKIN.)

Fringilla spinus, Linneus, Syst. Nat. i. p. 322 (1766).

The Siskin is the smallest Finch found in the Japanese Empire (wing from carpal joint about 2¾ inches). It is a small Greenfinch with a more slender bill. The male has a black crown. The flanks are streaked in the adult of both sexes as well as in the young.

Figures: Dresser, Birds of Europe, iii. pl. 169.

The Siskin appears to be only a winter visitor to the Japanese Islands. It is common in the woods near Hakodadi in autumn (Whitely, Ibis, 1867, p. 201), and in the Pryer collection there are eight examples from the neighbourhood of Yokohama. It appears in large flocks in autumn and winter in Central Hondo (Jouy, Proc. United States Nat. Mus. 1883, p. 297). Mr. Ringer obtained it at Nagasaki, where it was also procured by the Siebold expedition (Temminck and Schlegel, Fauna Japonica, Aves, p. 89); and it has been recorded from the central group of the Loo-Choo Islands (Stejneger, Proc. United States Nat. Mus. 1886, p. 651).

Westwards we find the Siskin breeding in the lower Amoor and wintering in South China; but beyond this range it is not known that the Siskin occurs, except west of the Ural Mountains, whence its range extends to the British Islands. It is highly improbable that the range of the Siskin is discontinuous, and it will most probably be found to extend to the mountain-ranges of Southern Siberia.

Eastern examples have been described as distinct under the name of *Chrysomitris dybowskii* (Taczanowski, Journ. Orn. 1876, p. 199), but I have failed to discover the alleged difference.

101. FRINGILLA LINARIA.
(MEALY REDPOLE.)

Fringilla linaria, Linneus, Syst. Nat. i. p. 322 (1766).

In the Mealy Redpole the forehead and the front half of the crown is crimson; the breast is also frequently suffused with rosy pink.

Figures: Dresser, Birds of Europe, iv. pl. 187.

The Mealy Redpole is probably only a winter visitor to the Japanese Islands. In the Swinhoe collection there are three examples from Hakodadi (Swinhoe, Ibis, 1874, p. 160), and in the Pryer collection there are four from Yokohama. Two examples are recorded from Central Hondo, obtained in November (Jouy, Proc. United States Nat. Mus. 1883, p. 297); and those procured by the Siebold expedition were probably obtained at Nagasaki (Temminck and Schlegel, Fauna Japonica, Aves, p. 89).

It is a circumpolar bird, breeding in high latitudes at or near the limit of forest-growth, and migrating irregularly southwards in cold weather, occasionally visiting the British Islands.

In all the Redpoles which I have seen from Japan the rump and upper tail-coverts are streaked with brown, and the bill is large.

102. FRINGILLA MONTIFRINGILLA.
(BRAMBLING.)

Fringilla montifringilla, Linneus, Syst. Nat. i. p. 318 (1766).

The Brambling is the only Japanese Finch with a white rump. In both sexes the throat and breast are chestnut-buff, but in the male the rest of the head and the back are black (edged with buff in the autumn), whilst in the female these parts are dark brown, edged with buff, at all seasons.

Figures: Dresser, Birds of Europe, iv. pl. 184.

The Brambling is a common winter visitor to all the Japanese Islands. It frequents the neighbourhood of Hakodadi during winter in small flocks (Whitely, Ibis, 1867, p. 201); and there are examples in the Paris Museum procured near Aomori, in the north of Hondo, by l'Abbé Fauric. There are six examples in the Pryer collection from Yokohama. The examples procured by the Siebold expedition were probably obtained at Nagasaki (Temminck and Schlegel, Fauna Japonica, Aves, p. 87); and it occurs on the central group of the Loo-Choo Islands (Stejneger, Zeitschr. ges. Orn. 1887, p. 176), and on the Bonin Islands on migration (Seebohm, Ibis, 1890, p. 101).

The breeding-range of the Brambling extends across the Palæarctic Region from Lapland to Kamtschatka, but in the British Islands, as in Japan, it is only known as a winter visitor.

103. FRINGILLA SINICA.
(CHINESE GREENFINCH.)

Fringilla sinica, Linneus, Syst. Nat. i. p. 321 (1766).

The Chinese Greenfinch is less than its European representative (wing from carpal joint 3·3 to 3·1 inches), and further differs from it in having the bases of the secondaries yellow on both webs.

It is rather less than the Japanese Greenfinch. The crown of the male is grey, and that of the female greyish brown.

Figures: Temminck and Schlegel, Fauna Japonica, Aves, pl. 49.

The Chinese Greenfinch is a resident in the Japanese Islands. It is very common in Yezzo (Whitely, Ibis, 1867, p. 202); there is a large series from Yokohama in the Pryer collection; and several examples from Nagasaki were presented by Mr. Ringer to the British Museum.

This species is very common throughout China, and a single example was obtained by Dr. Radde in the valley of the Amoor. It was also found by General Prjevalski in the north-east of Mongolia.

The rump and underparts are very slightly suffused with yellow in the female. In the young in first plumage most of the small feathers have dark central streaks. The three Greenfinches found in the Japanese Empire are very nearly allied, and may eventually prove to be only subspecifically distinct.

104. FRINGILLA KAWARAHIBA.
(JAPANESE GREENFINCH.)

Fringilla kawarahiba, Temminck, Planches Coloriées, no. 588. fig. 1 (1836).

The Japanese Greenfinch is on an average larger than the Chinese Greenfinch (wing from carpal joint 3·4 to 3·2 inches), but agrees with it in having the bases of the secondaries yellow on both webs.

It further differs in having the crown of the male brown instead of grey, and that of the female sandy brown instead of greyish brown.

Figures: Temminck and Schlegel, Fauna Japonica, Aves, pl. 48.

The Japanese Greenfinch is only known from the islands whose name it bears, but it is by no means a common bird. I have one example collected by Captain Blakiston at Hakodadi, where it is said to be a rare bird (Whitely, Ibis, 1867, p. 202), and there are several examples in the Paris Museum procured by l'Abbé Fauire near Aomori in Northern Hondo. In the Pryer collection there are two examples obtained by Mr. Snow from the Kurile Islands, and two others from Yokohama. In the British Museum there are two examples from Nagasaki presented by Mr. Ringer.

105. FRINGILLA KITTLITZI.
(BONIN-ISLAND GREENFINCH.)

Fringilla kittlitzi, Seebohm, Ibis, 1890, p. 101.

The Bonin-Island Greenfinch is about the size of the Japanese Greenfinch, but the crown and nape are olive (instead of brown, as in *F. kawarahiba*, or grey, as in *F. sinica*). It has also less yellow at the base of the tail-feathers than either of its allies.

The Bonin-Island Greenfinch was discovered by Kittlitz in 1828, but was not regarded by its discoverer as distinct from the European species. Mr. Holst obtained three examples on one of the Parry Islands, and two on one of the Bailly Islands, all of which are in my collection.

106. MONTIFRINGILLA BRUNNEINUCHA.
(JAPANESE SNOW-FINCH.)

Fringilla (*Linaria*) *brunneinucha*, Brandt, Bull. Sc. Acad. Imp. Sc. St. Pétersbourg, 1842, p. 252.

The Japanese Snow-Finch is about the size of a Bullfinch (wing from carpal joint 4¼ inches). The wing-coverts, rump, upper tail-coverts, and underparts are much suffused with rose-pink in the male, and slightly so in the female.

Figures: David and Oustalet, Oiseaux de la Chine, pl. 89.

The Japanese Snow-Finch is only a winter visitor to Japan, generally appearing in large flocks (Whitely, Ibis, 1867, p. 245). I have two examples collected by Captain Blakiston at Hakodadi (Swinhoe, Ibis, 1875, p. 450), and a third collected by Mr. Henson in the same locality. Mr. Fukushi obtained it in the Kurile Islands in July (Blakiston and Pryer, Trans. As. Soc. Japan, 1882, p. 174).

This is probably the southern limit of its breeding-range, which extends to Kamtschatka and Eastern Siberia. It winters in Northern China, as well as Japan. It sometimes appears in winter in Central Hondo in great numbers (Jouy, Proc. United States Nat. Mus. 1883, p. 296).

107. PYRRHULA GRISEIVENTRIS.
(ORIENTAL BULFINCH.)

Pyrrhula griseiventris, Lafresnaye, Rev. Zool. 1841, p. 240.

The Oriental Bulfinch differs from the Common Bulfinch in having the breast, belly, and flanks grey (sometimes slightly suffused with red) instead of bright brick-red.

Figures: Temminck and Schlegel, Fauna Japonica, Aves, pl. 53, sub nomine *Pyrrhula orientalis*.

The Oriental Bulfinch is a resident in Japan. I have two examples collected by Wossnesenski on the Kurile Islands in July, and there are two others in the British Museum from the same locality. These are so much paler than Japanese examples that they have been separated (Sharpe, Cat. Birds Brit. Mus. xii. p. 450, pl. xi.) as *Pyrrhula griseiventris kurilensis*. The typical form appears to be found both in Yezzo and near Yokohama, but in both localities together with others, in which the underparts are much suffused with rosy and the upper parts slightly so. If the latter be subspecifically distinct, they may be called *P. griseiventris rosacea*. Mr. Whitely, who got both forms near Hakodadi in March, says (Ibis, 1867, p. 203) that this species was very abundant, but disappeared before summer.

The range of this species is very restricted. The typical form has been found on the Island of Askold opposite Vladivostok, once or twice near Pekin, and once in the upper valley of the Ussuri, and the Roseate form on an island in the Bay of Okhotsk (Seebohm, Ibis, 1887, p. 101). It is quite possible that the roseate tint may be the result of eating some particular food, the Bulfinches being apparently specially sensitive to the influence of certain seeds.

There are eight males in the Pryer collection of the typical colour from Yokohama, and four of the roseate form, besides six females. There is one male in the Swinhoe collection, from Hakodadi, of the roseate form, besides three females. There are two males in the British Museum of the typical colour from Yokohama, but none of the roseate form. In the same collection there is only one male of the typical colour from Yezzo, and four of the roseate form. Mr. Jouy obtained five adult males in Tate-yama in winter, four of which were more or less roseate on the breast, and in two of them the back was also roseate (Jouy, Proc. United States Nat. Mus. 1883, p. 293). The two forms certainly intergrade, and the appearance of an example of the typical form from Yezzo throws considerable doubt on the distinctness of their geographical ranges.

108. PASSER MONTANUS.
(TREE-SPARROW.)

Fringilla montana, Linneus, Syst. Nat. i. p. 324 (1766).

The Tree-Sparrow is easily recognized by its chestnut-brown crown and nape, and by the large black patch in the middle of the white on the side of the head. The female resembles the male, but is slightly duller in colour.

Figures: Dresser, Birds of Europe, iii. pl. 178.

The Tree-Sparrow is the Common Sparrow of the towns and villages of Japan (Blakiston and Pryer, Ibis, 1878, p. 244), and was obtained probably at Nagasaki by the Siebold Expedition (Temminck and Schlegel, Fauna Japonica, Aves, p. 89). There is an example in the Swinhoe collection from Hakodadi (Swinhoe, Ibis, 1877, p. 145); and there are four in the Pryer collection from the neighbourhood of Yokohama, and nine from the central group of the Loo-Choo Islands.

The Tree-Sparrow is found throughout the Palæarctic Region, from the British Islands to Japan.

It is abundant everywhere near houses or towns in Central Hondo, breeding in the thatched roofs of the native houses (Jouy, Proc. United States Nat. Mus. 1883, p. 297). Eggs in the Pryer collection do not differ from those obtained in the British Islands.

Examples from the Loo-Choo Islands have been described as distinct under the name of *Passer saturatus* (Stejneger, Proc. United States Nat. Mus. 1885, p. 19), but I am unable to distinguish them from the European species.

The example described by Dr. Stejneger appears to have been in immature plumage. Examples collected by General Prjevalski at Lob Nor and other localities in Central Asia are so pale that they are almost worthy of being recognized as a desert form.

109. PASSER RUTILANS.
(RUSSET SPARROW.)

Fringilla rutilans, Temminck, Planches Coloriées, no. 588, fig. 2 (1836).

The Russet Sparrow has the crown and rump chestnut-red in the male, and brown in the female. The throat is black in the male, and there is a buff eye-stripe extending to the nape in the female.

Figures: Temminck and Schlegel, Fauna Japonica, Aves, pl. 50.

The Russet Sparrow is doubtless a resident in Japan. It is not uncommon in Yezzo, and is occasionally brought into the Yokohama market (Blakiston and Pryer, Ibis, 1878, p. 244). The example figured in the 'Fauna Japonica' as *Passer russatus* was probably obtained at Nagasaki. There are five examples in the Pryer collection from Yokohama.

This species has a very restricted range; it is found in Formosa and in the mountains of Central China, as far west as Moupin in Eastern Thibet.

110. EMBERIZA CIOPSIS.
(BONAPARTE'S JAPANESE BUNTING.)

Emberiza ciopsis, Bonaparte, Consp. Generum Avium, i. p. 466 (1850).

Bonaparte's Japanese Bunting has no trace of yellow on the underparts, and the rump and upper tail-coverts are uniform rich chestnut. The combination of these two characters prevents it being confounded with most of the other Buntings which are known to visit Japan. The fact that the nape is almost as conspicuously streaked as the crown will probably complete the diagnosis.

Figures: Temminck and Schlegel, Fauna Japonica, Aves, pl. 59 (male and female).

Bonaparte's Japanese Bunting is peculiar to Japan, and is the commonest Bunting on the islands. There is an example from Yezzo in the Swinhoe collection (Swinhoe, Ibis, 1874, p. 161); and there are four-and-twenty from Yokohama in the Pryer collection. Mr. Ringer obtained it at Nagasaki (Blakiston and Pryer, Trans. As. Soc. Japan, 1882, p. 168); and it was observed in abundance by the officers of the Perry Expedition at Simoda (Cassin, Exp. Am. Squad. China Seas and Japan, ii. p. 220). Dr. Henderson obtained it at Hakodadi in October 1857 (Cassin, Proc. Acad. Nat. Sc. Philad. 1858, p. 192).

It breeds in great abundance on Fuji-yama, making a nest on or near the ground of dried grass and leaves, lined with fine rootlets (Jouy, Proc. United States Nat. Mus. 1883, p. 298). Eggs in the Pryer collection closely resemble those of *Emberiza cia* and those of *Emberiza cioides*, being scrawled all over the larger end with fine hair-like streaks.

Bonaparte's Japanese Bunting is an island form of Brandt's Bunting, *Emberiza cioides*, and is possibly only subspecifically distinct from it. The adult male differs from that of its continental ally in having the ear-coverts nearly black instead of russet-brown. The female only differs from that of the continental species in having the throat and under tail-coverts more suffused with buff. Intermediate forms occur in Japan, but these may possibly be immature examples.

Brandt's Bunting is a resident in Eastern Siberia, and is the eastern representative of the Meadow-Bunting, *Emberiza cia*, a perfectly distinct species which ranges from Spain across Europe and Southern Siberia as far east as Lake Saissan, whence I have an example collected by General Prjevalski.

111. EMBERIZA YESSOENSIS.
(SWINHOE'S JAPANESE BUNTING.)

Schœnicola yessoensis, Swinhoe, Ibis, 1874, p. 161.

Swinhoe's Japanese Bunting combines two characters, *rump and upper tail-coverts uniform chestnut-buff* and *no trace of yellow on the underparts*. None of the other Buntings which are known to occur in Japan possess both these characters except *E. ciopsis*, *E. fucata*, and *E. rustica*. From the males of these three species and from both sexes of the last mentioned, the entire absence of white on the throat is a sufficient distinction. From the females of the two first-mentioned species the fact that the nape is scarcely streaked (in marked contrast to the conspicuously streaked crown) is a good distinction.

Figures: Seebohm, Ibis, 1879, pl. 1. fig. 2.

Swinhoe's Japanese Bunting is peculiar to Japan, and cannot be a very rare bird, as, in addition to the type in the Swinhoe collection obtained by Captain Blakiston near Hakodadi, there are eleven examples in the Pryer collection from the neighbourhood of Yokohama. Two other examples from Yezzo are in the Blakiston collection (Seebohm, Ibis, 1884, p. 42). It is principally known in the south as a winter visitor, but it breeds on Fuji-yama (Blakiston, Amended List of the Birds of Japan, p. 61).

112. EMBERIZA SCHŒNICLUS.
(REED-BUNTING.)

Emberiza schœniclus, Linneus, Syst. Nat. i. p. 311 (1766).

The male Reed-Bunting differs from all other Buntings known to visit Japan in having a white nuchal collar. The female closely resembles that of *Emberiza yessoensis*, but is rather larger and much greyer, especially on the rump and upper tail-coverts.

Figures : Dresser, Birds of Europe, iv. pl. 221 (females), pl. 222. fig. 1 (male); Swinhoe, Ibis, 1876, pl. viii. fig. 2 (male in autumn plumage of eastern race).

The Reed-Bunting is said to be only a summer visitor to Yezzo, and to the mountains in the more southerly Japanese Islands, descending to the plains in winter. There are two examples from Hakodadi (one of them the type of *Schœnicola pyrrhulina*) in the Swinhoe collection, and there are twenty from the Yokohama game-market in the Pryer collection.

The range of the Reed-Bunting extends eastwards from the British Islands, across Europe and Asia at least as far as the meridian of Calcutta. I found it common in the valley of the Yenesay, and General Prjevalski obtained it at Lob Nor. Taczanowski says that examples obtained by Dybowski in Kamtschatka are identical with the European bird, and it is probable that the range of this species across Siberia is continuous. At what point in its geographical distribution the bill begins to thicken is not known, but Radde remarks it in an example obtained by him near Tarei Nor.

The Eastern race may be regarded as subspecifically distinct under the name of *Emberiza schœniclus palustris*, on the ground that in the east the thick-billed birds are most numerous, whilst in the west the contrary is the case.

It is not known that the Reed-Bunting of Japan differs in the slightest particular from its representative in the British Islands, except in having a slightly thicker bill, and in having rather fewer dark stripes on the flanks. Both these characters are, however, very variable, and examples from Italy and Asia Minor may be found which are undistinguishable from examples from Japan. It is absolutely impossible to regard the two forms as specifically distinct, and it is quite as absurd to place them in different genera as it would be to separate the Siberian Nutcracker from the Japanese Nutcracker on the same grounds. In the dark ages of Ornithology there was a superstition that a variation in the shape of the bill was necessarily a generic character, but no student of Darwin's works can do otherwise than smile at such a theory.

113. EMBERIZA RUSTICA.
(RUSTIC BUNTING.)

Emberiza rustica, Pallas, Reise Russ. Reichs, iii. p. 698 (1776).

The Rustic Bunting differs from every other Bunting known to occur in Japan in having the breast and flanks broadly streaked with rich chestnut.

Figures: Temminck and Schlegel, Fauna Japonica, Aves, pl. 58 (male adult and immature); Dresser, Birds of Europe, iv. pl. 219 (male and female).

The Rustic Bunting breeds in Yezzo and winters in the more southerly of the Japanese Islands (Blakiston and Pryer, Ibis, 1878, p. 243). There is an example in the Swinhoe collection from Hakodadi (Swinhoe, Ibis, 1874, p. 161); and there is one in the Paris Museum procured near Aomori, in the north of Hondo, by l'Abbé Faurie. There are five examples in the Pryer collection from the neighbourhood of Yokohama.

The range of the Rustic Bunting during the breeding-season extends across the Arctic regions from Lapland to Kamtschatka. Its winter-quarters appear to be confined to China. It can only be regarded as an accidental visitor to the British Islands.

114. EMBERIZA FUCATA.
(GREY-HEADED BUNTING.)

Emberiza fucata, Pallas, Reise Russ. Reichs, iii. p. 698 (1776).

The Grey-headed Bunting differs from all the other Buntings which are known to occur in Japan in having the throat (white in the male and buff in the female) surrounded by bold black streaks.

Figures : Temminck and Schlegel, Fauna Japonica, Aves, pl. 57 (male and female); Gould, Birds of Asia, v. pl. 9.

The Grey-headed Bunting is a common winter visitor to the plains near Yokohama, retiring to the mountains and to Yezzo to breed (Blakiston and Pryer, Ibis, 1878, p. 212). There is an example in the Swinhoe collection from Hakodadi (Swinhoe, Ibis, 1874, p. 161), where it had been procured seventeen years previously by Dr. Henderson (Cassin, Proc. Acad. Nat. Sc. Philad. 1858, p. 192). There are nine examples in the Pryer collection from the neighbourhood of Yokohama, and Mr. Ringer has obtained it at Nagasaki.

Westward the range of this species extends, during the breeding-season, to Eastern Siberia and Northern China, and in winter to Southern China, Burma, and the plains of India. It is said to be a resident in the North-west Himalayas.

115. EMBERIZA SULPHURATA.
(SIEBOLD'S BUNTING.)

Emberiza sulphurata, Temminck and Schlegel, Fauna Japonica, Aves, p. 100 (1847).

Siebold's Bunting differs from all the other Buntings known to occur in Japan by its combination of the two characters, *chin, throat, and breast unstreaked yellow*, and *forehead, crown, and nape unstreaked olive-brown*.

Figures: Temminck and Schlegel, Fauna Japonica, Aves, pl. 60.

Siebold's Bunting is said to be a rare summer visitor to Yezzo (Whitely, Ibis, 1867, p. 203), but to be very common in the more southerly Japanese Islands. In the Pryer collection there are six examples from Yokohama, and I have a seventh collected by Mr. Heywood Jones on Fuji-yama.

Siebold's Bunting is said to leave Japan in autumn, and to winter in Formosa and Southern China.

It is the commonest Bunting on Fuji-yama in summer, and builds in the fork of a small bush, making its nest of grass, lined with horsehair or the seed-stalks of moss (Jouy, Proc. United States Nat. Mus. 1883, p. 299).

Eggs in the Pryer collection closely resemble a common variety of the eggs of the Garden-Warbler with dark spots.

116. EMBERIZA PERSONATA.

(TEMMINCK'S JAPANESE BUNTING.)

Emberiza personata, Temminck, Planches Coloriées, no. 580 (1835).

Temminck's Japanese Bunting combines two characters, *mantle russet-brown streaked with dark brown*, and *throat and breast yellow*, streaked with brown in the female, which no other Japanese Buntings possess, except the females of *Emberiza spodocephala* and *E. sulphurata*. The latter has an unstreaked yellow chin, throat, and breast. The male of *E. personata* has a black chin, and the female a streaked breast.

Figures: Temminck and Schlegel, Fauna Japonica, Aves, pl. 59 B.

Temminck's Japanese Bunting is peculiar to the Japanese Islands. It has been recorded from Eturop, the most southerly of the Kurile Islands (Blakiston and Pryer, Trans. As. Soc. Jap. 1882, p. 170). It is a summer visitor to Yezzo, and a few remain in that island during winter. In the more southerly Japanese Islands it is a resident. There is an example from Hakodadi collected by Captain Blakiston in the Swinhoe collection (Swinhoe, Ibis, 1874, p. 161); and I have two others from the same locality collected by Mr. Henson. There are eight examples from Yokohama in the Pryer collection, and I have three examples from Nagasaki collected by Mr. Ringer. It was observed in abundance by the officers of the Perry Expedition at Simoda (Cassin, Exp. Am. Squad. China Seas and Japan, ii. p. 221); and it is the only Bunting recorded from the Loo-Choo Islands (Seebohm, Ibis, 1887, p. 174). It breeds abundantly on Fuji-yama. The nest is placed on the ground or in a tussock of grass, and is made of dried grass, lined with fine roots and horsehair (Jouy, Proc. United States Nat. Mus. 1883, p. 298). Eggs in the Pryer collection resemble richly marked examples of those of the Ortolan Bunting.

Temminck's Japanese Bunting is an island form of the Blackfaced Bunting, *Emberiza spodocephala*, and is possibly only subspecifically distinct from it. The adult male differs from that of its continental ally in having the lower throat and breast yellow instead of olive-grey. The female only differs from that of the continental species in having rather less white on the outer tail-feathers, but this is a somewhat variable character. Some of the intermediate forms from China have been referred to a continental race of *E. personata* (Sharpe, Cat. Birds Brit. Mus. xii. p. 522).

117. EMBERIZA SPODOCEPHALA.
(BLACK-FACED BUNTING.)

Emberiza spodocephala, Pallas, Reise Russ. Reichs, iii. p. 698 (1776).

The male Black-faced Bunting differs from all the other Buntings that are known to occur in Japan in having a uniform olive-grey throat and breast. The female scarcely differs from that of *Emberiza personata*, except in having much more white on the outer tail-feathers.

Figures: Middendorff, Sibirische Reise, ii. pl. 13. figs. 5-8.

The claim of the Black-faced Bunting to be regarded as a Japanese bird rests upon a single example, a male, with grey breast, shot by Mr. Jouy in January near Tokio (Seebohm, Ibis, 1884, p. 182).

This species has a wide range, breeding in Siberia from the valley of the Yenesay eastwards, and wintering in the eastern Himalayas and China.

118. EMBERIZA ELEGANS.
(TEMMINCK'S YELLOW-BROWED BUNTING.)

Emberiza elegans, Temminck, Planches Coloriées, no. 583 (1835).

Temminck's Yellow-browed Bunting differs from all other Buntings known to occur in Japan in having a conspicuous yellow stripe over each eye, which is almost as bright in the female as in the male.

Figures: Temminck and Schlegel, Fauna Japonica, Aves, pl. 55; Gould, Birds of Asia, v. pl. 12.

Temminck's Yellow-browed Bunting is by no means a common bird in Japan. It has not hitherto been recorded from Yezzo, but there is an example in the Paris Museum procured by l'Abbé Fauire near Hakodadi. There are three examples in the Pryer collection from Yokohama, and I have two others obtained by Mr. Ringer near Nagasaki.

It is possibly a resident in Japan, but to Manchuria and the valley of the Amoor it is only a summer visitor, wintering in China.

119. EMBERIZA RUTILA.
(RUDDY BUNTING.)

Emberiza rutila, Pallas, Reise Russ. Reichs, iii. p. 698 (1776).

The Ruddy Bunting differs from all the other Buntings known to visit Japan by its combination of three characters: belly yellow, rump chestnut, mantle streaked.

Figures: Temminck and Schlegel, Fauna Japonica, Aves, pl. 56 B (male).

The claim of the Ruddy Bunting to be regarded as a Japanese bird rests solely on a single example figured in the 'Fauna Japonica.' It may be an accidental visitor on migration to the west of Japan, but no second example has been recorded.

It is an East-Asiatic species, breeding in Eastern Siberia and North China, and wintering in South China, Cochin China, and Burma.

120. EMBERIZA AUREOLA.
(YELLOW-BREASTED BUNTING.)

Emberiza aureola, Pallas, Reise Russ. Reichs, ii. p. 711 (1773).

The male of the Yellow-breasted Bunting is easily recognized by its uniform chestnut back; and the female is the only Bunting with yellow underparts (known to visit Japan), which has also the whole of the upper parts uniformly streaked.

Figures: Dresser, Birds of Europe, iv. pl. 218.

I have never seen an example of the Yellow-breasted Bunting from Japan, but it is occasionally found in Yezzo in summer (Blakiston and Pryer, Trans. As. Soc. Japan, 1882, p. 170). It has been obtained on the south-east coast of Yezzo in May (Blakiston, Chrysanthemum, 1882, p. 126), and has once occurred near Yokohama (Blakiston and Pryer, Ibis, 1878, p. 243).

This Bunting has a wide range across Northern Europe and Asia. It winters in China and Burma.

121. EMBERIZA VARIABILIS.
(GREY BUNTING.)

Emberiza variabilis, Temminck, Planches Coloriées, no. 583, fig. 2 (1835).

The Grey Bunting differs from every other Bunting known to visit Japan in having no white on any of the tail-feathers, a character common to both sexes.

Figures : Temminck and Schlegel, Fauna Japonica, Aves, pl. 56 (male and female).

The Grey Bunting is probably only a summer visitor to Yezzo, but it is a resident in the more southerly islands of Japan. I have an example collected by Captain Blakiston from Hakodadi (Swinhoe, Ibis, 1875, p. 450), and a second obtained by Mr. Henson from the same locality. There are six examples in the Pryer collection from the neighbourhood of Yokohama, and I have two obtained by Mr. Ringer near Nagasaki.

The Grey Bunting has a very restricted range. It has been three times recorded from Kamtschatka : in 1858 (Kittlitz, Denkwürdigkeiten, ii. p. 201), in 1881 (Dybowski, Journ. Orn. xxix. p. 184), and in 1885 (Stejneger, Orn. Expl. Comm. Isl. and Kamtschatka, p. 247) ; and once from the island of Askold (Taczanowski, Journ. Orn. 1881, p. 184).

It must be admitted that the Grey Bunting is a somewhat aberrant member of the genus, and it is not easy to trace its relationship to the other species. It has been recently placed (Sharpe, Cat. Birds Brit. Mus. xii. p. 566) in the genus *Fringillaria*, principally composed of African Buntings, which differ from the true Buntings in having no white on the outer tail-feathers. To this genus *Emberiza striolata* is also referred, a species which appears to be much nearer related to *E. cia* than to *E. variabilis*. The Grey Bunting appears to me to be more probably a *Spizella* allied to *S. atrigularis* from California, if it be not nearer allied to the typical Buntings.

The determination of the genera in the subfamily Fringillinæ is exceptionally difficult. On the whole, the pattern of colour and the variations due to age, sex, and season, appear to be of greater generic value than slight modifications of the form or size of the bill.

122. EMBERIZA **NIVALIS**.
(SNOW-BUNTING.)

Emberiza nivalis, Linneus, Syst. Nat. i. p. 308 (1766).

The Snow-Bunting differs from all the other Buntings which are known to occur in Japan in having the three outer tail-feathers on each side for the most part white.

Figures: Dresser, Birds of Europe, iv. pl. 225.

The Snow-Bunting is a rare or accidental visitor to Yezzo during winter. Only two examples are recorded (Blakiston and Pryer, Trans. As. Soc. Japan, 1882, p. 172), one of which I have had an opportunity of examining, thanks to the kindness of Captain Blakiston. I have also two examples, obtained by Mr. Snow, from the Kurile Islands, a locality which it has long been known to frequent (Pallas, Zoogr. Rosso-Asiat. ii. p. 33).

It is a circumpolar bird, breeding in the Arctic Regions above the limit of forest-growth, and wandering southwards in winter.

123. EMBERIZA **LAPPONICA**.
(LAPLAND BUNTING.)

Fringilla lapponica, Linneus, Syst. Nat. i. p. 317 (1766).

The Lapland Bunting differs from the other Buntings found in Japan in having the rump and upper tail-coverts grey, with nearly black centres to each feather. The male has a chestnut nape and a black throat.

Figures: Dresser, Birds of Europe, iv. pl. 223.

It is not known that the Lapland Bunting has occurred in Japan, but I have an example, an adult male in full breeding-plumage, obtained by Mr. Snow on the Kurile Islands.

The Lapland Bunting is a circumpolar bird, breeding on the tundras of both hemispheres beyond the limit of forest-growth, and wintering in great numbers in Eastern Mongolia and North China. It is a very rare winter visitor to the British Islands.

HIRUNDININÆ.

First primary obsolete; bill broad, flat, and notched; tertials not reaching beyond the middle of the wing.

The Swallows are an almost cosmopolitan group of birds, and number about eighty species, of which five are represented in the Japanese Empire.

124. HIRUNDO RUSTICA.
(CHIMNEY-SWALLOW.)

Hirundo rustica, Linneus, Syst. Nat. i. p. 343 (1766).

The Chimney-Swallow has a very deeply-forked tail. In the typical form the throat is chestnut, bounded below by a black band. In the Eastern race the chestnut extends below into the black band, which it divides in the middle.

Figures: Dresser, Birds of Europe, iii. pl. 160. fig. i. (typical race).

The Eastern race of the Barn-Swallow is a common summer visitor to all the Japanese Islands. There are several examples sent by Captain Blakiston from Hakodadi in the Swinhoe collection (Swinhoe, Ibis, 1874, p. 151); and there are four examples in the Pryer collection from Yokohama. The examples obtained by the Siebold Expedition were probably procured at Nagasaki (Temminck and Schlegel, Fauna Japonica, Aves, p. 31); and Mr. Holst procured a female on the Bonin Islands (Seebohm, Ibis, 1890, p. 102).

The breeding-range of the Barn-Swallow extends from the British Islands across Europe to Turkestan and West Siberia. Further east it ranges in a slightly modified form through Mongolia and the Himalayas, across China to Japan.

The Eastern race of the Chimney-Swallow differs from the Western race in having the black pectoral band almost interrupted in the middle by the chestnut of the throat. It was described as a distinct species as long ago as 1786, under the name of *Hirundo gutturalis* (Scopoli, Del. Flor. et Faun. Insubr. ii. p. 96); but as the two races completely intergrade, it can only be regarded as subspecifically distinct under the name of *Hirundo rustica gutturalis*.

The Barn-Swallows of Japan build in the native houses, where one or more little wooden shelves are placed for their accommodation, just inside the door on one of the rafters of the ceiling, and where

they are jealously guarded from molestation (Jouy, Proc. United
States Nat. Mus. 1883, p. 290). The eggs do not differ from those
of the European Barn-Swallow (Blakiston and Pryer, Trans. As. Soc.
Japan, 1882, p. 139).

In immature examples the upper parts are bronzed with green
instead of purple.

125. HIRUNDO JAVANICA.
(BUNGALOW-SWALLOW.)

Hirundo javanica, Sparrman, Mus. Carls. ii. pl. 100 (1789).

The Bungalow-Swallows closely resemble the Chimney-Swallows,
but are smaller, and the tail is only slightly forked. They further
differ from them in having no dark band across the breast below the
chestnut throat, and in having the white of the rest of the underparts replaced by brown.

Figures: Gould, Birds of Asia, i. pl. 32 (under the name of
Hypurolepis domicola).

A large race of the Bungalow-Swallow (wing from carpal joint
4·6 inches instead of 4·4 to 4·0) has been recorded from Okinawa-
shima, the largest island of the central group of the Loo-Choo
Islands. It is said to be green instead of blue on the upper parts,
but this is also the case with examples from Ceylon, Borneo, and
Lombock.

It has been described as a distinct species under the name of
Chelidon namiyei (Stejneger, Proc. United States Nat. Mus. 1886,
p. 646), but it can scarcely be regarded as more than a large race
of the Indian species under the name of *Hirundo javanica namiyei*.

126. HIRUNDO ALPESTRIS.
(MOSQUE-SWALLOW.)

Hirundo alpestris, Pallas, Reise Russ. Reichs, ii. p. 709 (1771).

The Mosque-Swallows closely resemble the Chimney-Swallows,
but they are easily recognized by the chestnut rump and the striped
underparts.

Figures: Temminck and Schlegel, Fauna Japonica, Aves, pl. 11
(under the name of *Hirundo alpestris japonica*).

The Japanese race of the Mosque-Swallow is a summer visitor to
the southern islands, but has not been recorded from Yezzo. There

are four skins in the Pryer collection from Yokohama. It builds a long bottle-shaped nest under the eaves of buildings, and lays six white eggs (Blakiston and Pryer, Ibis, 1878, p. 231, no. 173). The Mosque-Swallow of Asia Minor, *Hirundo rufula*, builds a similar nest in caves, and lays similar eggs. The Mosque-Swallow of South Africa, *Hirundo cucullata*, builds a similar nest, often under the verandahs of houses, and also lays white eggs.

The arrival of the Hume Collection in the British Museum has been of invaluable service to ornithology, and has made many groups, which were formerly in hopeless confusion, comparatively easy to arrange in a satisfactory manner. Of these the Asiatic Mosque-Swallows are a conspicuous example, and I find myself obliged to modify the opinions formerly published (Seebohm, Ibis, 1883, p. 167). The Japanese Mosque-Swallow belongs to the group in which the colour of the rump is uniform (not gradated). This group appears to contain four species, though it is very probable that some of them may hereafter be found to intergrade. Two of these species have *narrow* streaks on the underparts, whilst those on the rump are *almost or quite obsolete*. One of them is large (wing 5·2 to 4·9 inches), somewhat rufous on the underparts, and may be called *Hirundo alpestris*, breeding in South Siberia, and wintering in Mongolia and Thibet. The other is smaller (wing 4·5 to 4·2 inches), not quite so rufous on the underparts, and may be called *Hirundo erythropygia*, breeding in the Himalayas and wintering in the plains of India. The other two species have *broad* streaks on the underparts, whilst those on the rump, *though narrow, are very conspicuous*. One of the second pair is large (wing 5·4 to 4·9 inches), with little or no rufous on the underparts, and may be called *Hirundo striolata*, breeding from Assam across Southern China to Formosa, and in winter ranging as far south as Java. The other is smaller (wing 4·8 to 4·4 inches), more rufous on the underparts, and may be called *Hirundo nipalensis*, breeding in the Himalayas across North China to Japan, and wintering in Burma, Flores, and doubtless other islands of the Malay Archipelago. I am, however, of Mr. Sharpe's opinion (Sharpe, Cat. Birds Brit. Mus. x. p. 159), that these four forms intergrade, and can only be regarded as subspecies of *Hirundo alpestris*, in which case the Japan examples must be called *Hirundo alpestris nipalensis*. The extremes appear to be very distinct, but *H. alpestris* probably intergrades with *H. erythropygia*, which appears completely to intergrade with *H. nipalensis*, which again appears to intergrade with *H. striolata*.

127. CHELIDON DASYPUS.
(BLACK-CHINNED MARTIN.)

Chelidon dasypus, Bonaparte, Consp. Generum Avium, i. p. 343 (1850).

The Black-chinned Martin has a much less forked tail than the European House-Martin; and the black on the head descends farther below the eye to the upper part of the ear-coverts and the base of the chin.

Figures: Swinhoe, Ibis, 1874, pl. 7. fig. 1.

The Black-chinned Martin is the Japanese representative of our House-Martin, and is a very common summer visitor to all the islands, breeding on the cliffs and in the caves. It was first obtained in Japan by Captain Blakiston, and erroneously described as a new species under the name of *Chelidon blakistoni* (Swinhoe, Proc. Zool. Soc. 1862, p. 320). There are several examples from Hakodadi in the Swinhoe collection (Whitely, Ibis, 1867, p. 196), and there is a large series in the Pryer collection from Yokohama.

It spends its summers in Japan and winters in Borneo. It breeds in considerable numbers on the sides of an inaccessible cliff on Fuji-yama, above the limit of forest-growth (Jouy, Proc. United States Nat. Mus. 1883, p. 290). Eggs in the Pryer collection do not differ from those of the European House-Martin.

128. COTYLE RIPARIA.
(SAND-MARTIN.)

Hirundo riparia, Linnæus, Syst. Nat. i. p. 344 (1766).

The Sand-Martin is a smaller bird than any of the other Swallows of Japan. Its upper parts and a band across the breast are brown; the rest of the underparts are nearly white.

Figures: Dresser, Birds of Europe, iii. pl. 163.

The Sand-Martin is a summer visitor to Japan, but is nowhere very abundant (Blakiston and Pryer, Ibis, 1878, p. 234). I have four examples sent me by Captain Blakiston from Yezzo (Seebohm, Ibis, 1879, p. 30); and there are two examples in the Pryer collection from Yokohama.

The Sand-Martin is a circumpolar bird, breeding in the British Islands and across Europe and South Siberia to Japan, whence its range extends on the American continent as far east as Baffin's Bay.

Suborder II. *EURYLÆMI.*

Palate ægithognathous; young not passing through a complete downy stage; *flexor longus hallucis* leading to hallux after sending down a tendon to the *flexor perforans digitorum*, which leads to the three front digits.

There are about a dozen species of Broadbills, which are confined to the Oriental Region. They range as far as Borneo and Siam, but do not reach the Japanese Empire.

Suborder III. *TROCHILI.*

Young born helpless, and not passing through a complete downy stage; palate schizognathous; nasals holorhinal; front plantar leading to front toes, hind plantar leading to hallux.

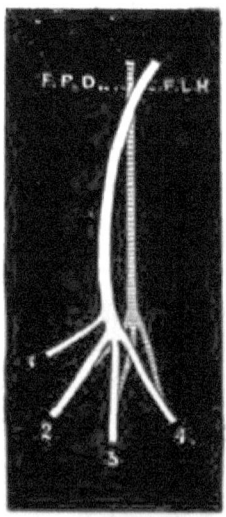

Deep plantar tendons of *Patagona gigas*.

There are about 400 species of Humming-birds, which are confined to the New World.

Suborder IV. SCANSORES.

Fourth digit reversed; front plantar leading to third digit only; spinal feather-tract well defined on the neck by lateral bare tracts, and continuing single on the upper back, but divided into two branches on the lower back.

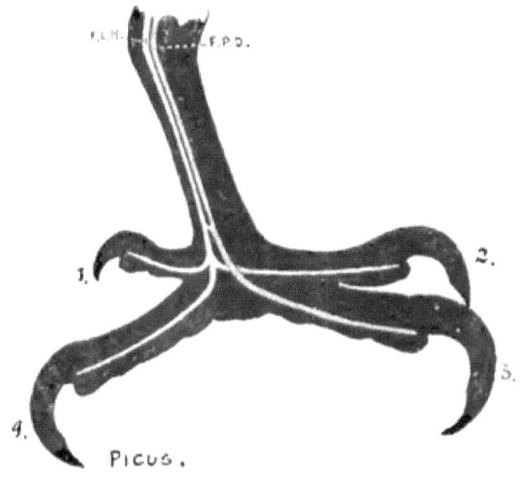

Deep plantar tendons of *Picus martius*.

The Scansores appear to be a natural group of birds consisting of half a dozen families, and rather more than 600 species. Three of these families—the *Rhamphastidæ* or Toucans, the *Galbulidæ* or Jacamars, and the *Bucconidæ* or Puff-birds—are exclusively Neotropical; a fourth, the *Capitonidæ* or Barbets, is not only Neotropical but also Ethiopian and Oriental; a fifth, the *Indicatoridæ* or Honey-Guides, is Ethiopian and Oriental; whilst the sixth, the *Picidæ* or Woodpeckers, is almost cosmopolitan.

The *Picidæ* is the only family belonging to the Scansores which is represented in the Japanese Empire.

The genera and subgenera of Japanese Woodpeckers may be diagnosed in the following manner:—

Distance between nasal grooves at nostrils one third the width of bill.	*Dryocopus.*	
Reversed toe shortest............	*Thripona.r.*	
Second primary at least 10 % longer than longest tail-feather.	*Gecinus* ..	Angle of mandible halfway between nostrils and tip of bill.
Reversed toe longest. Distance between nasal grooves at nostrils more than one half the width of bill	*Iyngipicus.* *Picus.* *Sapheopipo.*	First primary less than one third of second.

The only clue we possess to enable us to form an opinion of the relative value of these characters is their constancy at different ages. The characters at the left appear to be as much developed in the young as in the adult, whilst those on the right vary considerably with age. It would be very easy still further to multiply the subgenera, by the use of equally important characters; but the interests of science will probably be best served by restricting the genera in the following manner:—

Distance between nasal groove at nostrils at least one third the width of bill. Angle of mandible about halfway between eye and tip of bill	*Gecinus.* *Iyngipicus.* *Picus.*	Second primary at least 10 % longer than longest tail-feather.

129. GECINUS AWOKERA.
(JAPANESE GREEN WOODPECKER.)

Picus awokera, Temminck, Planches Coloriées, no. 585 (1836).

The Japanese Green Woodpecker may be easily distinguished from the Grey-headed Green Woodpecker by the red patch on the black malar stripe, which is also found in the male of *G. viridis*.

Figures: Temminck and Schlegel, Fauna Japonica, Aves, pl. 36 (male and female).

The Japanese Green Woodpecker is peculiar to Japan. It is not found in Yezzo, where its place is taken by the Grey-headed Green Woodpecker, but it is a resident in all the southern islands. There are examples in the Paris Museum procured in the north of Hondo by l'Abbé Fauire. There are eight examples in the Pryer collection from Yokohama; and Mr. Ringer has obtained it near Nagasaki (Blakiston, Am. List Birds of Japan, p. 46).

The Japanese Green Woodpecker is intermediate in the amount of black on the sides of its head between *G. viridis* and *G. canus*; and in the barring of the lower half of the underparts resembles the young of those two species and the adult of *G. vaillanti*. If the plumage of the young be regarded as an index to the plumage of recent ancestors, then we may assume that the Algerian Green Woodpecker and the Japanese Green Woodpecker have retained to a large extent the barring on the underparts characteristic of the common ancestor, and that the Green Woodpeckers of Europe, Siberia, Yezzo, and North China have more or less completely lost these bars in the adult.

130. GECINUS CANUS.
(GREY-HEADED GREEN WOODPECKER.)

Picus canus, Gmelin, Syst. Nat. i. p. 434 (1788).

The Grey-headed Green Woodpecker never has a red patch on the black malar stripe, and when adult has no dark bars on the underparts.

Figures: Dresser, Birds of Europe, v. pl. 95.

The Grey-headed Green Woodpecker is a resident in Yezzo, where the earliest recorded Japanese examples were taken in 1861 (Blakiston, Ibis, 1862, p. 325). It is unknown in Southern Japan, its place being taken by the Japanese Green Woodpecker. There are two examples in the Swinhoe collection from Hakodadi (Swinhoe, Ibis, 1875, p. 451); and there is an example in the Pryer collection from the same locality.

The range of the Grey-headed Green Woodpecker extends westwards from Yezzo and North China across Siberia into Europe; but although this species breeds in Scandinavia, Luxembourg, and Spain, it is not known to have occurred in the British Islands.

Dr. Stejneger regards the Grey-headed Green Woodpeckers from Japan as subspecifically distinct from those found on the mainland, under the name of *Picus canus yessoensis* (Stejneger, Proc. United States Nat. Mus. 1886, p. 106). He asserts that the head is much greener, that the underparts are paler, and that in the male the black streaks on the nape are longer than is the case with the typical form. I am unable to detect the slightest difference between European and Japanese examples. On the other hand, Siberian examples are sometimes so remarkably grey, and so devoid of streaks on the nape, that it seems quite possible that the *Picus canus perpallidus* of the same author may be recognized when sufficient material for comparison has been obtained.

131. PICUS MARTIUS.
(GREAT BLACK WOODPECKER.)

Picus martius, Linneus, Syst. Nat. i. p. 173 (1766).

The Great Black Woodpecker is a large bird (wing from carpal joint about 9 inches); and is black all over, with the crown and nape red in the male and the nape only in the female.

Figures: Dresser, Birds of Europe, v. pl. 274.

The Great Black Woodpecker is a resident in Yezzo, but does not occur south of the Straits of Tsugaru. It is common in the woods near Hakodadi (Blakiston, Ibis, 1862, p. 325). There are two examples in the Swinhoe collection (Swinhoe, Ibis, 1875, p. 451), one in the Pryer collection, and I have a fine example collected by Mr. Henson—all of them from Hakodadi.

It is doubtful whether the Great Black Woodpecker has ever occurred in the British Islands, but its range extends from Scandinavia across Europe and Southern Siberia to Japan.

132. PICUS RICHARDSI.
(TRISTRAM'S WOODPECKER.)

Dryocopus richardsi, Tristram, Proc. Zool. Soc. 1879, p. 386.

Tristram's Woodpecker is a large bird (wing from carpal joint nearly 10 inches). It is black, with the lower breast and belly, the lower back and rump, under wing-coverts and axillaries, and the base and tips of the primaries white.

Figures: Tristram, Proc. Zool. Soc. 1879, pl. 31.

Tristram's Woodpecker is only known from a single example, which was procured by Captain Richards on the island of Tsusima in the Straits of Corea. It appears to be most nearly related to *Picus feddeni* from the Burma peninsula, and to *Picus kulinowskii* from the Corean peninsula.

Picus richardsi.

These three Woodpeckers and half a dozen others form a compact little subgenus, to which the name of *Thriponax* has been applied. They agree with all the species of the genus *Picus* in the position of the nasal grooves, and of the angle of the mandible, as well as in the length of the tail; but they differ from *Dryocopus* and typical *Picus* in the comparative length of their toes. In typical *Picus* the reversed toe is the longest; in *Dryocopus* the reversed toe is equal in length to the middle toe; whilst in *Thriponax* (as in the genus *Gecinus*) the reversed toe is shorter than the middle toe. Somewhat the same relation of the subgenera of *Picus* to each other and to *Gecinus* is also observable in the distance between the nasal grooves.

In typical *Picus* this distance at the nostrils is more than half the width of the bill; in *Dryocopus* and *Thriponax* about a third; whilst in *Gecinus* it is less than a third.

133. PICUS NOGUCHII.
(PRYER'S WOODPECKER.)

Picus noguchii, Seebohm, Ibis, 1887, p. 178.

Pryer's Woodpecker is medium sized (wing from carpal joint 5¾ inches). It is black above and reddish brown below, and it has a few white spots on the primaries.

Figures: Seebohm, Ibis, 1887, pl. 7.

Picus noguchii.

Pryer's Woodpecker is only known from a single example obtained by Mr. Pryer's collectors on the largest island of the central group of the Loo-Choo Islands. It has a longer first primary than is usual in *Picus*, and the nostrils are also less concealed; but possibly both these characters may be affected by its extreme youth.

The reversed toe is too long and the tail is much too long for the genus *Blythipicus* or *Lepocestes*, as suggested by Dr. Stejneger (Zeitsch. ges. Orn. 1887, p. 172). I prefer to retain it in the genus *Picus*, under the subgeneric term of *Supheopipo*, as proposed by Mr. Hargitt—a subgenus which may possibly have to be abandoned when fully adult examples have been examined.

134. PICUS LEUCONOTUS.
(WHITE-BACKED WOODPECKER.)

Picus leuconotus, Bechstein, Naturg. Deutschl. ii. p. 1034 (1805).

The White-backed Woodpecker is a large bird (wing from carpal joint about 6 inches), and is easily distinguished from the other Japanese species by the crimson on its under tail-coverts extending also to the belly.

Figures: Dresser, Birds of Europe, v. pl. 279 (typical form).

The White-backed Woodpecker is a resident in all the Japanese Islands. The earliest record of its occurrence in Japan is that of an example procured in Yezzo in October 1861 (Blakiston, Ibis, 1862, p. 325). There are two examples in the Swinhoe collection from South Yezzo (Swinhoe, Ibis, 1875, p. 451); and I have an example collected by Mr. Henson at Hakodadi on the 13th of April. There are eighteen examples in the Pryer collection from Yokohama.

The range of the White-backed Woodpecker extends westwards from Japan across Siberia into Europe; but although it reaches Norway, Germany, and Spain, this species is not known to have occurred in the British Islands.

The White-backed Woodpecker is subject to much climatic variation. The Arctic race ranges from Russia across Southern Siberia to the mouth of the Amoor. The amount of white on the upper parts, especially on the tertials, is at least double that on examples of the typical form from Norway, and entitles it to rank as an excellent subspecies under the name of *Picus leuconotus cirris*. Dr. Stejneger has separated the race found in Southern Japan under the name of *Dryobates subcirris* (Stejneger, Proc. United States Nat. Mus. 1886, p. 113). In the amount of white on the upper parts they agree with examples from Yezzo and the Island of Askold in being intermediate between the Arctic and the typical form; but whilst the latter differ very slightly from the typical form in the colour of the underparts, the race which inhabits Southern Japan differs from all other races of this species in having the white confined to the throat, and the crimson on the belly much more developed and graduating on the breast into brownish buff. This race intergrades with the Yezzo race, and can only claim subspecific rank as *Picus leuconotus subcirris*.

135. PICUS NAMIYEI.
(STEJNEGER'S WOODPECKER.)

Dryobates namiyei, Stejneger, Proc. United States Nat. Mus. 1886, p. 116.

Stejneger's Woodpecker is rather less than the White-backed Woodpecker (wing from carpal joint 5¾ inches), which it very closely resembles in colour, except that the white is everywhere much reduced in extent.

Figures: Stejneger, Proc. United States Nat. Mus. 1886, pl. 2.

Stejneger's Woodpecker is only known from a single example in the Tokio Museum, which was obtained at Yamato, south-west of Osaka (about halfway between Nagasaki and Yokohama), and which has been examined and described by Dr. Stejneger (Blakiston and Pryer, Trans. As. Soc. Japan, 1882, p. 133). It is of the same size as *Picus leuconotus subcirris* (wing 5·75 inches), but in colour it comes nearest to *Picus insularis* from Formosa. The latter is a smaller bird (wing 5·4 to 5·2 inches), but has more white on the back and on the wings. Stejneger's Woodpecker agrees with *Picus insularis* in the colour of the underparts, which are much more streaked with black on the breast and flanks than in *Picus leuconotus*. It differs from *Picus insularis* in the upper parts in having broad instead of narrow black tips to the feathers of the lower back. Its specific rank is very doubtful, but until a series has been obtained it is impossible to say with which species it intergrades.

136. PICUS MAJOR.
(GREAT SPOTTED WOODPECKER.)

Picus major, Linnæus, Syst. Nat. i. p. 176 (1766).

The Great Spotted Woodpecker is a medium-sized species (wing from carpal joint about 5½ inches) with no white on the back except on the scapulars, and the crimson on the underparts scarcely extending above the under tail-coverts.

Figures: Dresser, Birds of Europe, v. pl. 275 (typical race).

The Great Spotted Woodpecker was first recorded as a Japanese bird on the authority of Mr. Heine, who obtained it at Hakodadi in May 1854, during the Perry Expedition (Cassin, Exp. Am. Squad. China Seas and Japan, ii. p. 222). Other examples were recorded from the same locality, obtained in October 1857 by Dr. Henderson during the cruise of the 'Portsmouth' (Cassin, Proc. Acad. Nat. Sc. Philad. 1858, p. 195), and the characters in which they differed from the European form were pointed out. It appears to be generally distributed in the Japanese Islands; I have three examples obtained by Mr. Snow in the Kurile Islands, and five examples obtained by Captain Blakiston in Yezzo, where it appears to be a resident, as the dates on the skins are February, March, May, and November. There are twelve examples in the Pryer collection from Yokohama.

The range of the Great Spotted Woodpecker extends westwards from Japan across Siberia and Europe to the British Islands. The variations in the plumage of this species are considerable and appear to be climatic. The arctic race extends across Lapland and Siberia, and may be called *Picus major cissa*. The throat, breast, and flanks are pure white, and the terminal half of the tertiaries is black. The typical form inhabiting Southern Scandinavia and Western Europe is an intermediate one, the extreme of the first character being found in the Caucasus: this race is called *Picus major poelzami*, and has the throat, breast, and flanks chocolate-brown. The extreme of the second character is found in Japan: this race is called *Picus major japonicus* (Seebohm, Ibis, 1883, p. 24), and has the tertials crossed by three broad white bands, only interrupted by a black shaft-line, one of the bands being nearly terminal.

The young in first plumage of the Japanese race differs so much from that of the west European that the two races may possibly prove to be specifically distinct. The young of our birds have nearly uniform buffish-white underparts, whilst those of the Japanese race are profusely streaked with black on the flanks, and more or less so on the throat and breast.

My examples from the Kurile Islands and from Yezzo are on an average whiter on the underparts than those from Southern Japan, but they do not differ from them in the amount of white on the tertials. They vary considerably in both respects in both localities. An example from Sakhalien, collected by Dr. Schrenck, leads, through an example from the Amoor, up to the arctic race.

Dr. Stejneger regards the Great Spotted Woodpeckers of Yezzo as specifically distinct from those of Hondo (Stejneger, Proc. United States Nat. Mus. 1886, p. 109) on the ground that the latter are darker on the underparts, and have much less white on the scapulars. Examples from the Kurile Islands and from Yezzo are on an average slightly whiter on the underparts than birds from Southern Japan, but in the amount of white on the scapulars they do not differ. The scapulars are always white with concealed black bases.

It is the commonest Woodpecker in the mountains of Central Hondo (Jouy, Proc. United States Nat. Mus. 1883, p. 307).

137. PICUS MINOR.
(LESSER SPOTTED WOODPECKER.)

Picus minor, Linnæus, Syst. Nat. i. p. 176 (1766).

The Lesser Spotted Woodpecker is a small species (wing from carpal joint about 3¾ inches), with the front part of the crown white (suffused with red in the male) and with the hinder part of the crown and the nape black.

Figures: Dresser, Birds of Europe, v. pl. 282.

The Lesser Spotted Woodpecker is a resident on Yezzo; but is not known to have occurred in Southern Japan. I have an example from Hakodadi, collected by Captain Blakiston (Seebohm, Ibis, 1879, p. 29) on the 11th of May.

The Lesser Spotted Woodpecker has a very wide range from the Azores and the British Islands into Algeria, and across Europe and Siberia to Kamtschatka and the north island of Japan. With such an extensive distribution it is not surprising that it may be subdivided into various climatic races. The arctic race extends across Lapland and Siberia, and may be called *Picus minor pipra*. The underparts, with the exception of a few dark streaks on the under tail-coverts, are pure white, and the black transverse bars on the lower back and rump are very obscurely indicated. If the typical form be that which occurs in Southern Scandinavia (which was presumably the one described by Linnæus), it scarcely differs from the race found in Japan, which is by no means white on the underparts, is streaked on the breast and flanks, as well as on the under tail-coverts, and is more barred on the lower back and rump than Siberian examples (Seebohm, Ibis, 1884, p. 36). In the British Islands and in Southern Europe all these characters are more pronounced, the extreme dark form, *Picus minor danfordi*, occurring in Asia Minor.

138. IYNGIPICUS KISUKI.
(TEMMINCK'S PIGMY WOODPECKER.)

Picus kisuki, Temminck. Planches Coloriées, text to no. 585 (1836).

Temminck's Pigmy Woodpecker may be best distinguished by its brown forehead and crown.

Figures: Temminck and Schlegel, Fauna Japonica, Aves, pl. 37.

Temminck's Pigmy Woodpecker is found on all the Japanese Islands, including the Loo-Choo Islands, and also occurs in Manchuria and on the Corean Peninsula. Even within this small range it is subject to considerable local variation. The typical form was originally described from the island of Kiusiu, whose name it bears, and whence I have examined three examples collected by Mr. Ringer at Nagasaki. On the main island and on Yezzo a larger and paler form occurs, *Iyngipicus kizuki seebohmi* (Hargitt, Ibis, 1884, p. 100), of which there is an example in the Swinhoe collection from Hakodadi (Swinhoe, Ibis, 1875, p. 451), and five examples in the Pryer collection from Yokohama. This is the form which has occurred on the Island of Askold and in the valley of the Ussuri, but examples from these localities are on an average larger than those from Japan. There are three examples in the Pryer collection from the Loo-Choo Islands, which are smaller and darker than the typical form, and to which I have given the name of *Iyngipicus kisuki nigrescens* (Seebohm, Ibis, 1887, p. 177). I have also an example of the typical form collected by Mr. Heywood Jones on Fuji-yama, two collected by Mr. Owston at Yokohama, one collected by Mr. Snow on the Kurile Islands, and one collected by Mons. Kalinowski in the valley of the Ussuri. They vary considerably in size, as the following measurements of the length of the wing from the carpal joint testify:—

	in.
Ussuri	3·55
Kuriles	3·5
Yezzo	3·4
Yokohama	3·45 to 3·2
Nagasaki	3·2
Loo-Choo Islands	3·2 to 3·1

All my Yokohama examples (eight), including a breeding female, agree in colour and markings with the skin from Yezzo, and not with that from Nagasaki.

There can be little doubt that Temminck's Pigmy Woodpecker is most nearly related to *I. pygmæus* and its climatic races, which range from the Himalayas across China, Hainan, and Formosa, to Askold and the valley of the Ussuri. It agrees with them in having black upper tail-coverts and central tail-feathers, but it differs from them in having the white superciliary stripe separated by a brown band from the white sides of the neck, and in having the nape and upper back brown like the crown instead of black. As no species of *Iyngipicus* is known to occur north of the Amoor, *I. kisuki* must be regarded as a tropical species which long ago emigrated from South China to Japan.

It almost invariably accompanies flocks of Tits, associating with them in perfect harmony, and uttering its warning cry *geed, geed*, as it moves from tree to tree (Jouy, Proc. United States Nat. Mus. 1883, p. 308).

139. IYNX TORQUILLA.
(WRYNECK.)

Yunx torquilla, Linneus, Syst. Nat. i. p. 172 (1766).

The Wryneck is a small bird (wing from carpal joint 3·5 to 3·2 inches). It is barred or spotted all over with dark brown or grey upon a white or buff ground.

Figures: Dresser, Birds of Europe, v. pl. 289.

The Wryneck is found in all the Japanese Islands, but is probably only a summer visitor to Yezzo. There are two examples in the Swinhoe collection from Hakodadi (Swinhoe, Ibis, 1874, p. 162), and one in the Pryer collection from Yokohama. Mr. Ringer has sent examples to the Norwich Museum obtained at Nagasaki, where those procured by the Siebold expedition were probably obtained (Temminck and Schlegel, Fauna Japonica, Aves, p. 75).

The breeding-range of the Wryneck extends from the British Islands across Europe and Southern Siberia to Japan. This species also breeds in the Himalayas.

Japanese examples have been described as distinct from the European and Asiatic species under the name of *Yunx japonica* (Bonaparte, Consp. Generum Avium, i. p. 112). It is alleged that they are smaller and paler than the typical form, but there does not seem to be any foundation for the statement.

Suborder V. *UPUPÆ*.

Plantars passerine; episternal process perforated to receive the feet of the coracoids; palate desmognathous.

The Upupæ consist of two small families—the Upupidæ (10 species), which are found in the Palæarctic, Ethiopian, and Oriental Regions; and the Irrisoridæ (12 species), only found in the Ethiopian Region.

Pterylosis of *Upupa epops* (upper parts).

Pterylosis of *Upupa epops* (under parts).

140. UPUPA EPOPS.
(HOOPOE.)

Upupa epops, Linnæus, Syst. Nat. i. p. 183 (1766).

The Hoopoe, with its long curved bill and its conspicuous crest, is too well known to need description.

Figures: Dresser, Birds of Europe, v. pl. 298.

The sole claim of the Hoopoe to be regarded as a Japanese bird rests upon a single example in the possession of Captain Blakiston, which was obtained off the south-east coast of Yezzo, and which he kindly sent me for examination (Seebohm, Ibis, 1884, p. 36).

The Hoopoe is not yet quite exterminated in the British Islands, and its breeding-range extends across Europe and Southern Siberia to the Himalayas and China.

Order TROGONES.

Feet heterodactyle; first and second digits directed backwards, third and fourth forwards; hind plantar (*flexor longus hallucis*) dividing into two tendons at the foot of the tarsus, leading to the two hind toes, front plantar (*flexor perforans digitorum*) also dividing into two tendons, leading to the two front toes.

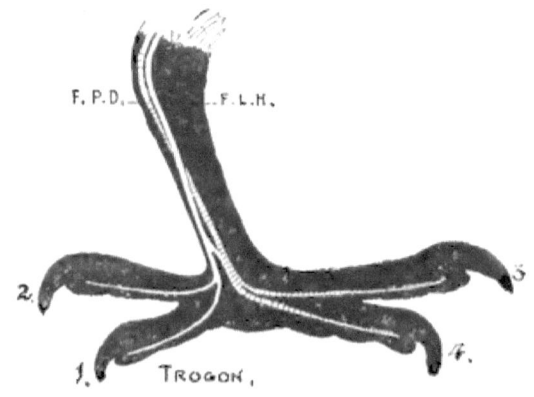

Suborder VI. *TROGONES*.

The Trogones consist of one suborder only, which comprises the Trogons, a group of about 60 species, distributed in the Neotropical, Ethiopian, and Oriental Regions, but not extending to Japan, though one species reaches the Philippine Islands.

Order COLUMBÆ.

Palate schizognathous; nasals schizorhinal; young born helpless.

Suborder VII. *COLUMBÆ.*

The Pigeons are so isolated a group of birds that the suborder may fairly claim ordinal rank. There are nearly 400 species contained in this almost cosmopolitan group, but with the exception of about a score which inhabit the Nearctic and Palæarctic Regions, they are all tropical or subtropical.

Most of the Japanese Pigeons belong to tropical genera.

141. COLUMBA LIVIA.
(BLUE ROCK-PIGEON.)

Columba livia, Brisson, Orn. i. p. 82 (1760); Bonnaterre, Table Encycl. et Méthod. i. p. 227 (1790).

The Rock-Pigeon of Japan is a very dark bird, and varies greatly in the colour of its belly and lower back. The former varies from dark grey to pale grey, and the latter from nearly black to nearly white.

The Rock-Dove of Japan, like that of Siberia, appears to be a feral bird. There are three examples in the Pryer collection from Yokohama. It is said to breed in a cave on Eno-sima (Blakiston and Pryer, Ibis, 1878, p. 227) ; and Captain Rodgers obtained examples on the Loo-Choo Islands (Cassin, Proc. Acad. Nat. Sc. Philad. 1862, p. 320).

142. TURTUR ORIENTALIS.
(EASTERN TURTLE-DOVE.)

Columba orientalis, Latham, Index Orn. ii. p. 606 (1790).

The Eastern Turtle-Dove is one of the medium-sized Japanese Columbæ (wing from carpal joint 7¾ to 7 inches). It may be easily recognized by the broad chestnut-brown margins of its dark-centred scapulars and tertials.

Figures: Temminck and Schlegel, Fauna Japonica, Aves, pl. 60 B (as *Columba gelastis*).

The Eastern Turtle-Dove has been observed on Eturop, one of the Kurile Islands (Blakiston and Pryer, Trans. As. Soc. Japan, 1882, p. 129), and is a summer visitor to Yezzo (Whitely, Ibis, 1867, p. 204). There is an example in the Swinhoe collection from Hakodadi (Swinhoe, Ibis, 1874, p. 162), whence it was procured by the Perry Expedition twenty years previously (Cassin, Exp. Am. Squad. China Seas and Japan, ii. p. 222). There are seven examples in the Pryer collection from Yokohama. Captain Rodgers obtained it on the Loo-Choo Islands (Cassin, Proc. Acad. Nat. Sc. Philad. 1862, p. 320); and there is an example in the Pryer collection from the central group of those islands (Seebohm, Ibis, 1887, p. 179).

Examples from the southern group of the Loo-Choo Islands have been described as distinct under the name of *Turtur stimpsoni* (Stejneger, Proc. United States Nat. Mus. 1887, p. 399). They are said to be deeper in colour than those from Japan. This is probably only individual variation, as the example in the Pryer collection from the central group of the Loo-Choo Islands is not so deep in colour as many of those from Japan, and agrees in every respect with an example from Hakodadi.

The Eastern Turtle-Dove is the eastern representative of our Common Turtle-Dove (*Turtur auritus*), and is as widely distributed in the temperate parts of the Eastern Palæarctic Region as its British ally is in the temperate parts of the Western Palæarctic Region.

In the Eastern species the under tail-coverts and the pale terminal band across the tail and the tips of the feathers on the sides of the neck are lavender-grey, instead of white or nearly so. The southern range of this species extends through China, Cochin China, and Burma, into India and Ceylon; but in Nepal, Turkestan, and Southwest Siberia the under tail-coverts and the bar across the tail are frequently almost as pale as in the Western form (which also reaches Turkestan), though the lavender-grey tips of the feathers on the sides of the neck are retained. This local race may be called *Turtur orientalis ferrago*.

It has been stated that the Japanese birds are larger than those from China and India, and ought therefore to be regarded as distinct under the name of *Turtur gelastes* of Temminck; but this generalization has been arrived at from the measurement of too small a series. The length of wing from carpal joint varies in twelve Japanese examples from 7·8 to 7·2 inches, and in twenty-two Indian and Chinese examples from 7·9 to 7·1 inches.

The Eastern Turtle-Dove has been recorded more than once as an accidental visitor to Scandinavia, and an example was shot at Scarborough in the autumn of 1889. In all these cases the birds were in first autumn plumage.

143. TURTUR RISORIUS.
(COMMON INDIAN DOVE.)

Columba risoria, Linnæus, Syst. Nat. i. p. 285 (1766).

The Common Indian Dove is less than the Eastern Turtle-Dove (wing from carpal joint 7 to 6½ inches). It may easily be distinguished from the other Japanese Columbæ by its uniform brownish-grey scapulars and tertials.

Figures: Jardine's Nat. Libr., Pigeons, pl. 17; Dresser, Birds of Europe, vii. pl. 464. fig. 2.

The Common Indian Dove is a summer visitor to Southern Japan, but has not been recorded from Yezzo (Swinhoe, Ibis, 1876, p. 334). There is an example in the Swinhoe collection from Yokohama (Swinhoe, Ibis, 1877, p. 145), and there are three in the Pryer collection from the same locality.

The Common Indian Dove has a wide distribution from Turkey and Asia Minor across India, Ceylon, Burma, and China, to Japan; and Schrenck records a single example from the Lower Amoor. It must be regarded as a tropical species whose summer range extends to the extreme south of the Palæarctic Region.

144. TURTUR HUMILIS.
(CHINESE RED DOVE.)

Columba humilis, Temminck, Planches Coloriées, nos. 258, 259 (1824).

The Chinese Red Dove is a small bird (wing from carpal joint about 5¼ inches). Its wing-coverts, scapulars, and tertials are vinous-red.

The Chinese Red Dove has very small claims to be regarded as a Japanese bird. An example was obtained by Mr. Owston from a dealer at Yokohama, who asserted that it had been shot in the neighbourhood (Seebohm, Ibis, 1884, p. 179). The skin is in my collection, and shows no marks of having been in confinement; on the contrary, it appears to have been shot in the wings.

The Chinese Red Dove is a resident in South China and Formosa, the Philippine Islands, and the Burma Peninsula, but is replaced in India by a very nearly allied species, *Turtur tranquebaricus*.

145. TRERON SIEBOLDI.
(JAPANESE GREEN PIGEON.)

Columba sieboldi, Temminck, Planches Coloriées, no. 549 (1835).

The Japanese Green Pigeon differs from its ally on the Loo-Choo Islands in the great extent of white on its belly, and in the yellowness of the green on its head and breast.

Figures: Temminck and Schlegel, Fauna Japonica, Aves, pl. 60 D.

The Japanese Green Pigeon is peculiar to Japan. It is a summer visitor to Yezzo, but in Southern Japan it is a resident. There is an example in the Swinhoe collection from Hakodadi (Whitely, Ibis, 1867, p. 201), and there are six examples in the Pryer collection from Yokohama. Mr. Ringer has obtained it at Nagasaki (Blakiston, Am. List Birds Japan, p. 44), whence he has sent an example to the Norwich Museum.

It is tolerably abundant on Fuji-yama, but exceedingly shy, and is very fond of feeding on wild cherries (Jouy, Proc. United States Nat. Mus. 1883, p. 314). In Yezzo it prefers the wooded bluffs near the sea, and frequently alights on the sandy shore (Blakiston and Pryer, Trans. As. Soc. Japan, 1882, p. 129). It has a long and varied coo.

The Japanese Green Pigeon is nearest related to *Treron sororia* from Formosa. It is doubtful whether the females of the two races are separable, but the males differ slightly in the colour of the mantle. In *T. sieboldi* the vinous red of the wing-coverts is distinctly traceable across the mantle, but in *T. sororia* the green of the mantle scarcely differs from that of the lower back, rump, and upper tail-coverts. In both races the tail is much graduated, the outer feathers being an inch shorter than the centre ones. These two Pigeons are, of course, island races of a continental species, which appears to be *T. sphenura*, a Himalayan bird ranging into Burma. This species scarcely differs from its Japanese offshoots in colour; it agrees with both its insular races in the shape and colour of its tail, which is much graduated and is crossed by a dark terminal band, but it differs from both of them, and from all the other species of *Treron* (except from the long-tailed *T. apicauda*), in having lost the curious sinuation on the inner web of the third primary, so characteristic of the other species in the genus.

146. TRERON PERMAGNA.
(LOO-CHOO GREEN PIGEON.)

Treron permagna, Stejneger, Proc. United States Nat. Mus. 1886, p. 637.

The Loo-Choo Green Pigeon is a larger bird than its Japanese ally; and the head, breast, and belly are a nearly uniform dark green.

The Loo-Choo Green Pigeon was described by Dr. Stejneger from an example obtained by Mr. Namiye on the island of Okinawa-Shima, one of the central group of the Loo-Choo Islands. It is so nearly allied to one of the Formosan Green Pigeons that its specific distinctness must be regarded as somewhat doubtful until a larger series is obtained. There are two examples, apparently male and female, in the Pryer collection.

Both the Formosan Green Pigeons are represented by allied races in the islands lying to the north. In Japan *T. sororia* is represented by *T. sieboldi*, and on the Loo-Choo Islands *T. formosana* is represented by *T. permagna*. The Loo-Choo Green Pigeon can only be regarded as a large race of its Formosan ally; the length of the wing in the former varying from 8·2 to 7·7 and that of the latter from 7·6 to 7·1 inches. The females of the two races scarcely differ in colour, but in the male of *T. formosana* in the Swinhoe collection (the type) the green of the crown is suffused with orange, which is not the case with the male of *T. permagna* in the Pryer collection. In both races the graduation of the tail is very slight (about half an inch), and the third primary is sinuated. It is very difficult to trace the affinities of these nearly allied races, but they do not seem to belong to the same group as the Hainan Green Pigeon, the species belonging to which are characterized by a broad pale terminal band across the under surface of the tail. They probably belong to the same stock as the other two races inhabiting nearly the same area.

147. CARPOPHAGA IANTHINA.
(JAPANESE FRUIT-PIGEON.)

Columba janthina, Temminck, Planches Coloriées, no. 503 (1830).

The Japanese Fruit-Pigeon is a large bird (wing from carpal joint 9 to 8½ inches). It is slaty brown, bronzed with reddish purple and green.

Figures: Temminck and Schlegel, Fauna Japonica, Aves, pl. 60 c.

The Japanese Fruit-Pigeon is peculiar to Japan and some of the neighbouring islands. There are two examples in the Pryer collection from Yokohama, and one from the central group of the Loo-Choo Islands. There is an example in the Norwich Museum obtained by Mr. Ringer near Nagasaki, and there are several examples in the British Museum from Nagasaki and Yokohama. There is an example in the Senckenberg Museum in Frankfort (labelled *Columba metallica*) obtained by Kittlitz on one of the Bonin Islands, and there is a second example in the St. Petersburg Museum from the same source. The latter has been made the type of a new species, *Ianthœnas nitens* (Stejneger, Proc. United States Nat. Mus. 1887, p. 421), on the ground that the head is brown instead of grey. The difference is doubtless due to abrasion. Amongst the examples in the British Museum from Yokohama and Nagasaki are several in which the slate-grey ground-colour has more or less faded to russet-brown, and the metallic purples and greens have become dull. The metallic colours are very deceptive. In typical examples the breast is green, very slightly suffused with pinkish purple when seen with the spectator's back to the light. The Bonin example in the Senckenberg Museum is the greenest I have seen. On the other hand, the Loo-Choo example in the Pryer collection has the breast-feathers pinkish purple with green bases, when seen in the position mentioned. These are probably individual differences unconnected with geographical distribution.

The genera of the Columbæ have never been satisfactorily diagnosed, and it is possible that this species and the two following do not belong to the genus *Carpophaga*. It has been stated (Garrod, Proc. Zool. Soc. 1875, p. 367) that an allied species differs from the species of that genus in two important particulars: it has no gall-bladder and it has a cæcum.

148. CARPOPHAGA VERSICOLOR.

(BONIN FRUIT-PIGEON.)

Columba versicolor, Kittlitz, Kupfertafeln zur Naturgeschichte der Vögel, p. 5 (1832).

The Bonin Fruit-Pigeon differs from its Japanese ally in having the breast pinkish purple like the crown, and in having a pale throat. It is larger than the Japanese species (wing from carpal joint 11 to 10 inches), and much paler both on the upper and under parts.

Figures: Kittlitz, Kupfertafeln zur Naturgeschichte der Vögel, pl. 5. fig. 2.

The Bonin Fruit-Pigeon was discovered on one of the Bonin group of islands in 1827 by Captain Beechey during the voyage of the 'Blossom;' but in consequence of the unreasonable delay in the completion of the part relating to the Mollusca, the results of the voyage were not published until 1839, when this interesting bird received the name of *Columba metallica* (Vigors, Zool. Captain Beechey's Voyage, p. 25). In the meantime two events happened which make the use of this name impossible. In 1828 F. H. von Kittlitz spent a fortnight on the Bonin Islands, and also discovered the Fruit-Pigeon which is peculiar to them, which he figured and described in 1832 under the name of *Columba versicolor*. But not only was Vigors's name antedated by that of Kittlitz in consequence of the provoking delay, but it was completely nullified by its independent application in 1835 to another species of Fruit-Pigeon from the island of Timor (Temminck, Planches Coloriées, no. 562). A third name, *Columba kitlizii*, was given to the Bonin species in the same year (Temminck, Planches Coloriées, page following text to no. 578). The opinion that this name was applied to the Japanese species (Schlegel, Mus. Pays-Bas, iv. *Columbæ*, p. 74) is manifestly erroneous.

A fourth name was given to it in 1858 (Kittlitz, Denkwürdigkeiten einer Reise nach dem Russischen Amerika, nach Mikronesien und durch Kamtschatka, ii. p. 175), when it was proposed to substitute the name of *Columba iris* for that of *Columba versicolor*.

There can be little doubt that Vigors's type of this species was once in the Museum of the Zoological Society (together with that of *Nycticorax crassirostris* and *Coccothraustes ferreirostris*); but there is no evidence that it was transferred to the British Museum when the collection belonging to the Zoological Society was dispersed.

There is an example in the St. Petersburg Museum, which was obtained by Kittlitz on the Bonin Islands, and which has recently been described in detail (Stejneger, Proc. United States Nat. Mus. 1887, p. 421); and there is a second example in the Senckenberg Museum in Frankfort from the same source. The latter is a large bird (wing from carpal joint 11 inches), and differs conspicuously from its Japanese ally in being much paler in colour. It has also a yellower bill and a much paler throat. The St. Petersburg skin is 10·1 in length of wing, and Vigors gives 10 inches. Mr. Holst obtained for me a male from Nakondo-Shima, one of the Parry Islands (Seebohm, Ibis, 1890, p. 103). It measures 10 inches in length of wing, and is much larger and paler than its Japanese ally. The bronze on the wing-coverts is green in all positions, and the ground-colour of the underparts is lavender instead of dark bluish grey.

149. CARPOPHAGA JOUYI.
(LOO-CHOO FRUIT-PIGEON.)

Ianthænas jouyi, Stejneger, American Naturalist, 1887, p. 583.

The Loo-Choo Fruit-Pigeon is larger than its Japanese ally (wing from carpal joint $10\frac{1}{4}$ to $9\frac{3}{4}$ inches). It principally differs in having a white crescent across the upper back.

The Loo-Choo Fruit-Pigeon was described by Dr. Stejneger from an example obtained by Mr. C. Tasaki on one of the islands whose name it bears. There are two examples in the Pryer collection, obtained from the central group of the Loo-Choo Islands, most probably by the same collector (Seebohm, Ibis, 1887, p. 179).

This fine and remarkably distinct species is doubtless a resident on the Loo-Choo Islands, whilst the Japanese Fruit-Pigeon (*Carpophaga ianthina*), which also occurs on this group, may prove to be only a winter visitor.

These three Fruit-Pigeons have close allies on the Philippine Islands, as well as on some of the islands in the Malay Archipelago, and must be regarded as of Tropical origin.

Order COCCYGES.

Palate desmognathous; basipterygoid processes absent; hallux present, and connected with the *flexor longus hallucis*, not with the *flexor perforans digitorum*, which leads to the second, third, and fourth digits. Young not passing through a complete downy stage. Spinal feather-tract well defined on the neck.

Suborder VIII. *MUSOPHAGI*.

Palate desmognathous; feet semi-zygodactyle; plantars galline; spinal feather-tract well defined on neck by lateral bare tracts, but with no interscapular fork.

There are about 20 species of Plantain-eaters, which are confined to the Ethiopian Region.

Suborder IX. *CUCULI*.

Palate desmognathous; basipterygoid processes absent; feet zygodactyle; plantars galline; oil-gland nude.

The Cuckoos are an almost cosmopolitan group of birds, and number nearly 200 species. They have been divided into three subfamilies (Beddard, Proc. Zool. Soc. 1885, p. 187), which are more properly regarded as families:—

	Cuculidæ.	Syrinx tracheo-bronchial.
Accessory femoro-caudal present.	*Phœnicophæidæ.*	
	Centropodidæ.	

The Cuculidæ is the only family of this suborder which is represented in Japan.

150. CUCULUS CANORUS.
(COMMON CUCKOO.)

Cuculus canorus, Linneus, Syst. Nat. i. p. 168 (1766).

Japanese examples of the Common Cuckoo appear to be absolutely similar to European examples. They completely intergrade with the Himalayan Cuckoo in size (wing from carpal joint 9 to 8 inches); and it is not known that they differ in any way in colour, except that in the adult Common Cuckoo there is no tendency for the tail to darken near the tip, and in the rufous stage there are no bars across the rump.

Figures : Dresser, Birds of Europe, v. pl. 299.

The Common Cuckoo has long been known to occur both on the Kurile Islands and in Japan (Pallas, Zoogr. Rosso-Asiat. i. p. 443). It is a summer visitor to all the Japanese Islands. There is an example in the Swinhoe collection from Hakodadi (Swinhoe, Ibis, 1875, p. 451), whence it was obtained by the Perry Expedition nearly twenty years previously (Cassin, Exp. Am. Squad. China Seas and Japan, ii. p. 222). There are fourteen examples in the Pryer collection from Yokohama.

The breeding-range of the Common Cuckoo extends from the British Islands across Europe and Southern Siberia to Japan.

151. CUCULUS INTERMEDIUS.
(HIMALAYAN CUCKOO.)

Cuculus intermedius, Vahl, Scrift. Nat. Selsk. iv. pt. i. p. 59 (1797).

The Himalayan Cuckoo is a small form (wing from carpal joint $7\frac{3}{4}$ to $6\frac{3}{4}$ inches) of the Common Cuckoo; but, having a totally different note (Seebohm, Ibis, 1878, p. 326), it is regarded as specifically distinct. The tail has a slight tendency to darken towards the tip, and in the rufous stage the rump is barred.

The Himalayan Cuckoo is a summer visitor to all the Japanese Islands. It is not uncommon in Yezzo (Blakiston and Pryer, Trans. As. Soc. Japan, 1882, p. 131), and there are three examples in the Pryer collection from Fuji-yama.

The breeding-range of the Himalayan Cuckoo extends westwards from Japan and China to the Himalayas, Mongolia, and Eastern Siberia, as far west as the valley of the Yenesay.

The Himalayan Cuckoo has been singularly unfortunate as regards its nomenclature. Most writers have called it *Cuculus himalayanus* (Vigors, Proc. Zool. Soc. 1831, p. 172); but there can be no question that the figure of this bird (Gould, Century of Birds from the Himalaya Mountains, pl. 54) represents the rufous phase of *Cuculus poliocephalus*. A still earlier name, dating from 1823, *Cuculus striatus* (Drapiez, Dict. Class. d'Hist. Nat. iv. p. 570), describes a Cuckoo from Java with a total length of "douze pouces," or $12\frac{3}{4}$ English inches, and has been applied by many writers to this species. This can only refer to a large example of the Common Cuckoo. The types of *Cuculus canoroides* (Salomon Müller, Land- en Volkenkunde, p. 235) are fortunately in the Leyden Museum (Schlegel, Mus. Pays-Bas, Cuculi, p. 9), and are said to vary from $7\frac{1}{4}$ to $8\frac{1}{2}$ English inches in length of wing. It is therefore a composite species (from Java, Sumatra, Borneo, and Timor), though some of the types are unquestionably referable to the Himalayan Cuckoo. This name dates from 1839, and there are plenty of later date to choose from :—

1843. *Cuculus saturatus* (Hodgson, Journ. As. Soc. Beng. 1843, p. 942).

1845. *Cuculus optatus* (Gould, Proc. Zool. Soc. 1845, p. 18).

1858. *Cuculus horsfieldi* (Moore, Cat. B. Mus. E. I. Co. ii. p. 703).

1862. *Cuculus canorinus* (Cabanis, Mus. Hein. iv. p. 35).

1863. *Cuculus kelungensis* (Swinhoe, Ibis, 1863, p. 394).

1865. *Cuculus monosyllabicus* (Swinhoe, Ibis, 1865, p. 545).

These names are, however, so very modern that in this exceptional case it may be the wisest course to rake up an ill-defined and forgotten name which dates from 1797. *Cuculus intermedius* was described in a Danish periodical, published in Copenhagen, from an example obtained at Travancore in Madras, and is said to be similar to the Common Cuckoo, but smaller. As there are three species which scarcely differ from each other except in size, and as the Himalayan Cuckoo happens to be the intermediate one, the name is singularly appropriate, though of course it does not fulfil the impossible demands of the ill-starred Stricklandian code.

152. CUCULUS POLIOCEPHALUS.
(LITTLE CUCKOO.)

Cuculus poliocephalus, Latham, Index Orn. i. p. 214 (1790).

The Little Cuckoo appears to be almost similar, both in form and colour, to the Common Cuckoo and the Himalayan Cuckoo, from which it scarcely differs except in size (wing from carpal joint $6\frac{1}{2}$ to $6\frac{1}{4}$ inches). Its note is quite different from that of either of its close allies.

Figures: Gould, Century of Birds from the Himalaya Mountains, pl. 54 (rufous phase).

The Little Cuckoo is a summer visitor to all the Japanese Islands (Blakiston, Am. List Birds of Japan, p. 13). There are seven examples in the Pryer collection from Yokohama.

The range of the Little Cuckoo extends westwards across China to India and Ceylon, and various parts of tropical Africa.

153. HIEROCOCCYX HYPERYTHRUS.
(AMOOR CUCKOO.)

Cuculus hyperythrus, Gould, Proc. Zool. Soc. 1856, p. 96.

The Amoor Cuckoo is the same size as the Common Cuckoo (wing from carpal joint 8 to $7\frac{1}{2}$ inches), but it differs in colour. In adults the breast is uniform vinaceous buff; in the young it is white, longitudinally striped with dark brown. The tail is always barred.

Figures: Schrenck, Reis. u. Forsch. Amur-Lande, i. pl. 10 (immature); Gould, Birds of Asia, vi. pl. 43 (adult).

The Amoor Cuckoo is a summer visitor to all the Japanese Islands (Blakiston and Pryer, Trans. As. Soc. Japan, 1882, p. 132). There are four examples in the Pryer collection from Fuji-yama, and I have two collected by Mr. Heywood Jones from the same locality (Seebohm, Ibis, 1879, p. 28).

It breeds in the valley of the Amoor as well as in Japan, and winters in South China and the Philippine Islands. It has two somewhat close allies—*Hierococcyx fugax,* which inhabits the Malay Peninsula and the adjacent islands of the Malay Archipelago, Sumatra, Java, Borneo, &c.; and *Hierococcyx nisicolor,* which inhabits the Himalayas and Burma.

Subclass CORACIIFORMES.

The Coraciiformes may be diagnosed by a single character. So far as is known they differ from every other bird in the arrangement of their deep plantar tendons. In all other birds the hallux (if it be present and important enough to have any connection with the deep

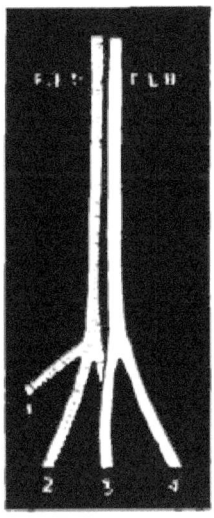

Deep plantar tendons of *Catharista atratus*.

plantar tendons) is connected with the *flexor longus hallucis* and not with the *flexor perforans digitorum*. All the Coraciiformes have a hallux, and in all of them it is connected with the *flexor perforans digitorum*, and not with the *flexor longus hallucis*. The two plantars are always coalesced (as they are in the Accipitres, Anseres, &c.), but may easily be separated by gently tearing them asunder; but in very many cases the tendon to the hallux branches off from the *flexor perforans digitorum* before the two deep plantars coalesce.

The subclass Coraciiformes contains two orders, one of which is represented in Japan, the other being confined to the American continent.

Order PICARIÆ.

The Picariæ differ from all other birds in combining the following two characters :—*Flexor perforans digitorum* leading to hallux; ambiens muscle absent.

The order Picariæ contains three suborders, two of which are represented in Japan.

Suborder X. *HALCYONES*.

Front plantar leading to hallux; spinal feather-tract well defined on the neck by lateral bare tracts, and continuing single down the upper back; vomer absent; palate desmognathous; no basipterygoid processes.

The Halcyones consist of four families. The *Coliidæ* are a very small family, containing half a dozen species, confined to the Ethiopian Region. The *Momotidæ*, with less than a score species, and the *Todidæ*, with about half a dozen, are confined to the Neotropical Region; but the *Alcedinidæ* contain nearly a hundred and fifty species, and, with the exception of the Arctic and Antarctic Regions, are found all over the world.

Three species are found in Japan, all of them apparently of tropical origin.

154. HALCYON COROMANDA.
(RUDDY KINGFISHER.)

Alcedo coromanda, Latham, Index Orn. i. p. 252 (1790).

The Ruddy Kingfisher is more or less rufous all over except a stripe down the centre of the rump and upper tail-coverts, which is white marked with blue.

Figures: Temminck and Schlegel, Fauna Japonica, Aves, pl. 39, under the name of *Alcedo* (*Halcyon*) *coromanda major*; Sharpe, Alcedinidæ, pl. 57.

The Ruddy Kingfisher is said to be only a summer visitor to Yezzo, but to be a resident in the other islands belonging to the Japanese group. In the Swinhoe collection there is an example collected by

Captain Blakiston at Hakodadi (Blakiston and Pryer, Ibis, 1878, p. 230); and in the Pryer collection there are two examples from Yokohama, and three from the central group of the Loo-Choo Islands (Seebohm, Ibis, 1887, p. 176). It has also occurred in the southern group of the Loo-Choo Islands (Stejneger, Proc. United States Nat. Mus. 1887, p. 403).

The Ruddy Kingfisher has a wide range. It occurs in Nepal and Sikkim, the Andaman Islands, Burma and the Malay Peninsula, Sumatra and Java, Borneo and Celebes, and in the Philippine Islands and Formosa. The fact that it has not been recorded from the continent of China is presumptive evidence that it found its way to Japan *viâ* the Loo-Choo Islands and Formosa.

Japanese examples vary in length of wing from 5·1 to 4·6 inches, and may possibly be on an average slightly larger than Indian skins, but scarcely sufficiently so to be regarded as subspecifically distinct. The alleged variations in colour and in the wing-formula do not appear to have any geographical significance.

155. CERYLE GUTTATA.
(ORIENTAL SPOTTED KINGFISHER.)

Alcedo guttatus, Vigors, Proc. Zool. Soc. 1830, p. 22.

The Oriental Spotted Kingfisher is the largest representative of its suborder, not only in Japan, but in the Asiatic continent. It has been asserted that Japanese examples are larger than Indian ones, but this is an error. In both countries the variation is the same (wing from carpal joint 7 to 7½ inches). In the male the breast and the sides of the neck are suffused with chestnut-buff, and the axillaries and under wing-coverts are white; and the dark spots across the breast are few and far between. In the female exactly the opposite is the case; the breast and the sides of the neck are white, profusely spotted with black, but the axillaries and under wing-coverts are chestnut-buff.

Figures: Temminck and Schlegel, Fauna Japonica, Aves, pl. 38 B; Sharpe, Alcedinidæ, pl. 18.

The Oriental Spotted Kingfisher is a resident in the southern islands of Japan; but in Yezzo it is said to be a partial migrant. There are two examples in the Swinhoe collection obtained at Hakodadi in January (Swinhoe, Ibis, 1875, p. 449), and there are three in the Pryer collection from Yokohama. The example figured in the 'Fauna Japonica' as *Alcedo lugubris* was probably obtained by Dr. Siebold at Nagasaki.

On Tate-yama it is found in the wildest mountain-streams and gorges and is exceedingly wary (Jouy, Proc. United States Nat. Mus. 1883, p. 310).

The range of the Oriental Spotted Kingfisher extends from Japan across China, Burma, and the Himalayan valleys as far west as Cashmere. As this species is not found in Siberia, nor in Formosa or the Philippine Islands, it is fair to assume that it reached Japan across China. In the Swinhoe collection there is an example from Ningpo, and l'Abbé David records it from various localities in Central China. In the Christiania Museum there is an example collected by Herr Baun at Puching in North Fokien.

156. ALCEDO ISPIDA.
(COMMON KINGFISHER.)

Alcedo ispida, Linneus, Syst. Nat. i. p. 179 (1766).

Japanese examples of the Common Kingfisher vary in length of wing from 2·8 to 3·0 inches, and in length of bill from 1·2 to 1·5 inches, and may be regarded as belonging to the Eastern race *Alcedo ispida bengalensis*.

Figures: Temminck and Schlegel, Fauna Japonica, Aves, pl. 38.

The Eastern form of the Common Kingfisher is generally distributed throughout the Japanese Islands. It is a summer visitor to Eturop (the most southerly of the Kurile Islands) and to Yezzo (Blakiston and Pryer, Trans. As. Soc. Japan, 1882, p. 136), but further south it is a resident. There is an example in the Swinhoe collection from Hakodadi (Swinhoe, Ibis, 1874, p. 152); there are eight examples in the Pryer collection from Yokohama, and Mr. Ringer has obtained it at Nagasaki. Capt. Rodgers procured it from the Loo-Choo Islands (Cassin, Proc. Acad. Nat. Sc. Philad. 1862, p. 318), and there are two examples in the Pryer collection from the central group of those islands (Seebohm, Ibis, 1887, p. 176).

Few species having so wide a range, and being migratory in so few localities, vary less than the Common Kingfisher. In the

western half of the Palæarctic Region the length of wing varies from
3·2 to 2·8 inches; whilst in the eastern half of that Region and in
the Oriental Region it varies from 2·9 to 2·6 inches. It is note-
worthy that the length of bill (which varies according to age from
1½ to 2 inches) is not known to present any geographical variation;
hence the Eastern form has relatively a slightly longer bill than its
Western representative. It is, however, impossible to recognize the
two forms as specifically distinct.

The range of the Common Kingfisher extends across the Palæarctic
Region from the British Islands to Japan, but does not reach further
north than about latitude 55°. To the south it includes the Canary
Islands, Egypt, India, China, and the islands of the Malay Archipelago.

The Eastern form was described as a distinct species as long ago
as 1788 under the name of *Alcedo bengalensis* (Gmelin, Syst. Nat. i.
p. 450), but it is scarcely probable that any one would claim specific
rank for it now.

Suborder XI. *CORACIÆ*.

Front plantar leading to hallux; spinal feather-tract well-defined
on the neck by lateral bare tracts, but dividing into two tracts on
the upper back; oil-gland nude.

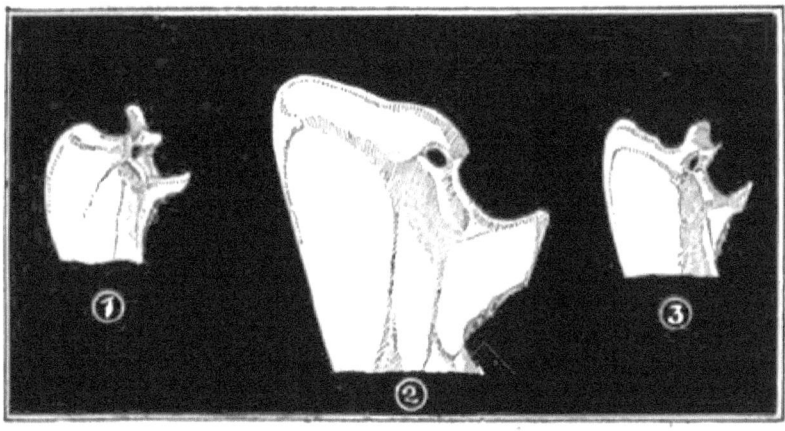

Front portion of sternum of (1) *Upupa epops*, (2) of *Buceros albirostris*,
(3) of *Merops apiaster*.

The Coraciæ consist of seven families. The *Meropidæ*, containing about thirty species, and the *Coraciidæ*, containing nearly a score species, inhabit the tropical and subtropical parts of the Old World. The *Leptosomidæ* contains only one species, which is peculiar to Madagascar. The *Podargidæ* may contain a score species, which are confined to the Oriental and Australian Regions. The *Steatornithidæ* contains only one species, which is peculiar to the Neotropical Region. The *Caprimulgidæ*, numbering a hundred species, and the *Cypselidæ*, numbering about seventy species, are cosmopolitan, except that they are not found in the Arctic or Antarctic regions.

Of these families the *Coraciidæ*, the *Caprimulgidæ*, and the *Cypselidæ* are represented in Japan.

157. CYPSELUS PACIFICUS.
(WHITE-RUMPED SWIFT.)

Hirundo pacifica, Latham, Index Orn. Suppl. p. lviii (1801).

The White-rumped Swift is slightly larger than the Common Swift (wing from carpal joint 6·5 to 7·6 inches), and is easily recognized by its white rump.

Figures: Jardine and Selby, Illustrations of Ornithology, iv. pl. 39; Gould, Birds of Australia, ii. pl. 11.

The White-rumped Swift was first procured in Japan by Captain Blakiston (Swinhoe, Ibis, 1876, p. 331), and has since been found to be a summer visitor to all the Japanese Islands. It has occurred on Eturop, the most southerly of the Kurile Islands (Blakiston and Pryer, Trans. As. Soc. Japan, 1882, p. 140); Captain Blakiston sent me a skin from Hakodadi (Seebohm, Ibis, 1879, p. 31); and there are seven skins in the Pryer collection from Yokohama.

The breeding-range of the White-rumped Swift extends eastwards from Japan across Southern Siberia as far west as Krasnoyarsk in the valley of the Yenesay, whence I have an example procured by Mr. Kibort in June, and as far south as the Lam-yit Islands (on the Chinese coast opposite North Formosa). It winters in the Burma Peninsula and in Australia.

Other white-rumped Swifts are found in the Oriental and Ethiopian Regions, but they are all much smaller birds.

158. CHÆTURA CAUDACUTA.

(NEEDLE-TAILED SWIFT.)

Hirundo caudacuta, Latham, Index Orn. Suppl. p. lvii (1801).

The Needle-tailed Swift is a large bird (wing from carpal joint 8 inches or more). Japanese examples have less white on the forehead than is usual in birds from Siberia, and approach the resident Nepalese species, *Chætura nudipes*, which has no white on the forehead or lores.

Figures: Dresser, Birds of Europe, iv. pl. 270; Gould, Birds of Australia, ii. pl. 10.

The Needle-tailed Swift is a common summer visitor to all the Japanese Islands. There are several examples in the Swinhoe collection from Hakodadi (Swinhoe, Ibis, 1875, p. 448), and there are four examples in the Pryer collection from Yokohama.

The Needle-tailed Swift is an accidental visitor to the British Islands, its breeding-range extending westwards from Japan across Northern China to South-eastern Siberia. It winters in Australia.

159. CAPRIMULGUS JOTAKA.

(JAPANESE GOATSUCKER.)

Caprimulgus jotaka, Temminck and Schlegel, Fauna Japonica, Aves, p. 37 (1847).

The Japanese Goatsucker differs from its British representative in several points, of which perhaps the most important are the spots on the tail-feathers of the male. In the British species the white spots are terminal, but they only occur on the two outer feathers on each side, leaving six central feathers without them. In the Japanese species the white spots are subterminal, but they occur on the four outer feathers on each side, leaving only two central feathers without them. In the plains of India and in Ceylon a paler and smaller form of the Japanese Goatsucker occurs, *C. indicus*, which may possibly be specifically distinct from it, the length of wing varying from 7 to 7·6 inches instead of from 8·2 to 8·8 inches.

Figures: Temminck and Schlegel, Fauna Japonica, Aves, pl. 12 (male), pl. 13 (female).

The Japanese Goatsucker is only entitled to its name on the ground that it was originally described from Japan. It occurs in

Yezzo (Whitely, Ibis, 1867, p. 195), and is common near Yokohama, if we may judge by the fact that there are twenty skins in the Pryer collection. It has also been obtained near Nagasaki by Mr. Ringer, who has presented an example from that locality to the Norwich Museum. To the north its range extends through Manchuria to the valley of the Amoor, but further west in Siberia its place is taken by the European species. To the south its range extends to South-east Mongolia, China, Cochin China, and Burma to Nepal.

It is probably only a summer visitor to Japan, breeding on the mountains, and passing through the plains near Yokohama in May and October (Blakiston and Pryer, Ibis, 1878, p. 231). It is abundant on Fuji-yama in summer, when its cry *chuck, chuck, chuck*, is constantly heard in the still evening air and sometimes before daybreak. In autumn it is said to be silent (Jouy, Proc. United States Nat. Mus. 1883, p. 310).

It lays two eggs on the ground (Blakiston and Pryer, Trans. As. Soc. Japan, 1882, p. 141). Examples in the Pryer collection exactly resemble the smaller varieties of the eggs of the European Goatsucker.

160. EURYSTOMUS ORIENTALIS.
(BROAD-BILLED ROLLER.)

Coracias orientalis, Linneus, Syst. Nat. i. p. 159 (1766).

The Broad-billed Roller is about the size of the European Roller (wing from carpal joint 8 inches), but it has a wider bill and a shorter tail. Its general colour is blue, violet on the wings and tail, and greenish on the body.

Figures: Daubenton, Planches Enluminées, no. 619.

The claim of the Broad-billed Roller to be regarded as a Japanese bird rests upon one example procured at Nagasaki in May 1879 (Blakiston and Pryer, Trans. As. Soc. Japan, 1882, p. 137), and a second obtained on the most southerly group of the Loo-Choo Islands (Stejneger, Proc. United States Nat. Mus. 1887, p. 402).

It inhabits most of the Oriental Region, and must be regarded as a tropical species which occasionally wanders as far as the valley of the Amoor.

Suborder XII. BUCEROTES.

Front plantar leading to hallux; no lateral bare tracts on the neck; no basipterygoid processes; episternal process perforated to receive the feet of the coracoids.

The Bucerotes consist of the family *Bucerotidæ*, containing about sixty species confined to the Ethiopian and Oriental Regions.

Order **MIMOGYPES**.

The Order Mimogypes contains only one Suborder, which is not represented in Japan.

Suborder XIII. *MIMOGYPES*.

The Pseudo-Vultures of America possess the following characters :—

Hallux present, and connected with the *flexor perforans digitorum*;

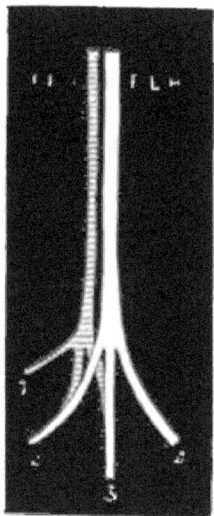

Deep plantar tendons of *Cathartes aura*.

spinal feather-tract not defined on the neck; basipterygoid processes present; young born helpless, but covered with down.

Of these four characters the combination either of the first three or of the last three is not known to occur in any other bird.

Subclass FALCONIFORMES.

The Falconiformes are supposed to be the only birds which combine the following characters:—

Young born helpless, and requiring to be fed by their parents in the nest for many days; young before acquiring feathers passing through a stage in which they are completely covered with down; hallux present, and connected with the *flexor longus hallucis* and not with the *flexor perforans digitorum*; spinal feather-tract well defined on the neck.

The Subclass Falconiformes contains two Orders, one only of which is represented in Japan.

Order PSITTACI.

The Order Psittaci only contains one Suborder.

Suborder XIV. *PSITTACI*.

The Parrots may be diagnosed as follows:—

Young born helpless and nearly naked; feet zygodactyle; spinal feather-tract well defined on the neck by lateral bare tracts, and forked on the upper back; oil-gland tufted or absent.

There are nearly 400 species of Parrots, which are all tropical or subtropical birds, and are distributed over both the Old and the New World. One species ranges into the southern portion of the Nearctic Region, but the suborder is unknown in the Palæarctic Region, including Japan, though a few species approach as near as South China.

The Parrots must be regarded as a very archaic group of birds, inasmuch as many if not all of them have opisthocœlous dorsal vertebræ.

Order **RAPTORES**.

The Raptores possess four characters which are not known to be combined in any other birds.

Young born helpless; young passing through a complete downy stage; hallux present, and connected with the *flexor longus hallucis* (not with the *flexor perforans digitorum*); spinal feather-tract well defined on the neck.

The Order Raptores contains three Suborders.

Suborder XV. *STRIGES*.

Young born helpless, but completely covered with down; oil-gland present, but nude; spinal feather-tract well defined on the neck.

The following alternative diagnoses are supposed to be equally exclusive :—

Basal phalanx of the third digit shortened almost to a cube; basipterygoid processes present.

Ambiens, accessory femoro-caudal, semitendinosus, and accessory semitendinosus muscles absent; basipterygoid processes present.

The distribution of the Owls is almost cosmopolitan. The number of species known is about 200.

There are eleven species of Owls which have been found in the Japanese Empire. Of these three belong to the genus *Strix*, in which the ear-conch is very large and protected by an operculum. One has been placed in the genus *Ninox* (scarcely separable from *Noctua*), in which the nostrils are placed in a projection formed by an inflation of the cere. Of the remaining seven, one belongs to the genus *Surnia*, which possesses neither of the characters already named, but has white or transversely barred underparts, longitudinal streaks on the underparts, and ear-tufts obsolete or nearly so. The remaining six have very conspicuous ear-tufts, and the broad longitudinal streaks on the underparts are more conspicuous than the narrow transverse bars. Two of them belong to the genus *Bubo*, which contains the large species (wing from carpal joint never less than 12 inches); and the remaining four to the genus *Scops*, which contains the small species (wing from carpal joint never more than 9 inches). It is not known that there are any structural

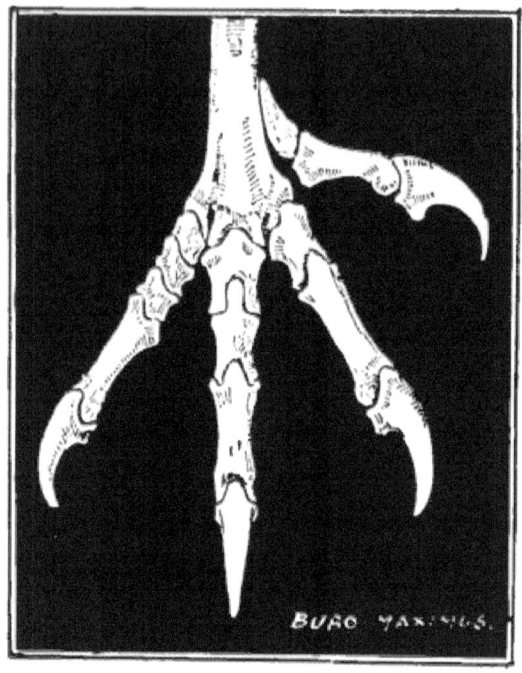

differences between the two last-named genera, which, like most other genera of Owls, are very unsatisfactory.

161. BUBO MAXIMUS.
(EAGLE-OWL.)

Bubo maximus, Gerini, Orn. Meth. Dig. i. p. 84 (1767).

The Eagle-Owl is very large (wing from carpal joint 20 to 18 inches). Its feet are densely feathered to the claws, and it has very conspicuous ear-tufts.

Figures: Dresser, Birds of Europe, v. pl. 315.

It is not known that the Eagle-Owl is found on any of the three or four large islands which may be regarded as continental Japan; but in the Norwich Museum there is an example (presented by

Mr. Ringer) which was shot on one of the Goto Islands, a group which lies only about fifty miles to the west of Nagasaki (Gurney, Ibis, 1886, p. 524).

The range of the Eagle-Owl extends from the British Islands, where it is now only an occasional visitor, across Europe and Asia to the confines of Japan.

162. BUBO BLAKISTONI.
(BLAKISTON'S EAGLE-OWL.)

Bubo blakistoni, Seebohm, Proc. Zool. Soc. 1883, p. 466; Seebohm, Ibis, 1884, p. 42.

Blakiston's Eagle-Owl is probably the largest Owl known (wing from carpal joint 22 inches). Its tarsus is feathered, but its feet are bare. It has very conspicuous ear-tufts.

Figures: Seebohm, Ibis, 1884, pl. 6.

Blakiston's Eagle-Owl is only known from the island of Yezzo, where it is a resident. It had long been confounded with the Eagle-Owl of Europe, *Bubo maximus*, until in 1883 Captain Blakiston sent an example to London for identification, when it was found to be an undescribed species. The type is in the British Museum, and I have a second example, to which I am indebted to the kindness of Captain Blakiston, in my collection. There is a fine example in the Norwich Museum, which also possesses a skeleton of this interesting species. On the 20th of January, 1887, two live specimens were presented to the Zoological Society by Mr. J. H. Leech, who procured them from Mr. Henson at Hakodadi. They came from the Lake district twenty miles north of that port (Sclater, Proc. Zool. Soc. 1887, p. 138).

The affinities of *Bubo blakistoni* have been supposed to be with the subgeneric group of Eagle-Owls which was called *Pseudoptynx* by Kaup (Gurney, Proc. Zool. Soc. 1887, p. 138). This group is represented by two species from the Philippine Islands, and is supposed to be characterized by the absence of feathering on the toes. It is highly improbable that a subgenus should be represented in Yezzo and the Philippine Islands and not in Formosa or Southern Japan. The feathering of the toes varies so much in the allied genus *Scops* that it can scarcely be regarded as of much taxonomic value; and it seems more probable that the nearest ally of Blakiston's Owl is *Bubo coromandus*, which has occurred in China (Seebohm, Ibis, 1884, p. 183).

163. SURNIA NYCTEA.
(SNOWY OWL.)

Strix nyctea, Linneus, Syst. Nat. i. p. 132 (1766).

The Snowy Owl is very large (wing from carpal joint 19 to 15½ inches). Its feet are densely feathered to the claws, but its ear-tufts are very small. It is white, more or less barred with brown.

Figures: Dresser, Birds of Europe, v. pl. 310 (adult), pl. 309 (young).

The claim of the Snowy Owl to be regarded as a Japanese bird rests upon a single example, which was brought alive into Hakodadi on the 29th of November, 1879, and was said to have been caught in the neighbourhood (Blakiston and Pryer, Trans. As. Soc. Japan, 1882, p. 177).

The Snowy Owl is a circumpolar species, breeding in the Arctic Region of both continents. It is a rare winter visitor to the British Islands.

164. STRIX URALENSIS.
(URAL OWL.)

Strix uralensis, Pallas, Reise Russ. Reichs, i. p. 445 (1771).

There are two forms of Ural Owl in Japan, the typical form and a dark tropical form, which may be called *Strix uralensis fuscescens*. The Ural Owl is a large bird (wing from carpal joint 15½ to 12½ inches), but it has a very rounded wing (1st and 10th primaries nearly of equal length). Its ear-conch is furnished with an operculum).

Figures: Dresser, Birds of Europe, v. pl. 307 (typical form); Temminck and Schlegel, Fauna Japonica, Aves, pl. 10 (tropical form, sub nomine *Strix fuscescens*, dating from 1845).

The typical form of the Ural Owl probably breeds in Yezzo (Whitely, Ibis, 1867, p. 194), and there is an example in the British Museum (formerly in my collection) which does not differ from pale examples from Europe (Seebohm, Ibis, 1879, p. 41). It was collected by Captain Blakiston at Hakodadi. There are ten examples in the Pryer collection from Yokohama, most of which may be regarded as typical *Strix uralensis*; but one of them is a typical *Strix uralensis fuscescens*, the *Strix rufescens* of the text of Temminck and Schlegel's 'Fauna Japonica,' Aves, p. 30. All the examples that I have seen from Nagasaki, one of which was presented by Mr. Ringer to the Norwich Museum, belong to this tropical form (Seebohm, Ibis, 1884, p. 183). It is very much darker, both above and below, than the typical form, all the pale markings are smaller, and the white is confined to the throat, and a few spots on the flanks and scapulars. The two centre tail-feathers, instead of being crossed by half a dozen pale bars, are uniform brown. Some of the Yokohama examples are, however, so intermediate that there can scarcely be a doubt that the two forms completely intergrade.

The range of the Ural Owl extends westwards through Siberia to Scandinavia, but it does not reach the British Islands.

Eggs in the Pryer collection measure 1·9 by 1·6 inches.

165. STRIX OTUS.
(LONG-EARED OWL.)

Strix otus, Linnæus, Syst. Nat. i. p. 132 (1766).

The Long-eared Owl measures from 12 to 11 inches in length of wing from carpal joint. Its ear-conch has an operculum, and its ear-tufts are conspicuous. Its first primary is nearly as long as the fourth, and the feathers of the underparts have narrow transverse bars as well as broad longitudinal stripes.

Figures: Dresser, Birds of Europe, v. pl. 303.

The Long-eared Owl is a resident in all the Japanese Islands. It is not a very common bird in Yezzo (Whitely, Ibis, 1867, p. 195), but Captain Blakiston has sent an example from Hakodadi (Seebohm, Ibis, 1879, p. 41); there are three skins in the Pryer collection from Yokohama; and Mr. Ringer has sent examples from Nagasaki (Blakiston, Am. List Birds of Japan, p. 65), which are now in the Norwich Museum.

The breeding-range of the Long-eared Owl extends from the British Islands across Europe and Southern Siberia to Japan.

166. STRIX BRACHYOTUS.
(SHORT-EARED OWL.)

Strix brachyotus, Forster, Phil. Trans. lxii. p. 384 (1772).

The Short-eared Owl measures from 13 to 12 inches in length of wing from carpal joint. Its ear-conch has an operculum, but its ear-tufts are small. Its first primary is nearly as long as the third, but the feathers of the underparts have no transverse bars, though most of them have conspicuous longitudinal stripes.

Figures: Dresser, Birds of Europe, v. pl. 304.

The Short-eared Owl is common to all the Japanese Islands, and is probably only a summer visitor to Yezzo, but a resident in the more southerly islands. Captain Blakiston has sent an example from Hakodadi (Seebohm, Ibis, 1879, p. 41), and it has been procured in Yezzo by native bird-catchers (Whitely, Ibis, 1867, p. 195). In the Pryer collection there are three examples from Yokohama, besides one from Yezzo; and Mr. Ringer has sent examples to the Norwich Museum procured near Nagasaki (Blakiston, Am. List Birds of Japan, p. 65).

The Short-eared Owl breeds in the British Islands, and may almost be regarded as cosmopolitan, its breeding-range comprising most of the temperate regions of the world.

167. NINOX SCUTULATA.
(BROWN OWLET.)

Strix scutulata, Raffles, Trans. Linn. Soc. xiii. p. 280 (1822).

The Brown Owlet is one of the smaller species (wing from carpal joint 9 to 8 inches). No other Japanese Owl has the projecting cere of this species. It is chocolate-brown, with barred wings and tail; the underparts are streaked with white, and the under tail-coverts are nearly all white.

Figures: Temminck and Schlegel, Fauna Japonica, Aves, pl. 9 B, sub nomine *Strix hirsuta japonica*.

The Brown Owlet is very doubtfully recorded from Yezzo (Blakiston and Pryer, Trans. As. Soc. Japan, 1882, p. 177), but it is not uncommon in summer near Yokohama and Nagasaki. There are eight examples in the Pryer collection from Yokohama, and two from the central group of the Loo-Choo Islands (Seebohm, Ibis, 1887, p. 174). It has also been obtained in the southern group of the Loo-Choo Islands (Stejneger, Proc. United States Nat. Mus. 1887, p. 401). The examples obtained by the Siebold expedition were probably procured at Nagasaki.

The Brown Owlet is found throughout the Oriental Region as well as in Japan. It has been subdivided into various species or subspecies, but it is very doubtful whether any of them can be defined geographically.

168. SCOPS SEMITORQUES.
(FEATHERED-TOED SCOPS OWL.)

Otus semitorques, Temminck and Schlegel, Fauna Japonica, Aves, p. 25 (1845).

The Feathered-toed Scops Owl is one of the smaller species (wing from carpal joint 7 to 6 inches). It has a conspicuous broad pale band on the hind neck.

Figures: Temminck and Schlegel, Fauna Japonica, Aves, pl. 8.

The Feathered-toed Scops Owl is found in all the Japanese Islands. In the Swinhoe collection there are many examples from Yezzo (Swinhoe, Ibis, 1875, p. 448); and in the Paris Museum there are examples from Hirosaki in the north of Hondo, procured by l'Abbé Fauire. There is a fine series in the Pryer collection from Yokohama; and in the Norwich Museum as well as in the British Museum there are examples presented by Mr. Ringer from Nagasaki; but the example recorded by Mr. Pryer from the Loo-Choo Islands (Seebohm, Ibis, 1887, p. 174) proved upon examination to belong to the following species.

169. SCOPS ELEGANS.
(CASSIN'S SCOPS OWL.)

Ephialtes elegans, Cassin, Proc. Acad. Nat. Sc. Philad. 1852, p. 185.

Cassin's Scops Owl is a giant race of *Scops japonicus*, measuring $6\frac{1}{2}$ to $6\frac{3}{4}$ inches in length of wing from carpal joint, and having the tarsus bare for a short distance above the base of the toes, as may be seen in the woodcut on page 58 of the British Museum Catalogue of Striges.

I only know of the existence of five examples of Cassin's Scops Owl. The type in the Philadelphia Museum was caught on boardship a few miles west of the Loo-Choo Islands; a second example in the British Museum (erroneously described in the Catalogue, vol. ii. p. 56, as *Scops japonicus*) was obtained by Captain St. John at Nagasaki; a third, in the Educational Museum of Tokio, was procured on Okinawa-Shima, one of the Loo-Choo Islands, and is recorded under the name of *Megascops elegans* (Stejneger, Proc. United States Nat. Mus. 1886, p. 639); and the Pryer collection contains the fourth example, also from the central group of the Loo-Choo Islands (Seebohm, Ibis, 1888, p. 232). The fifth example is that of a very young bird in the Smithsonian Institution, and was collected by Mr. Tasaki on one of the northerly islands of the Loo-Choo group (Stejneger, Proc. United States Nat. Mus. 1887, p. 401).

170. SCOPS SCOPS.
(SCOPS OWL.)

Strix scops, Linnæus, Syst. Nat. i. p. 132 (1766).

Japanese examples of the Scops Owl appear all to belong to the small dark race of this species, which may perhaps only be entitled to be regarded as subspecifically distinct, under the name of *Scops scops japonicus*. It is slightly smaller than the typical form (wing from carpal joint 5¾ to 5¼ inches), and decidedly darker and browner. The ear-tufts are well developed; the tarsus is feathered, but the feet are bare.

Figures: Dresser, Birds of Europe, v. pl. 314 (typical form); Temminck and Schlegel, Fauna Japonica, Aves, pl. 9 (eastern form).

The Scops Owl is said to be rather common in Japan (Blakiston and Pryer, Ibis, 1878, p. 247), but it is rare in collections. When the second volume of the Catalogue of Birds in the British Museum was published there were no Japanese examples in the National Collection, and only two have since been added—one from the Tweeddale collection (brown phase) from Yokohama, and a second from the Swinhoe collection (rufous phase) from Hakodadi (Swinhoe, Ibis, 1875, p. 448, no. 71). There are only two examples in the

Pryer collection from Yokohama (one in the brown, and the other in the rufous phase). Mr. Ringer has obtained an example (very rufous) from Nagasaki (Blakiston and Pryer, Trans. As. Soc. Japan, 1882, p. 178).

The Japanese race of this species was originally described as *Otus scops japonicus* (Temminck and Schlegel, Fauna Japonica, Aves, p. 27); and the Chinese race was, about twenty years afterwards, described as *Scops stictonotus* (Sharpe, Cat. Birds Brit. Mus. ii. p. 54). These races cannot, however, be regarded as distinct, and the range of the subspecies extends beyond China to Nepal and Siam. The typical form is larger (wing from carpal joint $6\frac{1}{2}$ to $5\frac{3}{4}$ inches), and has the dark stripes both above and below more conspicuous. There is little difference in colour between the rufous phases of the two races, but the grey phase of the typical form is represented by a brown phase in the eastern race.

171. SCOPS PRYERI.
(PRYER'S SCOPS OWL.)

Scops pryeri, Gurney, Ibis, 1889, p. 302.

Pryer's Scops Owl is the largest species of Scops Owl found in the Japanese Empire (wing from carpal joint $7\frac{1}{4}$ inches). The feathering of the tarsus, like that of *S. scops*, extends to the base of the toes, but not beyond; it is consequently more than that of *S. elegans*, but less than that of *S. semitorques*. It agrees with the last-mentioned species in having a short first primary (equal to the 9th or 10th), but differs from it in having the pale band on the hind neck almost obsolete.

Mr. J. H. Gurney informed me that he "thinks *Scops pryeri* is nearest allied to *S. leucospilus*" (from Batchian and Gilolo, figured on plate 6 of Sharpe's Catalogue), "*S. morotensis*" (from the Moluccas, figured on plate 7 of the same work), "and *S. bouruensis*" (from Bouru, also figured on plate 7), "a group in which *S. magicus*" (from Ceram and Amboyna, figured on plate 5) "ought perhaps to be included."

Pryer's Scops Owl is only known from two examples, an adult in the Norwich Museum and an immature example in the Pryer collection. Both specimens were procured on one of the islands of the central group of the Loo-Choo chain.

Suborder XVI. *ACCIPITRES*.

Young born helpless, but completely covered with down; no basipterygoid processes; spinal feather-tract well defined on the neck; hallux present, and connected with the *flexor longus hallucis*, and not with the *flexor perforans digitorum*, the two tendons bound together by a fibrous vinculum; dorsal vertebræ heterocœlous.

The Birds of Prey may be regarded as cosmopolitan in their distribution. They number about 350 species, of which 20 have occurred in the Japanese Empire.

The Japanese genera of Accipitres may be divided into three groups founded upon the peculiarities of the covering of their tarsi; but until the osteology of the Birds of Prey has been examined, it is impossible to say how far these groups are natural ones.

FALCONINÆ.—Lower half of tarsus reticulated all round.

Bill deeply notched .... *Falco* ....⎫ First primary between the third
                                     Pandion ..⎭ and fifth.

                                       Butaster.

Lores covered with small feathers,⎫ *Pernis.*
    not hairs ....⎭

AQUILINÆ.—Tarsus scutellated or feathered in front; reticulated or feathered at sides and back.

Third, fourth, and fifth primaries⎧ *Milvus.* ⎫
    longest ....⎨ *Haliaetus.*⎬ Tarsus scutellated in front.
                                 ⎩ *Aquila.* ⎫
Tarsus feathered in front to the ⎨   ⎬ Tail not forked.
    toes ....⎩ *Spizaetus.*⎭

ACCIPITRINÆ.—Tarsus scutellated at back and almost always in front; reticulated at sides.

                              Buteo. ⎧ Tarsus less than a fourth of wing,
Carpal joint to tip of shortest ⎧ *Circus.* ⎨ and less than half of first primary.
    primary much less than length ⎨   ⎩
    of tail ....⎩ *Accipiter.*

FALCONINÆ.

172. FALCO GYRFALCO.
(JER-FALCON.)

Falco gyrfalco, Linneus, Syst. Nat. i. p. 130 (1766).

The Jer-Falcon is a large bird (wing from carpal joint 16½ to 14 inches). It differs from the Peregrine in having the outer toe no longer than the inner, and in having the general colour of the tail not darkened towards the tip. It is not known which of the various races of the Jer-Falcon occasionally strays as far as Japan.

Figures: Dresser, Birds of Europe, vi. pls. 367 to 371.

Mr. Henson informs me that he procured an example of one of the various races of Jer-Falcon at Hakodadi.

The Jer-Falcon is a circumpolar species, varying considerably in different parts of its range.

173. FALCO PEREGRINUS.
(PEREGRINE FALCON.)

Falco peregrinus, Tunstall, Orn. Brit. p. 1 (1771).

The Peregrine is a small Jer-Falcon (wing from carpal joint 15 to 12 inches). It differs from that species in having the tail darkening towards the tip.

Figures: Dresser, Birds of Europe, vi. pl. 372.

The Peregrine was recognized by Pallas as one of the birds found by Steller in the Kurile Islands (Pallas, Zoogr. Rosso-Asiat. i. p. 326), and has recently been found by Mr. Snow to be very common there in summer (Blakiston and Pryer, Trans. As. Soc. Japan, 1882, p. 185); but it is a resident in the more southerly Japanese Islands. There are examples from Hakodadi both in the Norwich Museum and in the British Museum (Whitely, Ibis, 1867, p. 194), and there are four examples from Yokohama in the Pryer collection. The examples obtained by the Siebold Expedition (Temminck and Schlegel, Fauna Japonica, Aves, p. 1) were probably procured near Nagasaki. It has been recorded from the most southerly group of the Loo-Choo Islands (Stejneger, Proc. United States Nat. Mus. 1887, p. 401).

The Peregrine is a circumpolar species, and breeds in the British Islands as well as in Japan.

American ornithologists regard the Nearctic Peregrine Falcons as subspecifically distinct from those found in the Palæarctic Region under the name of *Falco peregrinus anatum*. They allege that in the Nearctic species the breast of the adult bird is generally unspotted. They originally described East-Asiatic examples as an intermediate race under the name of *Falco peregrinus orientalis*; but Dr. Stejneger and Mr. Ridgway appear to have abandoned this position, and now regard Japanese and American examples as identical. Mr. Gurney did not recognize the Japanese birds as distinct from those of Europe.

174. FALCO SUBBUTEO.
(HOBBY.)

Falco subbuteo, Linneus, Syst. Nat. i. p. 127 (1766).

The Hobby is a miniature Peregrine (wing from carpal joint 11 to 10 inches). It is easily distinguished when adult by its chestnut thighs, and at all ages by the absence of bars on its under tail-coverts and central tail-feathers.

Figures: Dresser, Birds of Europe, vi. pl. 379 (male and female adult), pl. 387 (young in first plumage and in down).

The Hobby is tolerably abundant in Yezzo (Seebohm, Ibis, 1879, p. 42); but, strange to say, it has not been recorded from Southern Japan.

The range of the Hobby extends from the British Islands across Europe and Siberia to Japan.

175. FALCO ÆSALON.
(MERLIN.)

Falco æsalon, Tunstall, Orn. Brit. p. 1 (1771).

The Merlin is one of the smallest Falcons (wing from carpal joint 9 to 8 inches). The adult male Merlin (like the adult male Kestrel) has a blue-grey tail, crossed by a broad subterminal black band; but the blue-grey extends to the wing-coverts, scapulars, tertials, and interscapulars. The female and immature male closely resemble those of the Kestrel, but may be distinguished by having seven instead of nine dark bars across the tail, which is also less rounded.

Figures: Dresser, Birds of Europe, iv. pl. 380 (male and female adult), pl. 381 (very old female and young in first plumage).

The Merlin is a common resident in all the Japanese Islands. There are several examples in the Swinhoe collection from Hakodadi, and there are ten examples in the Pryer collection from Yokohama, whence the first occurrence of this species in Japan was recorded (Swinhoe, Ibis, 1877, p. 144).

The range of the Merlin extends from the British Islands across Europe and Siberia to Japan.

176. FALCO TINNUNCULUS.
(KESTREL.)

Falco tinnunculus, Linneus, Syst. Nat. i. p. 127 (1766).

The Japanese Kestrel belongs to the eastern race of the dark form of the Kestrel; and may be regarded as subspecifically distinct, under the name (dating from 1845) of *Falco tinnunculus japonicus* (Temminck and Schlegel, Fauna Japonica, Aves, p. 2).

The Kestrel varies in length of wing from $10\frac{1}{2}$ to $9\frac{1}{4}$ inches. It is very closely allied to the Merlin, the adult males of both species having a blue-grey tail, crossed by a broad subterminal black band; but in the Kestrel the wing-coverts, scapulars, tertials, and interscapulars are chestnut, barred with black. The females and immature males are much more difficult to determine, but in the Kestrel there are nine (instead of seven) dark bars across the tail, and the outer feathers are more than an inch (instead of less than half an inch) shorter than the central ones. The outer toe is also more nearly equal to the inner toe than it is in the Merlin.

Figures: Temminck and Schlegel, Fauna Japonica, Aves, pl. 1 (immature female), pl. 1 B (adult male).

The Japanese Kestrel is a common resident in the southern islands of Japan, but is not known to have occurred in Yezzo. Mr. Ringer has sent examples to the Norwich Museum procured at Nagasaki

(Blakiston and Pryer, Trans. As. Soc. Japan, 1882, p. 184), and there are seven examples in the Pryer collection from Yokohama.

The range of the Kestrel extends from the British Islands across Europe and Siberia to Japan; but examples from Japan, Formosa, and South China differ somewhat from the typical race (Gurney, Ibis, 1881, p. 462). They constitute a dark richly coloured local race of the same species which inhabits the British Islands, and only differs in size from the smaller dark race which breeds on the islands off the coast of West Africa (Cape Verd, Canaries, and Madeira).

Eggs in the Pryer collection resemble those of the common form.

177. PANDION HALIAETUS.
(OSPREY.)

Falco haliaetus, Linneus, Syst. Nat. i. p. 129 (1766).

The Osprey varies in length of wing (from carpal joint) from 21 to 16 inches. It is brown above and white below, with brown

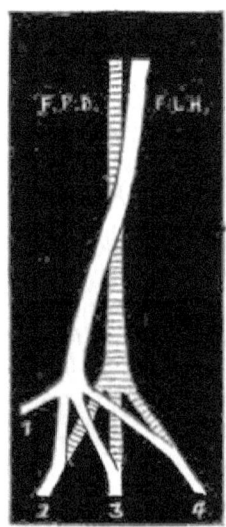

Deep plantar tendons of *Pandion haliaetus*.

streaks on the breast and white streaks on the crown and nape. Immature birds have most of the feathers of the upper parts more or less margined with white.

Figures: Dresser, Birds of Europe, vi. pl. 386 (adult), pl. 387 (young in first plumage).

The Osprey is a resident in all the Japanese Islands. Captain Blakiston has sent an example obtained at Hakodadi in October (Seebohm, Ibis, 1884, p. 183), and there are five examples in the Pryer collection from Yokohama. In the Norwich Museum there are several examples, which were brought by Mr. Ringer from Nagasaki, where those procured by the Siebold Expedition were probably obtained (Temminck and Schlegel, Fauna Japonica, Aves, p. 13).

The Osprey is a circumpolar species, and has not yet been quite exterminated in the British Islands. As it is not known to breed in India or Burma, it probably emigrated to Japan from Siberia.

178. BUTASTER INDICUS.
(JAVAN BUZZARD.)

Falco indicus, Gmelin, Syst. Nat. i. p. 264 (1788).

The Javan Buzzard varies in length of wing from $13\frac{1}{2}$ to $12\frac{1}{4}$ inches. It is brown above, more or less barred with dark brown on the wings and tail; and white, barred with brown, below.

Figures: Temminck, Planches Coloriées, no. 325 (adult); Temminck and Schlegel, Fauna Japonica, Aves, pl. 7 B (immature).

The Javan Buzzard is a very common resident in Southern Japan, but it has not been recorded from Yezzo. There are four examples in the Pryer collection from Yokohama, and two from the central group of the Loo-Choo Islands. There are several examples, including a nestling, in the Norwich Museum, which were brought by Mr. Ringer from Nagasaki (Blakiston and Pryer, Trans. As. Soc. Japan, 1882, p. 183), where the example figured in the 'Fauna Japonica' as *Buteo pyrrhogenys*, and described as *Buteo polyogenys*, was probably procured.

The range of the Javan Buzzard extends from Japan, the Philippine Islands, and Celebes, across China, Borneo, and Java to the Malay Peninsula.

Allied species occur in India and North-east Africa, consequently the Japan Buzzard must be regarded as one of the Tropical contributions to the Avifauna of Japan.

179. PERNIS APIVORUS.
(HONEY-BUZZARD.)

Falco apivorus, Linnæus, Syst. Nat. i. p. 130 (1766).

The Honey-Buzzard varies in length of wing (from carpal joint) from 17½ to 15 inches. It varies greatly in colour, but the tail always has four broad dark bands across it. It is easily recognized by the small feathers on the lores.

Figures: Dresser, Birds of Europe, vi. pl. 365 (male and female adult), pls. 364, 366 (immature).

The Honey-Buzzard appears to be a very rare bird in Japan. There is one example in the Leyden Museum, probably from Nagasaki, obtained during the Siebold Expedition (Temminck and Schlegel, Fauna Japonica, Aves, p. 24); and Captain Blakiston has recorded in the 'Chrysanthemum' the capture of a fine male (without crest) by Mr. Jouy at Chinsenji, in Tokio, during August.

The range of the Honey-Buzzard during the breeding-season extends from the British Islands across Europe and Southern Siberia to Japan.

AQUILINÆ.

180. MILVUS ATER.
(BLACK KITE.)

Falco ater, Gmelin, Syst. Nat. i. p. 262 (1788).

Japanese examples of the Black Kite belong to the large Siberian race, which was described in 1845 as *Milvus melanotis* (Temminck and Schlegel, Fauna Japonica, Aves, p. 14), and which may fairly claim to be regarded as subspecifically distinct, under the name of *Milvus ater melanotis*.

The Siberian race of the Black Kite is a large form (wing from carpal joint 21½ to 18½ inches). It is easily recognized amongst other Japanese birds of prey by its forked tail.

Figures: Temminck and Schlegel, Fauna Japonica, Aves, pl. 5 (brown phase), pl. 5 B (rufous phase).

The Siberian Black Kite is a very common resident in Japan. It

is very numerous during the fishing-season at Eturop, the most
southerly of the Kurile Islands (Blakiston and Pryer, Trans. As. Soc.
Japan, 1882, p. 181); and there is an example in the Swinhoe collection from Hakodadi (Swinhoe, Ibis, 1874, p. 150), whence it had
been obtained twenty years earlier by the Perry Expedition (Cassin,
Exp. Am. Squad. China Seas and Japan, ii. p. 219). There are
three examples in the Pryer collection from Yokohama, and one
in the Norwich Museum procured by Mr. Ringer at Nagasaki.

The range of the Black Kite extends from Western Europe to
Eastern Siberia and Japan; but examples from the latter districts
differ slightly from those inhabiting Europe. Some of the Siberian
birds appear to winter in India.

The Siberian form of the Black Kite is slightly larger than the
western form, and has the white at the base of the outer primaries
extending below the under wing-coverts, but that on the margins of
the feathers of the head is confined to the forehead.

Eggs in the Pryer collection are on an average larger than those
of the European form.

181. HALIAETUS ALBICILLA.
(WHITE-TAILED EAGLE.)

Vultur albicilla, Linneus, Syst. Nat. i. p. 123 (1766).

The White-tailed Eagle is a large bird (wing from carpal joint
29 to 26 inches). The white tail is only characteristic of adult
birds, but the absence of feathers on the lower half of the tarsus,
combined with the large size, prevent it being confused with any
other Japanese bird except Steller's Sea-Eagle, which has a very
cuneiform tail consisting of 14 (instead of 12) feathers.

Figures: Dresser, Birds of Europe, v. pl. 318 (adult), pl. 317
(young in first plumage).

The White-tailed Eagle is a common resident on all the Japanese
coasts (Blakiston and Pryer, Trans. As. Soc. Japan, 1882, p. 180).
There are two examples in the Pryer collection from Yokohama; and
there are others in the Norwich Museum brought by Mr. Ringer from
Nagasaki, where the example obtained by the Siebold Expedition was
probably procured (Temminck and Schlegel, Fauna Japonica, Aves,
p. 12).

The breeding-range of the White-tailed Eagle extends from the British Islands across Europe and Siberia to Japan.

As it is only a winter visitor to South China, it must be regarded as a Palæarctic species which probably emigrated to Japan from Siberia.

182. HALIAETUS PELAGICUS.
(STELLER'S SEA-EAGLE.)

Aquila pelagica, Pallas, Zoogr. Rosso-Asiat. i. p. 343 (1826).

Steller's Sea-Eagle may always be recognized by its wedge-shaped tail, consisting of 14 (instead of 12) feathers. It is about the same size as the White-tailed Eagle, but has a larger bill.

Figures: Temminck and Schlegel, Fauna Japonica, Aves, pl. 4 (young); Temminck, Planches Coloriées, no. 489 (adult).

Steller's Sea-Eagle is a frequent winter visitor from its breeding-grounds in Kamtschatka to the Japanese Islands, but is more often seen in Yezzo than further south (Blakiston and Pryer, Trans. As. Soc. Japan, 1882, p. 180). Mr. Henson has sent several examples from Hakodadi, three of which are in the British Museum. There are two examples in the Pryer collection from Yezzo (Seebohm, Ibis, 1884, p. 183), and one from the central group of the Loo-Choo Islands (Seebohm, Ibis, 1888, p. 232), the latter having been caught exhausted in a paddy-field.

The breeding-range of Steller's Sea-Eagle is probably confined to the shores of the Sea of Okhotsk.

183. AQUILA CHRYSAETUS.
(GOLDEN EAGLE.)

Falco chrysaetus, Linneus, Syst. Nat. i. p. 125 (1766).

The Golden Eagle is not quite so large as the Sea-Eagle (wing from carpal joint 28 to 22 inches). Its tarsi feathered to the toes, the absence of bars on the thighs and the rest of the underparts, and its large size, prevent it from being confounded with any other Japanese bird of prey.

Figures: Dresser, Birds of Europe, v. pl. 345.

Several examples of the Golden Eagle have been obtained in Southern Japan, but it has not yet been recorded from Yezzo (Blakiston and Pryer, Ibis, 1878, p. 247). There is a fine example in the Pryer collection from the game-market in Yokohama (Seebohm, Ibis, 1884, p. 43).

The range of the Golden Eagle extends from the British Islands across Europe and Siberia to Japan. As it is not known to occur in South China, it must be regarded as a Palæarctic species which has emigrated to Japan from Siberia.

184. AQUILA LAGOPUS.
(ROUGH-LEGGED BUZZARD-EAGLE.)

Falco lagopus, Gmelin, Syst. Nat. i. p. 260 (1788).

The Rough-legged Buzzard-Eagle is much smaller than any other Japanese bird of prey which has the tarsus feathered to the toes (wing from carpal joint 19 to 17 inches). A narrow strip at the back of the tarsus is free from feathers, a peculiarity also found in the nestling of the Golden Eagle.

Figures: Dresser, Birds of Europe, v. pls. 334, 335.

The Rough-legged Buzzard-Eagle is a rare winter visitor to Japan. Two examples have been procured at Hakodadi (Seebohm, Ibis, 1884, p. 43), but it has not yet been recorded from Southern Japan.

The Rough-legged Buzzard-Eagle is a rare winter visitor to the British Islands, as well as to Japan. Its breeding-range extends from the Atlantic to the Pacific, and across Bering Straits into Alaska.

185. SPIZAETUS NIPALENSIS.
(INDIAN CRESTED EAGLE.)

Nisaetus nipalensis, Hodgson, Journ. As. Soc. Beng. 1836, p. 229.

The Indian Crested Eagle is a large bird (wing from carpal joint 20 to 18 inches). It has the tarsus feathered to the toes. The flanks and thighs are always barred.

Figures: Temminck and Schlegel, Fauna Japonica, Aves, pl. 3 (immature), sub nomine *Spizaetus orientalis*.

The Indian Crested Eagle is a resident on Hondo, and wanders in winter as far as Yezzo (Blakiston and Pryer, Trans. As. Soc. Japan, 1882, p. 181). An immature bird from Japan moulted in the Zoological Gardens in London into the adult plumage. There is also an adult example in the Pryer collection from Fuji-yama.

The breeding-range of this species extends from the Himalayas, and the mountains of Southern India and Ceylon, across Southern China to Formosa and Japan.

As it has not been recorded from North China or Siberia, it must be regarded as a Tropical species which has emigrated to Japan from the south.

ACCIPITRINÆ.

186. BUTEO HEMILASIUS.
(SIBERIAN BUZZARD.)

Buteo hemilasius, Temminck and Schlegel, Fauna Japonica, Aves, p. 18 (1845).

The Siberian Buzzard is a large bird (wing from carpal joint 20 to 18 inches). The tarsus is scutellated at the back, but reticulated in front.

Figures: Temminck and Schlegel, Fauna Japonica, Aves, pl. 7.

The Siberian Buzzard appears to be only an accidental visitor to Japan. The Siebold Expedition obtained a single example at Nagasaki, but no second specimen has been recorded from any of the Japanese Islands.

This species breeds in Dauria and winters in Mongolia and North China.

187. BUTEO VULGARIS.
(COMMON BUZZARD.)

Buteo vulgaris, Leach, Syst. Cat. M. & Birds Brit. Mus. p. 10 (1816).

Japanese examples of the Buzzard belong to the Eastern race of this species, in which the upper half of the tarsus is plumed. It was originally described from Nepal as *Buteo plumipes* (Hodgson, Proc.

Zool. Soc. 1845, p. 37); and in the same year from Japan as *Falco buteo japonicus* and *Buteo japonicus*, in the text, and as *Buteo vulgaris japonicus* on the plates (Temminck and Schlegel, Fauna Japonica, Aves, p. 16). It is a fairly distinct race (though it appears to intergrade with the Western race), and is entitled to the name of *Buteo vulgaris plumipes*.

The Eastern race of the Common Buzzard varies in length of wing (from carpal joint) from $16\frac{1}{2}$ to $13\frac{1}{2}$ inches. The tenth primary exceeds the primary-coverts by about $1\frac{1}{2}$ inches; in *Butaster indicus* by $2\frac{1}{2}$ inches.

Figures: Temminck and Schlegel, Fauna Japonica, Aves, pls. 6, 6 B.

The Japanese race of the Common Buzzard is probably only a summer visitor to the Kurile Islands and to Yezzo, but a resident in Southern Japan. There are several examples from Hakodadi in the Swinhoe collection, and there are seven examples in the Pryer collection from Yokohama. Mr. Ringer has obtained it at Nagasaki (Blakiston and Pryer, Trans. As. Soc. Japan, 1882, p. 182), whence he has sent many examples to the Norwich Museum; and Mr. Holst procured it on Peel Island, one of the central group of the Bonins (Seebohm, Ibis, 1890, p. 102).

The range of the Common Buzzard extends from the British Islands across Europe, Central Asia, and Southern Siberia to Japan. Examples from Europe and Turkestan differ slightly from those found in Eastern Siberia, China, and Japan.

The Eastern form of the Common Buzzard is said to be always distinguishable from the Western form by the greater extent to which the tarsus is feathered. Adult birds are said, further, to differ in having uniform brown tails without bars.

188. CIRCUS CYANEUS.
(HEN-HARRIER.)

Falco cyaneus, Linneus, Syst. Nat. i. p. 126 (1766).

The Hen-Harrier is smaller than the Marsh-Harrier (wing from carpal joint $15\frac{1}{2}$ to $13\frac{1}{2}$ inches). Adult males are easily recognized by the pale bluish-grey throat and breast, but females and young males are very close to those of the Eastern Marsh-Harrier. In *C. æruginosus* the 1st primary is an inch or more longer than the 7th; in *C. cyaneus* they are nearly equal.

Figures: Dresser, Birds of Europe, v. pl. 329 (male and female adult).

The Hen-Harrier is a summer visitor to the Kurile Islands and to Yezzo, but a winter visitor to Southern Japan (Blakiston and Pryer, Trans. As. Soc. Japan, 1882, p. 185). There is an example in the Swinhoe collection from Hakodadi (Swinhoe, Ibis, 1875, p. 448), and there are eight examples in the Pryer collection from Yokohama. The example obtained by the collectors of the Siebold Expedition, and erroneously identified as *Circus uliginosus*, was doubtless procured at Nagasaki (Temminck and Schlegel, Fauna Japonica, Aves, p. 9).

The breeding-range of the Hen-Harrier extends from the British Islands across North Europe and Siberia to Japan.

189. CIRCUS ÆRUGINOSUS.
(MARSH-HARRIER.)

Circus æruginosus, Linneus, Syst. Nat. i. p. 130 (1766).

The Marsh-Harrier is on an average a somewhat larger bird than the Hen-Harrier (large females 17 inches in length of wing from carpal joint). The 1st and 7th primaries are nearly equal in length.

Figures: Dresser, Birds of Europe, v. pl. 326 (Western form); pl. 327 (intermediate form); Swinhoe, Ibis, 1863, pl. 5 (Eastern form).

It is impossible to determine whether the Marsh-Harriers of Japan belong to the Eastern or to the Western form of that species, or to both. The male of the Eastern form, *Circus spilonotus* (Kaup, Contr. Orn. 1850, p. 59), when fully adult has the underparts white, streaked on the throat and breast with black; whilst in the Western form the throat and breast are buff streaked with brown, and the rest of the underparts are chestnut. The females of the Eastern form are said to have broad bands across the tail-feathers, but otherwise to resemble those of the Western form, which never has a banded tail in either sex or at any age. I have never seen an adult male (with lavender-grey tail) of either form from Japan; but immature males and females with uniform brown tails (presumably *C. æruginosus*), and females with barred tails (probably *C. spilonotus*), are represented in the Pryer collection from Yokohama. The existence of intermediate forms between the two races, of which at least five examples have been recorded (Gurney, Diurnal Birds of Prey, p. 115), appears to prove that they are only subspecifically distinct, and that probably the immature examples of the two forms are indistinguishable.

The Marsh-Harrier is probably a summer visitor to all the Japanese Islands, whence it was first procured by Captain Blakiston from Awomori on Hondo opposite Hakodadi (Swinhoe, Ibis, 1877, p. 144). There are several examples in the Swinhoe collection from Hakodadi, and there are four examples in the Pryer collection from Yokohama.

The breeding-range of the Marsh-Harrier extends from the British Islands across Europe and Southern Siberia to Japan.

190. ACCIPITER PALUMBARIUS.
(GOSHAWK.)

Falco palumbarius, Linneus, Syst. Nat. i. p. 130 (1766).

The Goshawk measures from 14 to 12 inches in length of wing from carpal joint. In the adult male the upper parts are slate-grey, and the underparts white barred with slate-grey. In the female and young male the upper parts are brown with pale markings, and the underparts pale rufous streaked with brown.

Figures: Dresser, Birds of Europe, v. pl. 354 (female adult, and young in first plumage).

The Goshawk is a resident in all the Japanese Islands. There is an example in the Hakodadi Museum obtained in Yezzo (Blakiston and Pryer, Ibis, 1878, p. 218), and there are five examples in the Pryer collection from Yokohama. There is an example in the Paris Museum procured by l'Abbé Fauire near Awomori in the north of Hondo.

The range of the Goshawk extends from the British Islands across Europe and Siberia to Japan.

191. ACCIPITER NISUS.
(COMMON SPARROW-HAWK.)

Falco nisus, Linneus, Syst. Nat. i. p. 130 (1766).

The Common Sparrow-Hawk varies in length of wing (from carpal joint) from $9\frac{1}{2}$ to $7\frac{1}{2}$ inches. The feathers of the throat appear each of them always to have a dark shaft-streak.

Figures: Dresser, Birds of Europe, v. pls. 355, 356, 357, 358.

The Common Sparrow-Hawk is a resident in all the Japanese Islands. It has frequently been recorded from Yezzo (Whitely, Ibis, 1867, p. 194), and there are fifteen examples in the Pryer collection from Yokohama. It is also common at Nagasaki (Blakiston and Pryer, Trans. As. Soc. Japan, 1882, p. 183), where the examples obtained by the Siebold Expedition were probably obtained (Temminck and Schlegel, Fauna Japonica, Aves, p. 4), and whence a large series has been sent by Mr. Ringer to the Norwich Museum.

The range of the Common Sparrow-Hawk extends from the British Islands across Europe and Siberia to Japan.

192. ACCIPITER GULARIS.
(CHINESE SPARROW-HAWK.)

Astur (Nisus) gularis, Temminck and Schlegel, Fauna Japonica, Aves, p. 5 (1845).

The Chinese Sparrow-Hawk is smaller than the Common species (wing from carpal joint $8\frac{1}{2}$ to $6\frac{1}{2}$ inches). It has a line of black streaks down the centre of the throat, which in the female and young male is separated from the moustachial streaks by an unstreaked longitudinal band.

Figures: Temminck and Schlegel, Fauna Japonica, Aves, pl. 2; Gurney, Ibis, 1863, pl. 11.

The Chinese Sparrow-Hawk was first described in 1845 by Temminck and Schlegel, in the 'Fauna Japonica.' In 1847 it was redescribed from Malacca under the name of *Accipiter nisoides* (Blyth, Journ. As. Soc. Beng. xvi. p. 727), and again in 1863 from China as *Accipiter stevensoni* (Gurney, Ibis, 1863, p. 417).

It is found in all the Japanese Islands. The Perry Expedition obtained examples at Hakodadi (Cassin, Exp. Am. Squad. China Seas and Japan, ii. p. 219), and several examples have been since procured in Yezzo (Blakiston and Pryer, Trans. As. Soc. Japan, 1882, p. 184). There are five examples in the Pryer collection from Fuji-yama, near Yokohama, and there are others in the Norwich Museum obtained by Mr. Ringer at Nagasaki (Gurney, Diurnal Birds of Prey, p. 165).

It is probably only a summer visitor to Japan and China, wintering in the islands of the Malay Archipelago and in the Burma peninsula. It is very common on migration in October and November in Central Hondo (Jouy, Proc. United States Nat. Mus. 1883, p. 312).

Suborder XVII. *SERPENTARII*.

The Secretary Bird may be diagnosed as follows :—

Palate desmognathous; basipterygoid processes present; spinal feather-tract well defined on the neck; deep plantar tendons galline; oil-gland tufted.

This suborder contains only one species, which is only found in the Ethiopian Region.

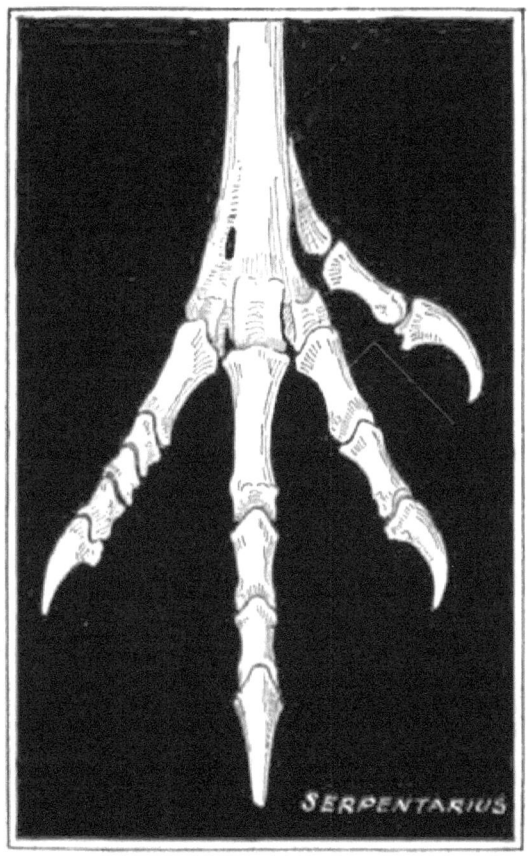

Subclass ANSERIFORMES.

The Anseriformes appear to form a natural group of birds which may be diagnosed as follows:—

Palate desmognathous; spinal feather-tract not defined on the neck (either coalesced with the ventral feather-tracts, or replaced by a spinal bare tract); front plantar not leading to the hallux.

The Subclass Anseriformes contains two Orders.

Order PELECANO-HERODIONES.

The Pelicans, the Herons, and their allies possess, of course, the three characters which diagnose the Subclass to which they belong, and in addition possess the following character, which distinguishes them from the Lamellirostres:—the young are born helpless, and require to be fed in the nest by their parents for many days.

The Order Pelecano-Herodiones contains three Suborders: Steganopodes, Herodiones, and Plataleæ, each of which is represented in the Japanese Empire.

Suborder XVIII. *STEGANOPODES.*

Palate desmognathous; no bare tracts on neck; mandible not produced and recurved behind its articulation with the quadrate; no basipterygoid processes; hallux united to second digit by a web.

The number of species which comprise the Steganopodes probably does not much exceed 50. They generally breed in large colonies, which are distributed in the tropical and temperate regions of both hemispheres. They may be grouped in families, which are easily diagnosed by well-marked osteological characters.

Eight species are found within the Japanese Empire.

193. PHALACROCORAX CARBO.
(COMMON CORMORANT.)

Pelecanus carbo, Linnæus, Syst. Nat. i. p. 216 (1766).

Head of *Phalacrocorax carbo*. ½ natural size.

In the Common Cormorant the bare space on each side of the throat extends behind the gape; and in adult birds the gorget is white, and the scapulars and wing-coverts are bronzy brown margined with black.

Figures: Dresser, Birds of Europe, vi. pl. 388.

The Common Cormorant appears to be a resident in the Southern Japanese Islands (Blakiston and Pryer, Ibis, 1878, p. 216), but it has been so much confused with Temminck's Cormorant that its exact range is difficult to determine. There are no skins in the Swinhoe collection from Hakodadi, but there are two in the Pryer collection from Yokohama. The Perry Expedition found it very common in the Bay of Yedo (Cassin, Exp. Am. Squad. China Seas and Japan, ii. p. 234), and the Siebold Expedition obtained it at Nagasaki (Temminck and Schlegel, Fauna Japonica, Aves, p. 129).

The breeding-range of the Common Cormorant extends from the British Islands across Europe and both Northern and Southern Asia to Japan. It also extends to Australia and the Atlantic coast of North America; but on the Pacific coast of the American continent the Common Cormorant appears to be crowded out by other species, some of which range as far west as Japan.

194. PHALACROCORAX CAPILLATUS.
(TEMMINCK'S CORMORANT.)

Carbo filamentosus vel *capillatus*, Temminck and Schlegel, Fauna Japonica, Aves, p. 129 (1847).

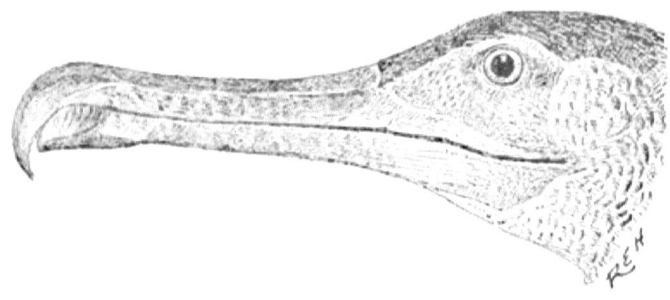

Head of *Phalacrocorax capillatus*. ⅔ natural size.

In Temminck's Cormorant the bare space on each side of the throat does not extend behind the gape; the gorget is profusely streaked with greenish black, and the scapulars and wing-coverts are bronzy green narrowly margined with black.

Figures: Temminck and Schlegel, Fauna Japonica, Aves, pl. 83 (adult), pl. 83 B (young).

Temminck's Cormorant is a resident in East China and Japan (Seebohm, Ibis, 1885, p. 271). There are two examples from Amoy in the Swinhoe collection, one dated February, the other April; there are also two examples from Hakodadi in the same collection procured in winter. I have a third example from Hakodadi collected by Mr. Henson on the 22nd of February, and there is one in the Pryer collection from Sarushima. It was originally described from Nagasaki.

Temminck's Cormorant is a very distinct species. It is a Cormorant, having fourteen tail-feathers, but in some respects it resembles a Shag; the scapulars and wing-coverts are bronzy green (not bronzy brown as in the Common Cormorant). The gorget of the latter species is white, but that of Temminck's Cormorant is profusely streaked with greenish black. Temminck's Cormorant is slightly the larger bird, and immature examples of the two species may be distinguished by the difference in the shape of the bare space on the throat: in Temminck's Cormorant the margin of the feathering extends from the gape at a right angle to the line of the commissure, and meets the margin of the other side at an acute angle, considerably in front of the gape (Seebohm, Ibis, 1885, p. 270).

195. PHALACROCORAX PELAGICUS.
(RESPLENDENT SHAG.)

Phalacrocorax pelagicus, Pallas, Zoogr. Rosso-Asiat. ii. p. 303 (1826).

Head of *Phalacrocorax pelagicus*. ⅔ natural size.

The Shags have only ten tail-feathers, and the feathers of the back and scapulars are not margined with black. The Resplendent Shag has two crests when adult, and the forehead is always feathered to the base of the bill.

Figures: Temminck and Schlegel, Fauna Japonica, Aves, pl. 84 (adult), pl. 84 a (young); misnamed *Carbo bicristatus*.

The Resplendent Shag breeds on the Kurile Islands, and is common on the coast of Yezzo during summer (Whitely, Ibis, 1867, p. 211) and probably in winter also. Great numbers visit Tokio Bay in winter, but leave for the north in the spring. I have an example collected by Mr. Snow on the Kurile Islands, and there are two examples in the Swinhoe collection from Hakodadi, one of them collected by Captain Blakiston in winter (Swinhoe, Ibis, 1874, p. 164). I have also two examples from Hakodadi collected by Mr. Henson on the 22nd of February; and there are two in the Pryer collection from Yokohama. There is also a fine example in the Norwich Museum sent by Mr. Ringer from Nagasaki.

These examples agree in their measurements with an example from Kamtschatka and with all the examples from China in the Swinhoe collection, including the type of *Phalacrocorax æolus* (Swinhoe, Ibis, 1867, p. 395). They average:—wing 11 inches, tail 6 inches, bill from frontal feathers 2 inches, tarsus 2¼ inches. They are all feathered on the forehead to the base of the bill.

The breeding-range of the Resplendent Shag extends from the Kurile Islands and Kamtschatka up to Norton Sound, and across the Aleutian chain to the south coast of Alaska as far south-east as Sitka.

196. PHALACROCORAX BICRISTATUS.
(BARE-FACED SHAG.)

Phalacrocorax bicristatus, Pallas, Zoogr. Rosso-Asiat. ii. p. 301 (1826).

Head of *Phalacrocorax bicristatus*. ⅔ natural size.

In the Bare-faced Shag, when adult, the forehead and a considerable space round the eye is orange-red and bare of feathers. Young birds are scarcely distinguishable from the young of the Resplendent Shag.

Figures: Baird, Brewer, and Ridgway, Water-Birds N. Amer. ii. p. 163 (coloured woodcut of head).

The Bare-faced Shag was found on the Kurile Islands by Steller, though he states that it is rarer there than in Kamtschatka. I have only one example, a female, which has only partially completed its moult into adult plumage, that I can refer to this species. It was collected by Mr. Snow on the Kurile Islands. The crests are well developed, the white plumes on the thighs are appearing, but the wings are in full moult. The forehead is bare of feathers for some distance, and the feathering on the side of the lower mandible runs down in nearly a straight line.

Its range is said to be confined to the North Pacific, where it is supposed to breed on the coasts of Alaska, Kamtschatka, and the intervening islands.

197. SULA LEUCOGASTRA.
(BOOBY GANNET.)

Pelecanus leucogaster, Boddaert, Tabl. Pl. Enl. p. 57 (1783).

The Booby Gannet or Common Booby has always pale yellow feet. When adult it is brown all over, except the underparts below the breast, which are white. Immature specimens are brown all over, paler on the head, neck, and underparts.

Figures: Gould, Birds of Australia, vii. pl. 78.

The Booby Gannet breeds on the Bonin Islands (Blakiston and Pryer, Trans. As. Soc. Japan, 1882, p. 102), and a single example was brought from Japan by the Siebold Expedition (Temminck and Schlegel, Fauna Japonica, Aves, p. 131). Captain Rodgers also found it on the Bonin Islands, and brought home an example (with pale yellow feet) from the Eastern Sea, between the Loo-Choo Islands and Formosa, which is recorded under the name of *Sula fiber* (Cassin, Proc. Acad. Nat. Sc. Philad. 1862, p. 325). I have a skin from Peel Island, and an egg from Long Island in the central Bonin group, both collected by Mr. Holst (Seebohm, Ibis, 1890, p. 107). There is an example in the Norwich Museum sent by Mr. Ringer from Nagasaki. There is an example from Formosa in the Swinhoe collection, and there can be little doubt that *Sula sinicadvena* (Swinhoe, Ibis, 1865, p. 109), from the coast of China, west of the Loo-Choo Islands, must be referred to this species. It has a very wide range, southwards to Australia, westwards across the Indian and Atlantic Oceans to the West Indies, and eastwards across the Pacific Ocean.

198. SULA PISCATRIX.
(RED-FOOTED BOOBY.)

Pelecanus piscator, Linneus, Syst. Nat. i. p. 217 (1766).

The Red-footed Booby has coral-red feet at all ages. When adult it is white with brown quills; but immature birds are brown all over, slightly paler on the underparts.

Figures: Gould, Birds of Australia, vii. pl. 79.

The Red-footed Booby has been admitted to the Japanese fauna on the authority of an example in the possession of Mr. Whitely, said to have been collected by Mr. Abel A. J. Gower while Consul in Japan (Blakiston, Amended List of the Birds of Japan, p. 34). I have an example of this species which was caught at sea by Mr. Snow between Japan and the Kruzenstern Rocks, which lie about sixty degrees due east of Formosa. It has occurred on the Philippine Islands (Walden, Trans. Zool. Soc. ix. p. 246), and has a very wide range, westwards across the Indian Ocean, southwards to Australia, and eastwards across the Pacific Ocean.

199. PHAETON RUBRICAUDA.
(RED-TAILED TROPIC-BIRD.)

Phaeton rubricauda, Boddaert, Tabl. Pl. Enl. p. 57 (1783).

The Red-tailed Tropic-bird is a white bird with a yellow bill and two long red feathers in the tail.

Figures: Gould, Birds of Australia, vii. pl. 73.

Mr. Holst writes that there is a bunch of the tail-feathers of the Red-tailed Tropic-bird in the Tokio Museum labelled Bonin Islands; and he was told on the Parry Islands that a white bird with a red tail was common there at certain seasons (Seebohm, Ibis, 1890, p. 107). I have a skin which was procured by Mr. Snow in the spring of 1883 on the Kruzenstern Rocks, about forty degrees to the east of the Bonin Islands.

The Red-tailed Tropic-bird frequents the Indian and Pacific Oceans, principally within the tropics.

200. FREGATA MINOR.
(LESSER FRIGATE-BIRD.)

Pelecanus minor, Gmelin, Syst. Nat. i. p. 572.

The Frigate-birds look like small Cormorants with deeply forked tails.

Figures: Gould, Birds of Australia, vii. pl. 72.

The Lesser Frigate-bird has been once shot at Hakodadi by Consul Quin in October (Seebohm, Ibis, 1884, p. 33).

It frequents the Pacific Ocean, principally within the tropics.

Suborder XIX. *HERODIONES*.

Palate desmognathous; mandible not produced and recurved behind its articulation with the quadrate; spinal bare space extending halfway or more up the neck.

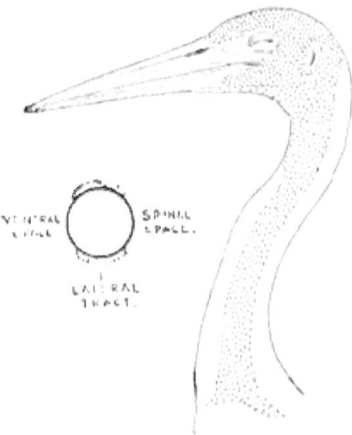

Pterylosis of neck of *Ardea cinerea*.

The Herodiones comprise three families. The *Ardeidæ*, containing about eighty species, are almost cosmopolitan, but they are not found in the arctic or antarctic regions. The *Scopidæ* contains only one species, which is peculiar to the Ethiopian Region. The *Ciconiidæ* consist of a score species, and are nearly as cosmopolitan as the *Ardeidæ*.

Of the 15 Japanese species belonging to the suborder Herodiones, 1 only is peculiar to Japan during the breeding-season; 3 breed in the Eastern Palæarctic Region; 5 breed both in the Palæarctic and Oriental Regions, one of which breeds also in the Nearctic Region; whilst 6 may be regarded as exclusively tropical, breeding in the Oriental Region.

Genus ARDEA.—The typical Herons differ from the Night-Herons (*Nycticorax*) and agree with the Bitterns (*Botaurus*) in having the whole of the front of the tarsus covered with wide transverse plates; but they agree with *Nycticorax* and differ from *Botaurus* in having twelve tail-feathers, and in having the inner toe shorter than the outer. They differ from both these genera in having the tibia bare of feathers for a greater distance than the length of the hind toe without the claw, but this character is subject to much individual variation. In *Ardea garzetta* it varies from 1·5 to 2·5 inches, being sometimes longer and sometimes shorter than the length of the inner toe without the claw. In *Ardea coromanda* it is always shorter than the inner toe, but never so short as the hind toe without the claw.

The genus *Ardea* may be divided into few or many subgenera according to the caprice of the systematist. The Japanese species consist of one typical Heron, five Egrets, and a Reef-Heron; but I know of no generic characters to distinguish one group from another.

201. ARDEA CINEREA.
(HERON.)

Ardea cinerea, Linnæus, Syst. Nat. i. p. 236 (1766).

The Common Heron is a large grey-backed species like the Purple Heron (which is very likely to occur in Japan), but may be distinguished from it at all ages by the colour of its forehead and crown, which is slate-grey in young in first plumage, moulting to white in the adult, whilst that of the Purple Heron is russet-brown in the young, moulting to black in the adult.

Figures: Dresser, Birds of Europe, iv. pl. 395.

The Common Heron is somewhat sparingly distributed in all the Japanese Islands. There is an example in the Swinhoe collection from Hakodadi (Swinhoe, Ibis, 1876, p. 335), and there are four examples in the Pryer collection from Yokohama. It also occurs in Nagasaki (Blakiston and Pryer, Trans. As. Soc. Japan, 1882, p. 118) and the Loo-Choo Islands (Cassin, Exp. Am. Squad. China Seas and Japan, ii. p. 244). The examples obtained by Dr. Siebold were doubtless procured at Nagasaki (Temminck and Schlegel, Fauna Japonica, Aves, p. 114).

The breeding-range of the Common Heron extends from the British Islands, across Europe and Southern Siberia to Japan. It also breeds in India. Chinese and Japanese examples do not appear to differ from European ones, but Dybowski states that Siberian examples have more developed nuptial plumes and redder feet (Taczanowski, Journ. Orn. 1874, p. 333).

202. ARDEA ALBA.
(GREAT WHITE EGRET.)

Ardea alba, Linnæus, Syst. Nat. i. p. 239 (1766).

The Great White Egret has no nuptial plumes on the head or breast, but in breeding-dress they are well developed on the scapulars. The bill is black in summer and yellow in winter. It is the largest of the Japanese White Egrets (wing from carpal joint 18 to $13\frac{1}{2}$ inches).

Figures: Dresser, Birds of Europe, vi. pl. 398 (Western race); Gray and Hardwicke, Ill. Ind. Zool. ii. pl. 49 (Eastern race, described as *Ardea modesta*).

Both races of the Great White Egret appear to visit Japan, the Eastern race as a common summer visitor, the Western race as a more or less accidental winter visitor. The two races only differ in size. The length of the wing (from carpal joint) of the Western or typical race, *Ardea alba*, varies from 18 to $15\frac{1}{2}$ inches; that of the Eastern race, *Ardea alba modesta*, from $15\frac{1}{2}$ to $13\frac{1}{2}$ inches.

The Eastern race of the Great White Egret is a summer visitor to all the Japanese Islands, arriving in Tokio Bay in April. It has been seen on Eturop, the most southerly of the Kurile Islands (Blakiston and Pryer, Trans. As. Soc. Japan, 1882, p. 118); and there is an example in the Swinhoe collection from Hakodadi (Swinhoe, Ibis, 1876, p. 335). There is an example in the Pryer collection from Tokio; and it has been collected by Mr. Ringer at Nagasaki.

It is to this race that the Great White Heron procured by Dr. Siebold doubtless at Nagasaki (Temminck and Schlegel, Fauna Japonica, Aves, p. 114) must be referred; and also that procured by Captain Rodgers on the Loo-Choo Islands (Cassin, Proc. Acad. Nat. Sc. Philad. 1862, p. 321), which has lately been examined in the Philadelphia Museum (Stejneger, Zeitschr. ges. Orn. 1887, p. 170).

The Eastern race of the Great White Heron breeds in Southern Siberia, and in India, Burma, and China.

A large form of the Great White Egret, with a length of wing varying from $16\frac{1}{2}$ to $17\frac{1}{2}$ inches, has occurred several times in Japan. Dr. Stejneger gives the measurements of one example from Yezzo and two from Tokio; and Captain Blakiston mentions three others from Yokohama, all obtained in winter, and all with yellow bills. I have not had an opportunity of examining any of these skins, but feel little doubt as to the species to which they should be referred. They agree apparently with the winter plumage of the Western form of *A. alba*, and it is scarcely possible that they can be examples of *A. egretta* from America, or of *A. syrmatophora* from Australia and New Zealand. The length of the bill ($4\frac{3}{4}$ to 5 inches from frontal feathers) and of the tarsus ($6\frac{1}{2}$ to 8 inches) appears to be too great for either the American or Australian species, so that the evidence seems to be strongly in favour of regarding these large Japanese Egrets as examples of the western race of *Ardea alba*, which have wandered eastwards in winter. So far as I know, none of the other Egrets named ever have the tarsus as much as 7 inches long.

203. ARDEA INTERMEDIA.
(PLUMED EGRET.)

Ardea intermedia, Wagler, Isis, 1829, p. 659.

The Plumed Egret is fairly entitled to the name given it by Gould. In breeding-plumage the dorsal plumes frequently extend six inches or more beyond the tail, and the pectoral plumes are often six inches long, and disintegrated like those of the scapulars. The combination of these two characters with the absence of nuchal plumes is found in no other Japanese Heron. The bill in summer is always more or less dark at the point and yellow at the base of both mandibles. In winter the bill is entirely yellow, a character sufficient to distinguish it from *A. garzetta*. Its length of wing from the carpal joint, which varies from 11½ to 12½ inches, distinguishes it from *A. coromanda* and *A. eulophotes*, in which the wing varies from 9 to 10 inches; and from *A. alba modesta* or *A. alba*, in which the wing varies from 13½ to 18 inches. The length of bill from the frontal feathers varies from 2¾ to 3½ inches, which is more than that of *A. coromanda*, and less than that of *A. alba modesta*. The Plumed Egret is also remarkable for its very long toes, the longest measuring more than 3 inches without the claw.

Figures: Temminck and Schlegel, Fauna Japonica, Aves, pl. 69 (erroneously named *Ardea egrettoides*, which is a synonym of *Ardea alba*); Gould, Birds of Australia, vi. pl. 57 (erroneously described as a new species under the name of *Herodias plumiferus*).

The Plumed Egret is a summer visitor to all the Japanese Islands, probably remaining to winter in the south. There are eight examples in winter dress from Yokohama in the Pryer collection, and it has been found in Nagasaki and Yezzo in summer dress (Blakiston and Pryer, Trans. As. Soc. Japan, 1882, p. 119).

This Egret has a very wide range. It is said to be a resident throughout the Ethiopian and Oriental Regions, and the Austro-Malayan and Australian Subregions.

The Plumed Egret is said to vary in the amount of black which the bill acquires during the breeding-season. In the Swinhoe collection is an example from Canton in full summer dress with the bill yellow, except for about half an inch at the tip, which is dark brown. An example from Yokohama has the terminal two thirds of the bill brown, whilst those from India and Ceylon are described as black in summer. It is possible that the eastern birds are subspecifically distinct, and that they should bear the name of *Ardea intermedia plumifera*.

204. ARDEA GARZETTA.

(LITTLE EGRET.)

Ardea garzetta, Linnæus, Syst. Nat. i. p. 237 (1766).

The Little Egret has a black bill both summer and winter. The length of wing from carpal joint varies from 10 to 11 inches, which is rather more than that of *A. coromanda*, and rather less than that of *A. intermedia* (neither of which ever has a black bill), and very much less than that of either the eastern or western forms of *A. alba* (which have a black bill in summer, and a yellow bill in winter). Although the Little Egret is a smaller bird than *A. intermedia*, it has a longer bill. Measured from the frontal feathers the bill of the smaller species varies from $3\frac{1}{2}$ to 3 inches, whilst that of the larger species varies from 3 to $2\frac{3}{4}$ inches.

Figures: Dresser, Birds of Europe, vi. pl. 399.

The Little Egret has not been obtained in Yezzo; but it is a resident in Southern Japan. There are eight examples in the Pryer collection from Yokohama (Seebohm, Ibis, 1879, p. 27). The examples procured by Dr. Siebold were doubtless obtained near Nagasaki (Temminck and Schlegel, Fauna Japonica, Aves, p. 115).

The Little Egret is only an accidental visitor to the British Islands, but its breeding-range extends from South Europe across Persia, India, Burma, and China to Japan.

205. ARDEA COROMANDA.
(EASTERN BUFF-BACKED HERON.)

Cancroma coromanda, Boddaert, Tabl. Pl. Enl. p. 54 (1783).

The Eastern representative of the Buff-backed Heron appears to be specifically distinct from its Western ally, and it is supposed that the ranges of the two species do not coalesce. The Eastern Buff-backed Heron differs from *A. bubulcus* in being rather larger in its size, and somewhat more brilliant in the colour of its nuptial plumes. In breeding-dress the chestnut-buff plumes on the head, nape, breast, and scapulars distinguish it from the other Japanese species. In winter plumage the length of the wing from carpal joint, which varies from 9 to 10 inches, distinguishes it from every other Japanese White Egret except perhaps from very small examples of *A. garzetta*. The colour of its bill, which is yellow at all seasons of the year, prevents it from being confused with *A. garzetta*, in which the bill is always black. There is, however, another small white Egret which is found in China (and possibly in the Malay Archipelago and Australia, if *A. immaculata* be the same species), and which may probably occasionally visit Japan, which is about the same size in length of wing. *A. eulophotes* may be recognized in summer by its white occipital crest. In winter it may be known by its longer bill, which measures from the frontal feathers $3\frac{1}{4}$ to $2\frac{3}{8}$ inches, instead of only from $2\frac{1}{2}$ to $2\frac{1}{4}$ inches; and by its shorter middle toe, which measures, without the claw, only from $2\frac{1}{4}$ to $2\frac{3}{8}$ instead of from $2\frac{3}{4}$ to $2\frac{3}{4}$ inches; so that in *A. eulophotes* the bill is longer than the middle toe, but in *A. coromanda* it is shorter.

Figures: D'Aubenton, Planches Enluminées, no. 910.

The Eastern Buff-backed Heron is a summer visitor to Southern Japan, but it has not been recorded from Yezzo. There are four examples from Tokio in winter plumage in the Pryer collection (Seebohm, Ibis, 1884, p. 35), and I have one in full summer dress from Sakai, in the south-west of the main island, collected by Mr. Owston. It has been sent by Mr. Ringer from Nagasaki, where the examples obtained by Dr. Siebold were doubtless also procured (Temminck and Schlegel, Fauna Japonica, Aves, p. 115).

It is a tropical species, inhabiting India, Ceylon, the Burma Peninsula, Cochin China, Southern China, Java, Borneo, Celebes, and the Philippine Islands.

206. ARDEA JUGULARIS.
(EASTERN REEF-HERON.)

Ardea jugularis, Wagler, Syst. Av. p. 214 (1827).

The grey phase of the Eastern Reef-Heron may be known by the nearly uniform slate-grey colour of the plumage; the white phase is of the same length of wing as *Ardea intermedia*, but the bill from frontal feathers ($3\frac{5}{8}$ to $3\frac{3}{8}$ inches) is longer instead of shorter than the middle toe and claw ($2\frac{1}{8}$ to $2\frac{5}{8}$ inches).

Figures: Gould, Birds of Australia, vi. pl. 60 (grey form), pl. 61 (white form).

The synonymy of the Reef-Herons is in the greatest confusion in consequence of there being two forms, one pure white, and the other slate-grey with a white line down the chin and upper throat. The white form appears to be the rarer of the two, but they are generally found together, and occasionally produce piebald examples, presumably by interbreeding. The Japanese birds belong to the Eastern species, which is said to range from the Andaman Islands eastwards on the coasts of Burma, the islands of the Malay Archipelago, to the coasts of Australia and New Zealand, and to some of the Pacific Islands. It is probable that the name of *Ardea sacra* (Gmelin, Syst. Nat. i. p. 640) may apply to the white form; but in the absence of proof it is wisest to adopt Wagler's name for the grey form, and *Ardea jugularis greyi* (Gray, List Spec. Birds Coll. Brit. Mus. iii. p. 80) for the white form. From Ceylon westwards along the coasts of Africa to the Gulf of Guinea a nearly allied species, *Ardea gularis*, is found, with much more white on the throat and a much more elongated crest.

The Eastern Reef-Heron is found as far north as Southern Japan, where it was obtained by Dr. Siebold (Schlegel, Mus. Pays-Bas, v. pt. 4, p. 28), though it was not included in the 'Fauna Japonica.' Mr. Ringer gave me two skins from Tsu-sima, an island in the Straits of Corea, and he also procured it from the Goto Islands a little further south (Blakiston and Pryer, Trans. As. Soc. Japan, 1882, p. 120). Captain Rodgers obtained both the grey form and the white form on the shores of the Loo-Choo Islands (Cassin, Proc. Acad. Nat. Sc. Philad. 1862, p. 321); and there are two examples of the grey form in the Pryer collection from the same locality. There are also examples of both the grey form and the white form from the central group of the Loo-Choo Islands in the museum of the Smithsonian Institution at Washington (Stejneger, Proc. United States Nat. Mus. 1887, pp. 301-303). The grey forms from the Corean Straits and from the Loo-Choo Islands are described by Stejneger as a new species under the name of *Demiegretta ringeri*.

In none of my examples is the crown or the occipital crest lighter in colour than is the case with typical examples in summer plumage, as stated by Dr. Stejneger of birds from the same localities. They all agree in having a narrow white line on the chin and upper throat, though this is sometimes obsolete or nearly so. Examples from the Pacific are on an average larger, darker and browner on the under-parts than those from the Bay of Bengal.

Genus NYCTICORAX.—The Night-Herons differ from the Herons (*Ardea*) and from the Bitterns (*Botaurus*) in having the lower portion of the tarsus reticulated instead of scutellated in front. They further differ from *Ardea*, and resemble *Botaurus*, in having the tibia feathered almost to the joint; and they further differ from

Botaurus and resemble *Ardea* in having twelve tail-feathers, and in having the inner toe shorter than the outer.

The genus *Nycticorax* may be divided into several subgenera—Night-Herons (typical *Nycticorax*), Mangrove-Herons (*Butorides*), and Squacco Herons (*Ardeola*): the two former with grey (instead of white) quills; and the two latter with straight and somewhat slender (instead of decurved and stout) bills.

207. NYCTICORAX NYCTICORAX.
(NIGHT-HERON.)

Ardea nycticorax, Linnæus, Syst. Nat. i. p. 235 (1766).

The Night-Heron is a medium-sized species (wing from carpal joint 10½ to 11 inches). It has grey quills, a stout decurved bill, and uniform pale grey axillaries.

Figures: Dresser, Birds of Europe, vi. pl. 402.

The Night-Heron is a common summer visitor to Southern Japan, but is not known to have occurred in Yezzo. In the Pryer collection there is a large series of both adult and immature examples from Yokohama. The examples procured by the Siebold Expedition were doubtless obtained near Nagasaki (Temminck and Schlegel, Fauna Japonica, Aves, p. 116); and there is one in the Norwich Museum sent by Mr. Ringer from the same locality. It was included by Pryer in the list of birds obtained by Namiye on the central group of the Loo-Choo Islands (Seebohm, Ibis, 1887, p. 181), and the specimen was identified by Dr. Stejneger (Proc. United States Nat. Mus. 1887, p. 296).

The breeding-range of the Night-Heron does not reach the British Islands (where this species is only known as a rare visitor), but it extends across Southern Europe to Persia, India, Burma, China, and Japan. It also breeds on the American continent.

208. NYCTICORAX CRASSIROSTRIS.
(BONIN NIGHT-HERON.)

Nycticorax crassirostris, Vigors, Zool. Captain Beechey's Voyage, p. 27 (1839).

The Bonin Night-Heron has a white superciliary stripe and white axillaries, and is otherwise similar to *N. caledonica*, except that the height of the bill at the nostrils varies from 1·0 to ·9 (instead of from ·9 to ·8) inches.

The Bonin Night-Heron was discovered by Kittlitz in 1828, but he identified it with the Australian species; and when he recorded it in 1833 (Kittlitz, Kupfertafeln zur Naturgeschichte der Vögel, pt. iii. p. 27) he called it *Ardea caledonica*. It had, however, been discovered in 1827 by Captain Beechey during the voyage of the 'Blossom,' but the zoological results of this voyage were not published until 1839, when the Bonin Night-Heron was named *Nycticorax crassirostris*. Vigors's type was placed in the Museum of the Zoological Society, and was transferred to the British Museum, where it now is, all statements (Walden, Trans. Zool. Soc. ix. p. 238) to the contrary notwithstanding.

The only other example known to exist is one in my collection, which was procured by Mr. Holst on Nakoudo-Shima, one of the Parry Islands (Seebohm, Ibis, 1890, p. 106).

209. NYCTICORAX GOISAGI.
(JAPANESE NIGHT-HERON.)

Nycticorax goisagi, Temminck, Planches Coloriées, no. 582 (1835).

The Japanese Night-Heron is a medium-sized species (wing from carpal joint 10 to 10½ inches). It has dark grey quills tipped with chestnut, a stout decurved bill, and barred axillaries.

Figures: Temminck and Schlegel, Fauna Japonica, Aves, pl. 70.

The Japanese Night-Heron, or, as it is sometimes called, the Japanese Tiger-Bittern, is peculiar to Japan and Formosa. There are six examples in the Pryer collection from Yokohama; and Mr. Dresser has seven examples procured in the same locality by Mr. Owston. There are examples in the Tweeddale collection and in the Norwich Museum obtained by Mr. Ringer from Nagasaki, whence those obtained by Dr. Siebold were probably also procured. It is very closely allied to the Malayan Night-Heron, *Nycticorax melanolophus*, a species which ranges from Southern India (Bourdillon, Stray Feathers, vii. p. 525), Ceylon, the Nicobar Islands (Hume, Stray Feathers, ii. p. 312), the Malay Peninsula, Sumatra, Java, Borneo, the Philippine Islands, to Formosa (Büttikofer, Notes from the Leyden Museum, 1887, p. 81), but it has not been obtained in China. The two species are so nearly allied that Swinhoe regarded them as identical; but there seems good grounds for believing them to be distinct.

The Malay species always has a dark crest (with a central and subterminal white spot on each feather in the young in first plumage); there is much white on the axillaries, under wing-coverts, and on the tips of the primaries.

It is not known that the Japanese species ever has a dark crest (though two of my specimens are very dark and grey on the forehead and crown); the pale bars across the axillaries and under wing-coverts are buff instead of white, and in only one example is there any white on the tips of the primaries, except on the first. Young in first plumage are unknown.

It has been stated that the shape of the bill is different in the two species, but this is not the case, except that in the Malayan species the bill is slightly longer than in the Japanese bird.

Both species occur on Formosa, the Malayan species as a breeding bird, and the Japanese species probably as a winter visitor, as the examples without the black crest in the Swinhoe collection were obtained in March, a fact which caused Swinhoe to regard them as the winter plumage of the Malayan species.

There can be no manner of doubt that this species has 12 and not merely 10 tail-feathers as has been stated (Reichenow, Journ. Orn. 1877, p. 216).

210. NYCTICORAX JAVANICUS.
(AUSTRALIAN MANGROVE-HERON.)

Ardea javanica, Horsfield, Trans. Linn. Soc. 1821, p. 190.

The Australian Mangrove-Heron is a small species (wing from carpal joint 7·5 to 8·2 inches). It has grey quills, a straight bill, and grey axillaries.

Figures: Gould, Birds of Australia, vi. pl. 66 (under the name of *Ardetta macrorhyncha*), pl. 67 (under the name of *Ardetta stagnatilis*).

The Australian Mangrove-Heron is a summer visitor to Japan. Captain Blakiston obtained it at Hakodadi (Blakiston and Pryer, Trans. As. Soc. Japan, 1882, p. 120); and Mr. Ringer has sent examples to the Norwich Museum procured at Nagasaki (Seebohm,

Ibis, 1884, p. 35), where those obtained by the Siebold Expedition (recorded erroneously under the name of *Ardea scapularis*) were doubtless also procured (Temminck and Schlegel, Fauna Japonica, Aves, p. 116).

The Australian Mangrove-Heron was originally described from North Australia, under the name of *Ardetta stagnatilis* (Gould, Proc. Zool. Soc. 1847, p. 221); it was afterwards redescribed from the east coast of Australia, under the name of *Ardetta macrorhyncha* (Gould, Proc. Zool. Soc. 1848, p. 39). It is probable that the *Ardea patruelis*, described in the same year from Tahiti (Peale, Zool. U. S. Expl. Exp. p. 216), must be referred to this race; and there can be no doubt that the names *Ardea (Butorides) virescens* var. *amurensis* (Schrenck, Reis. Forsch. Amur-Lande, i. p. 441) and *Butorides schrenckii* (Bogdanow, Consp. Av. Imp. Ross. i. p. 115) do belong to it. This race breeds in the valley of the Amoor and probably in Japan, and winters in Formosa and South China. It is said also to breed in Australia. It may be known as *Nycticorax javanicus stagnatilis*. It cannot be regarded as more than subspecifically distinct from the typical *Nycticorax javanicus*, as it only differs from it in size (wing 8·2 to 7·5 inches). Indian examples are smaller (wing 7·2 to 6·5 inches). The typical form ranges from India and Ceylon, across the Burma Peninsula to the Malay Archipelago and South China. Probably both forms occur and completely intergrade in the tropics.

211. NYCTICORAX PRASINOSCELES.
(CHINESE SQUACCO HERON.)

Ardeola prasinosceles, Swinhoe, Ibis, 1860, p. 64.

The Squacco Herons form a group of half a dozen small species, in which the wing from carpal joint only measures from 8 to 9 inches. The Chinese Squacco Heron is white, with the head, nape, and sides of the neck chestnut, and with the disintegrated feathers of the back and breast greenish black in the adult. In immature birds the chestnut is replaced by brown streaked with buff, the breast is white streaked with brown, the back and the tertials are brown, the wing-coverts are pale brown, and the scapulars are obscurely streaked with buff.

The sole claim of the Chinese Squacco Heron to be regarded as a Japanese bird rests upon a single immature example procured by Captain Blakiston at Hakodadi on the 12th of October, 1879, and presented by him to the museum of the Smithsonian Institution in Washington (Seebohm, Ibis, 1884, p. 35).

The Chinese Squacco Heron is a resident in South China and Cochin China. It is said to be a summer visitor to Central China (David and Oustalet, Ois. Chine, p. 443). It has once occurred in Manchuria (Taczanowski, Bull. Soc. Zool. France, 1886, p. 309), and is also recorded from Independent Burma, Tenasserim, and the Malay Peninsula.

It is possible that the *Crabier de Malac* (D'Aubenton, Planches Enluminées, plate 911) may be intended to represent an immature example of this species, in which case the names *Cancroma leucoptera* (Boddaert, Table Pl. Enl. p. 54) and *Ardea malaccensis* (Gmelin, Syst. Nat. i. p. 643) must be added to its synonymy. The name of *Buphus bacchus*, dating from 1857 (Bonaparte, Consp. Generum Avium, ii. p. 127), belongs, without doubt, to it.

Genus BOTAURUS.—The Bitterns differ from the Herons (*Ardea*) and from the Night-Herons (*Nycticorax*) in having only ten instead of twelve tail-feathers, and in having the inner toe longer than the outer. They further differ from *Nycticorax* and resemble *Ardea* in having the whole of the front of the tarsus covered with wide transverse plates. They further differ from *Ardea* and resemble *Nycticorax* in having the tibia feathered almost to the joint.

212. BOTAURUS STELLARIS.
(BITTERN.)

Ardea stellaris, Linneus, Syst. Nat. i. p. 239 (1766).

The Common Bittern is a large bird (wing from carpal joint more than 12 inches) and it has barred primaries.

Figures: Dresser, Birds of Europe, vi. pl. 403.

The Bittern is found in all the Japanese Islands, and is probably a summer visitor to Yezzo, and a resident in the islands further south. There is an example in the Swinhoe collection from Hakodadi

(Swinhoe, Ibis, 1875, p. 455), and there are three examples in the Pryer collection from Yokohama. Mr. Ringer has also procured it at Nagasaki, where the examples obtained by the Siebold Expedition were also doubtless procured (Temminck and Schlegel, Fauna Japonica, Aves, p. 116).

The breeding-range of the Bittern extends from the British Islands across Europe and Asia, both north and south of Mongolia, to Japan.

213. BOTAURUS SINENSIS.
(ORIENTAL LITTLE BITTERN.)

Ardea sinensis, Gmelin, Syst. Nat. i. p. 642 (1788).

The Oriental Little Bittern is a small bird (wing from carpal joint 5 to $5\frac{1}{4}$ inches), with white or buff axillaries, and the tibiae completely feathered to the joint.

Figures: Gray and Hardwicke, Ill. Ind. Zool. i. pl. 66. fig. 2.

The Oriental Little Bittern or Yellow Bittern is found in all the Japanese Islands. There are numerous examples from Yokohama in the Pryer collection, and Captain Blakiston sent me one from Hakodadi. There are two examples in the Norwich Museum sent by Mr. Ringer from Nagasaki; they do not differ from Chinese examples in the Swinhoe collection (Seebohm, Ibis, 1879, p. 27).

It is a tropical species, resident in India, Ceylon, the Malay Peninsula, many of the Islands of the Malay Archipelago, and China.

214. BOTAURUS EURHYTHMA.
(SCHRENCK'S LITTLE BITTERN.)

Ardetta eurhythma, Swinhoe, Ibis, 1873, p. 73.

Schrenck's Little Bittern is a small bird (wing from carpal joint 5 to $5\frac{1}{2}$ inches), with grey axillaries, and the tibia not quite feathered to the joint.

Figures: Swinhoe, Ibis, 1873, pl. 2 (adult); Schrenck, Reis. Forsch. Amur-Lande, i. pl. 13. fig. 3 (young).

Schrenck's Little Bittern has occurred several times in Yezzo (Blakiston and Pryer, Trans. Asiat. Soc. Japan, 1882, p. 118). There is an example in the Swinhoe collection from Hakodadi (Swinhoe, Ibis, 1876, p. 335), and there are two examples from Yokohama in the Pryer collection. They do not differ from Chinese specimens.

It breeds in the valley of the Amoor and in North China, migrating in autumn to Japan and South China.

215. CICONIA BOYCIANA.
(JAPANESE STORK.)

Ciconia boyciana, Swinhoe, Proc. Zool. Soc. 1873, p. 512.

The Japanese Stork is a very large bird (wing from carpal joint 27 inches), and is white with black scapulars and black quills, except that the outer webs of some of the primaries are partly hoary white.

Figures: Sclater, Proc. Zool. Soc. 1874, pl. 1.

The Japanese Stork is said to be a resident in Japan. There are two examples in the Pryer collection from Yokohama. It was originally described by Swinhoe from a pair of living examples in the grounds of the British Consulate at Shanghai, which had been brought from Yokohama; and Mr. Sclater's figure was drawn from a second pair brought by Swinhoe from Japan. Dybowski found it in some numbers in the valley of the Ussuri (Taczanowski, Proc. Zool. Soc. 1874, p. 307); and Mr. Jouy obtained three examples in the Corea (Stejneger, Proc. United States Nat. Mus. 1887, p. 286). It is to be seen sailing on its immense spread of wings over the plains near Yokohama (Blakiston and Pryer, Ibis, 1878, p. 224).

The Japanese Stork appears to be nearest allied to the Maguari Stork, *Ciconia maguari*, a species which inhabits the pampas of South America. It is larger than its Neotropical ally (wing from carpal joint 27 instead of 20 inches); but it resembles it in having the bare skin in front of the eye red; its bill is black instead of horn-colour, and its upper tail-coverts white instead of black. There can, however, be little doubt that its real affinities are with the White Stork, *Ciconia alba*, the range of which extends from Spain across Europe and Western Asia to Eastern Turkestan, where it has been recorded as far east as Yarkand. The White Stork is intermediate in size (wing about 23 inches); the distribution of black and white on the plumage is the same as that of the Japanese Stork, but the bill is red, and the bare skin in front of the eye is black.

Suborder XX. *PLATALEÆ*.

The Spoonbills and Ibises form a connecting-link between the Anserine and the Ardeine groups of birds; but in some respects they closely resemble the Limicoline group. They appear to be easily diagnosed by two characters. They agree with the Anseres and the Herodiones in having the maxillo-palatines completely fused with each other, and with the Limicolæ in being schizorhinal in the bifurcation of their nasals. They further agree with the Anseres and the Limicolæ in having the mandible produced behind its articulation with the quadrate and recurved.

There are only about 30 species of Plataleæ, which are distributed throughout the tropical and subtropical parts of the world.

Of the four Japanese species which belong to this suborder one is only known to breed in Japan, one is confined to Japan and East China, one is Palæarctic and Oriental, whilst the fourth is only Oriental and Japanese.

216. PLATALEA LEUCORODIA.
(COMMON SPOONBILL.)

Platalea leucorodia, Linneus, Syst. Nat. i. p. 231 (1766).

The Common Spoonbill is rather larger than Swinhoe's Black-faced Spoonbill; the naked skin on the face is yellow; and the throat is bare in the middle for a much greater distance than it is on the sides. The feathering on the forehead and sides of the head reaches far in front of the eye.

Figures: Temminck and Schlegel, Fauna Japonica, Aves, pl. 75, under the name of *Platalea major*.

The Common Spoonbill is a somewhat rare bird in Japan, especially in Yezzo. There is an example in the Swinhoe collection obtained by Captain Blakiston at Hakodadi, where it had been previously procured on the 13th of October (Whitely, Ibis, 1867, p. 204), and another from Yokohama obtained in November.

The range of the Common Spoonbill extends from the British Islands (where it formerly bred) across Europe and Southern Siberia to Japan. To these countries it is only a summer visitor, but it is a resident in North Africa, India, and Ceylon. It is a winter visitor to Formosa and South China.

It has been stated that the Eastern race of the Common Spoonbill has a longer culmen than the Western race, and ought therefore to be regarded as subspecifically distinct (Ogilvie-Grant, Ibis, 1889, p. 41). I have been unable to find the slightest evidence of the truth of this statement. The extreme length of bill attained by old birds seems to be 9 inches, and this appears to be the case throughout the range of the species from Spain to Japan. It is true that Hume states (Stray Feathers, i. p. 256) that in Scinde and the Punjab they sometimes attain a length of 9·7 inches; but as no example in the Hume collection measures more than 9 inches, it is probable that this is a misprint for 9·1 inches. If it be not a printer's error, it only proves that there may be a large-billed race of the

Side of head and throat of *Platalea leucorodia*. ⅔ natural size.

Common Spoonbill which is a resident in the Oriental Region, perhaps only in the western half of the Oriental Region, and it still leaves the Japanese Spoonbills absolutely indistinguishable from those of Holland.

The alleged difference in the extent of the bare space on the throat between Eastern and Western examples does not appear to have any geographical significance, but to be due either to individual variations or to difference of age.

217. PLATALEA MINOR.
(SWINHOE'S BLACK-FACED SPOONBILL.)

Platalea minor, Temminck and Schlegel, Fauna Japonica, Aves, p. 120 (1847).

Swinhoe's Black-faced Spoonbill is rather smaller than the Common Spoonbill; the naked skin on the face is black; and the feathering scarcely extends in front of the eyes either on the forehead or on the sides of the head. The bare space on the throat is, on the other hand, much less extensive than in the European species.

Figures: Temminck and Schlegel, Fauna Japonica, Aves, pl. 76 (bare skin on face coloured wrong).

The type of Swinhoe's Black-faced Spoonbill has a black face, though it is neither described nor figured as such in the 'Fauna Japonica.' In all probability it was procured at Nagasaki, whence there is an example in the Christiania Museum, collected by Mr. Petersen in December (Stejneger, Proc. United States Nat. Mus.

Side of head and throat of *Platalea minor*. ⅔ natural size.

1887, p. 283). There is also a head in the collection of Canon Tristram at Durham, obtained by Lieutenant Gunn on North Goto Island, west of Nagasaki (Ogilvie-Grant, Ibis, 1889, p. 57).

So far as is known, Swinhoe's Black-faced Spoonbill is peculiar to Formosa and the extreme south of Japan. It is very nearly allied to *Platalea melanorhyncha* from Australia and Timor, which combines the black face of *P. minor* with the naked throat of *P. leucorodia*.

218. IBIS NIPPON.
(JAPANESE CRESTED IBIS.)

Ibis nippon, Temminck, Planches Coloriées, no. 551 (1835).

The Japanese Crested Ibis is a large species (wing from carpal joint 16 inches); it is white when adult (grey on the head, crest, neck, and back when immature), with the fore part and sides of the head naked and scarlet, and the quills suffused with scarlet.

Figures: Temminck and Schlegel, Fauna Japonica, Aves, pl. 71.

The Japanese Crested Ibis breeds in Yezzo, but probably migrates southwards in autumn. In Southern Japan it is a resident (Blakiston and Pryer, Ibis, 1878, p. 223). There is an example in the Swinhoe collection from Hakodadi (Swinhoe, Ibis, 1875, p. 455), and there is one in the Pryer collection from Yokohama. It is not peculiar to Japan, but breeds also at Ningpo, and occurs in winter in the Corean Peninsula, and on the coasts of South China and Hainan.

It does not seem to be closely related to any of the other Ibises, and it is impossible to guess which is its nearest ally.

219. IBIS MELANOCEPHALA.
(WHITE IBIS.)

Tantalus melanocephalus, Latham, Index Orn. ii. p. 709 (1790).

The White Ibis is a large species (wing from carpal joint 13 to 15 inches). It is white with grey disintegrated tertials in adult breeding plumage. The head and neck are naked and black.

Figures: Temminck, Planches Coloriées, no. 481 (1829).

The White Ibis is probably a summer visitor to Southern Japan, but has not been recorded from Yezzo (Blakiston and Pryer, Ibis, 1878, p. 223). There is an example in the Swinhoe collection and one in the Pryer collection, both from Yokohama (Seebohm, Ibis, 1884, p. 35).

It is a tropical species resident throughout the Oriental Region (Seebohm, Ibis, 1888, p. 437), and is represented by close allies in the Ethiopian and Australian Regions.

Order LAMELLIROSTRES.

So far as is known the Lamellirostres are the only birds which combine the following characters :—

Young born covered with down, and able to run and feed themselves in a few hours; palate directly desmognathous; spinal feather-tract not defined on the neck.

The Order Lamellirostres contains three Suborders, only one of which is represented in Japan.

Suborder XXI. *PHŒNICOPTERI.*

Palate desmognathous; basipterygoid processes absent or very rudimentary; nasals holorhinal; mandible much produced and recurved behind its articulation with the quadrate.

The number of known species of Flamingo does not reach a dozen, but they have a very wide range. They are found in the Neotropical and Ethiopian Regions, in the south of the Palæarctic Region, and in the Oriental Region as far east as Calcutta. The Common Flamingo is said to have occurred in the British Islands and on Lake Baikal, but neither statement rests on satisfactory evidence. It is somewhat remarkable that the range of the Phœnicopteri does not extend to China or Japan.

Suborder XXII. *ANSERES.*

The Ducks, Geese, and Swans possess a character which they share with the Gallinæ. They have basipterygoid processes on the rostrum of the basisphenoid which articulate with the pterygoids as near as possible to the palatines. They differ from the Gallinæ in many important characters, though they resemble them in having the mandible produced and recurved behind its articulation with the quadrate. They are desmognathous, and there is only one notch on each side of the posterior margin of the sternum, and that is a very shallow one compared with the deep clefts in the sternum of the Gallinæ. The

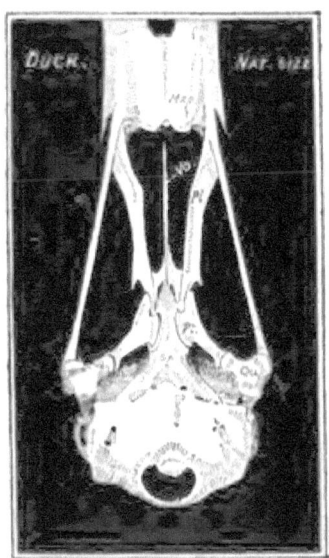

Skull of *Anas boschas*.

episternal processes are small, and do not meet over the feet of the coracoids.

The Anseres are cosmopolitan, and consist of nearly two hundred species, of which 37 have been recorded from the Japanese Empire.

220. CYGNUS MUSICUS.
(HOOPER SWAN.)

Cygnus musicus, Bechstein, Naturg. Deutschl. iii. p. 830 (1809).

The Hooper may be recognized by the distribution of the yellow and black on its bill. The yellow at the base of the mandible extends in front of the nostrils, whilst the black only reaches halfway to the gape.

Figures: Dresser, Birds of Europe, vi. pl. 419. fig. 4 (head).

The Hooper Swan is a common winter visitor to Yezzo. There is an example in the Swinhoe collection procured by Captain Blakiston at Hakodadi (Swinhoe, Ibis, 1875, p. 456). Messrs. Blakiston and

Pryer state that it is occasionally obtained in Tokio Bay in winter, but there is no skin in the Pryer collection. The example procured by the Siebold Expedition was probably obtained at Nagasaki (Temminck and Schlegel, Fauna Japonica, Aves, p. 125).

The breeding-range of the Hooper extends from the Atlantic to the Pacific, and it is not known that examples from the Eastern Palæarctic Region differ in any way from those obtained in the western part of that Region.

221. CYGNUS BEWICKI.
(BEWICK'S SWAN.)

Cygnus bewickii, Yarrell, Trans. Linn. Soc. xvi. p. 453 (1833).

In Bewick's Swan the yellow on the mandible does not reach so far forward as the nostrils, whilst the black extends backwards to the gape.

Figures: Gould, Birds of Great Britain, v. pl. 10; Dresser, Birds of Europe, vi. pl. 419. fig. 3 (head).

Bewick's Swan is a winter visitor to the Japanese Islands. It was first included in the Japanese list on the authority of a specimen in the Tokio Educational Museum (Blakiston and Pryer, Ibis, 1878, p. 212). There is an example in the Pryer collection from Tokio Bay.

Bewick's Swan breeds in the high north both of the Eastern and the Western Palæarctic Region, and winters in the British as well as the Japanese Seas. It is not known to vary in any way within its range.

222. ANSER CYGNOIDES.
(CHINESE GOOSE.)

Anser cygnoides, Gmelin, Syst. Nat. i. p. 502 (1788).

The Chinese Goose has a black bill, and a dark brown band down the back of the neck.

Figure: Temminck and Schlegel, Fauna Japonica, Aves, pl. 81.

The Chinese Goose appears to be a resident in Japan, and has been recorded from the Kurile Islands (Pallas, Zoogr. Rosso-Asiat. ii. p. 219). There is an example from Yokohama in the Blakiston collection (Seebohm, Ibis, 1884, p. 32).

It breeds in Eastern Siberia, and winters in China.

223. ANSER SEGETUM.
(BEAN-GOOSE.)

Anser segetum, Gmelin, Syst. Nat. i. p. 512 (1788).

The Bean-Goose has yellow legs, and an orange bill with a dark base and a dark nail.

Figures: Dresser, Birds of Europe, vi. pl. 412 (typical form).

The Eastern race of the Bean-Goose is a common winter visitor to the Japanese Islands. There is an example in the Swinhoe collection procured by Captain Blakiston at Hakodadi in October (Swinhoe, Ibis, 1875, p. 456), and there is another in the Pryer collection from Tokio Bay. The former of these appears to be the first recorded from Japan, unless we admit that the *Anser vulgaris* of Pallas (Zoogr. Rosso-Asiat. ii. p. 223), of which the Kurile Island and Japanese local names are quoted, refers to this species.

The Eastern form of the Bean-Goose completely intergrades with the Western form, and is consequently regarded as only subspecifically distinct from it. The range of the species extends from the Atlantic to the Pacific.

The Eastern form of the Bean-Goose is especially remarkable for its large size, and for the browner (not so grey) colour of its head. Examples from Japan measure from the frontal feathers to the tip of the beak 2·5, 2·7, and 2·9 inches. Western examples vary in this respect from 1·7 to 2·5 inches, and Eastern examples from 2·4 to 3·4 inches.

The Eastern form is known as *Anser segetum serrirostris* (Swinhoe, Proc. Zool. Soc. 1871, p. 417).

ANSER BRACHYRHYNCHUS.
(PINK-FOOTED GOOSE.)

Anser brachyrhynchus, Baillon, Mém. Soc. roy. d'ém. d'Abbeville, 1833, p. 74.

The Pink-footed Goose very closely resembles the Bean-Goose, but the yellow on the bill and feet is replaced by pink.

Figures: Dresser, Birds of Europe, vi. pl. 413.

The Pink-footed Goose was admitted to the Japanese fauna on the authority of a female obtained in October at Hakodadi by Captain Blakiston (Swinhoe, Ibis, 1875, p. 456). Unfortunately this example cannot be found in the Swinhoe collection, and some doubt attaches to the correctness of the identification.

224. ANSER ALBIFRONS.
(WHITE-FRONTED GOOSE.)

Branta albifrons, Scopoli, Ann. I. Hist. Nat. p. 69 (1769).

The White-fronted Goose is supposed to vary in length of bill from frontal feathers from 2·35 to 1·6 inches. The white on the forehead does not reach as far back as the eye. The legs are yellow and the bill pale.

Figures: Dresser, Birds of Europe, vi. pl. 414.

The White-fronted Goose is a common winter visitor to the southern islands of Japan, and passes along the coasts of Yezzo on migration in spring and autumn. There is a male in the Swinhoe collection procured in April at Hakodadi by Captain Blakiston (Swinhoe, Ibis, 1875, p. 456), and there is an example in the Pryer collection from Tokio Bay. The examples obtained by the Siebold Expedition were probably procured at Nagasaki (Temminck and Schlegel, Fauna Japonica, Aves, p. 125).

The two adult examples of the White-fronted Goose from Japan measure respectively 1·75 and 1·9 inches from frontal feathers to tip of beak. If *Anser gambeli* be regarded as a distinct race the latter would belong to it. The smaller bird is paler in colour and has a white patch on the chin, but there is no reason to suppose that American examples differ from European ones in any way. An example from Brighton measures 2·05 inches from frontal feathers to tip of beak.

The occurrence of the Pink-footed Goose on the Japanese coasts requires authentication. Captain Blakiston sent me the skin of a Goose which he obtained at Hakodadi in October (Seebohm, Ibis, 1882, p. 369).

There is no trace of black on the breast or belly.

The brown of the head joins the bill without any trace of black or white.

The whole bill is pale except the nail, which is nearly black.

It is possible that this may be an example of a Pink-footed Goose, but, in the absence of the black base to the bill, I am inclined to regard it as the young in first plumage of the White-fronted Goose, of which an example in the Swinhoe collection, dated Shanghai, 8 March, has only a narrow margin of white feathers at the base of the bill.

The Pink-footed Goose is only known with certainty to breed on Spitzbergen, but its breeding-range probably extends to Iceland and Franz-Josef Land. Its occurrence in Japan, if confirmed, is probably only accidental.

225. ANSER MINUTUS.
(LESSER WHITE-FRONTED GOOSE.)

Anser minutus, Naumann, Vög. Deutschl. xi. p. 364 (1842).

The Lesser White-fronted Goose is supposed to vary in length of bill from 1·6 to 1·15 inches. The white on the forehead extends as far back as the eye. The legs are yellow and the bill pale.

Figures: Naumann, Vögel Deutschlands, pl. 290.

The Lesser White-fronted Goose is a winter visitor to the Japanese Islands, and has been obtained both in Yezzo and in Southern Japan (Blakiston and Pryer, Ibis, 1878, p. 212). There is an example in the Swinhoe collection obtained in the Yokohama market in January (Seebohm, Ibis, 1879, p. 22), and there is a second in the Pryer collection from the same locality.

It breeds in the tundras of Siberia and Northern Europe.

226. ANSER HYPERBOREUS.
(SNOW-GOOSE.)

Anser hyperboreus, Pallas, Spicil. Zool. vi. p. 25 (1769).

The Snow-Goose, when adult, is white with black primaries (shading into grey at the base) and grey primary-coverts. The bill is light red, and the legs dark red. Young birds have the head, neck, back, and breast grey, and the bill and legs brown.

Figures: Dresser, Birds of Europe, vi. pl. 417.

American ornithologists (Ridgway, Man. North-Am. Birds, p. 115) admit two races of Snow-Geese: the typical form *Anser hyperboreus*, or Lesser Snow-Goose (wing from carpal joint 14½ to 17 inches); and a larger race *Anser hyperboreus nivalis*, or Greater Snow-Goose (wing from carpal joint 17¼ to 17½ inches). In the former the bill from the frontal feathers is said to vary from 1·9 to 2·3 inches, and in the latter from 2·5 to 2·7 inches.

Both races appear to be winter visitors to the Japanese coasts, occurring in large flocks (Blakiston and Pryer, Trans. As. Soc. Japan, 1882, p. 95); and the small race was recorded from Japan as long ago as 1840 (Temminck, Man. d'Orn. iv. p. 516).

There are two examples in the Pryer collection from Tokio Bay which undoubtedly belong to the smaller race; they measure 15¾ and 16¼ inches in length of wing from carpal joint, and 2 inches in length of bill from frontal feathers. On the other hand, two examples in the Blakiston collection, from the same locality, measure 17 and 17¾ inches in length of wing (Seebohm, Ibis, 1884, p. 32), and might almost be regarded as belonging to the larger race. The example procured by the Siebold Expedition was doubtless obtained at Nagasaki (Temminck and Schlegel, Fauna Japonica, Aves, p. 125).

The Snow-Geese breed in Arctic America, and possibly in Eastern Siberia, and occasionally appear in winter on the British coasts.

227. ANSER HUTCHINSI.
(HUTCHINS' BERNACLE GOOSE.)

Anser hutchinsii, Swainson and Richardson, Faun. Bor.-Amer. ii. p. 470 (1831).

Hutchins' Bernacle Goose is a small dark race of the Canada Goose, and is said to have fewer tail-feathers (14 to 16 instead of 18 to 20). Both races of the Canada Goose differ from the Bernacle Goose, which visits the British Islands, in having the black on the crown extending also over the forehead, but in not having the black of the throat reaching the breast.

Figures: Cassin, Birds of California &c. pl. 45; Baird, Brewer, and Ridgway, Water-Birds N. Amer. ii. p. 458 (coloured woodcut of head).

The occurrence of a species of Bernacle Goose in Japan was recorded as long ago as 1840 (Temminck, Man. d'Orn. iv. p. 520); but Captain Blakiston was probably the first to discover that the Japanese bird was one of the races of the Canada Goose, and quite distinct from the species of Bernacle Goose which winters on our coasts (Blakiston and Pryer, Ibis, 1878, p. 212, no. 27).

Hutchins' Bernacle Goose breeds on the Commander Islands (Stejneger, Orn. Expl. Comm. Isl. and Kamtschatka, p. 148) and on the Kuriles (Blakiston and Pryer, Trans. As. Soc. Japan, 1882, p. 96). It is a winter visitor to the coasts of the Japanese Islands; Captain Blakiston sent me an example procured at Hakodadi in November (Seebohm, Ibis, 1882, p. 369); and there is an example in the Pryer collection from Kadsusa in Tokio Bay.

The breeding-range of Hutchins' Bernacle Goose extends from the Kurile Islands across Arctic America to Hudson's Bay.

228. ANSER NIGRICANS.
(PACIFIC BRENT GOOSE.)

Anser nigricans, Lawrence, Ann. Lyc. New York, 1846, p. 171.

The Pacific Brent Goose only differs from the typical or dark-bellied race of the European Brent Goose in having the white crescentic markings on each side of the neck meeting in front, and in having the line of demarcation between the black breast and the dark belly rather more obscure.

Figures: Baird, Brewer, and Ridgway, Water-Birds N. Amer. ii. p. 472 (woodcuts).

The Pacific Brent Goose is a winter visitor to the Japanese coasts (Blakiston and Pryer, Ibis, 1878, p. 212). Mr. Henson has sent skins from Hakodadi; there are three examples in the Pryer collection from Tokio Bay, and Mr. Ringer has procured it near Nagasaki (Seebohm, Ibis, 1884, p. 32).

It is not known that the Pacific Brent Goose breeds in the Old World, but on the American continent it is a summer visitor to Alaska and Arctic North America as far east as Franklin Bay.

229. DENDROCYGNA JAVANICA.
(INDIAN WHISTLING TEAL.)

Anas javanica, Horsfield, Trans. Linn. Soc. 1821, p. 199.

The Indian Whistling Teal has a brown crown, with no stripe down the hind neck. The underparts vary from buff to chestnut, but are unstriated.

Figures: Hume and Marshall, Game Birds of India, Burmah, and Ceylon, iii. pl. 15.

There are three examples of the Indian Whistling Teal in the Pryer collection from the central group of the Loo-Choo Islands, and an example was obtained by Mr. Nishi on one of the most southerly group of the same chain (Stejneger, Proc. United States Nat. Mus. 1887, p. 397).

It is a tropical species, breeding in India, Ceylon, the Burma Peninsula, Java, and in the Nicobar and the Andaman Islands.

The occurrence of this species in the Loo-Choo Islands is very remarkable. An allied form with a spotted breast, *Dendrocygna vagans*, occurs in the Philippine Islands, Celebes, Timor, and North Australia; but the genus appears to be unrepresented in China.

230. TADORNA CORNUTA.
(COMMON SHELDRAKE.)

Anas cornuta, S. G. Gmelin, Reise Russl. ii. p. 185 (1774).

The Common Sheldrake has a greenish-black head and neck, and white lower back, rump, upper tail-coverts, sides of belly, and flanks, all these parts being chestnut in the Ruddy Sheldrake, except the rump and upper tail-coverts, which are black.

Figures: Dresser, Birds of Europe, vi. pl. 420.

The Common Sheldrake is not known to have occurred in Yezzo, but is not uncommon in Southern Japan, where it is probably a resident. There is an example in the Pryer collection from Tokio Bay, and Mr. Ringer has sent one to the Norwich Museum from Nagasaki (Seebohm, Ibis, 1884, p. 175), where the examples obtained by the Siebold Expedition were doubtless also procured (Temminck and Schlegel, Fauna Japonica, Aves, p. 128).

The breeding-range of the Common Sheldrake extends from the British Islands, across Europe and South Siberia to Japan. The mean temperature of July is so much less in Yezzo than it is in Dauria, that it is quite possible that this bird does not visit Yezzo, though it breeds ten degrees further north in Siberia.

231. TADORNA RUTILA.
(RUDDY SHELDRAKE.)

Anas rutila, Pallas, Nov. Com. Petrop. xiv. p. 579 (1770).

The Ruddy Sheldrake is chestnut above and below, except the rump, upper tail-coverts, tail, and quills, which are nearly black.

Figures: Dresser, Birds of Europe, vi. pl. 421.

The Ruddy Sheldrake must be a very rare bird in Japan, and confined to the extreme south, since neither Captain Blakiston nor Mr. Pryer were able to procure examples. On the other hand, several examples are said to have been procured by the Siebold Expedition, presumably near Nagasaki (Temminck and Schlegel, Fauna Japonica, Aves, p. 128). It is probably only an accidental visitor on migration.

The breeding-range of the Ruddy Sheldrake extends across Europe and Southern Siberia from the Atlantic to the Pacific.

232. ANAS STREPERA.
(GADWALL.)

Anas strepera, Linneus, Syst. Nat. i. p. 200 (1766).

The outer webs of the 9th, 10th, and 11th secondaries are nearly white in the Gadwall.

Figures: Dresser, Birds of Europe, vi. pl. 424.

The Gadwall has not yet been recorded from Yezzo, and appears to be a somewhat rare winter visitor to Southern Japan. There are nine examples in the Pryer collection from the Yokohama game-market; and it was met with, probably near Nagasaki, by the Siebold Expedition (Temminck and Schlegel, Fauna Japonica, Aves, p. 128).

The Gadwall is a circumpolar species, breeding in the subarctic regions of both continents. It can scarcely be supposed to reach Southern Japan without passing along the coasts of Yezzo.

233. ANAS CLYPEATA.
(SHOVELLER.)

Anas clypeata, Linneus, Syst. Nat. i. p. 200 (1766).

The Shoveller may always be recognized by its spoon-shaped bill, which is twice as wide near the tip as it is at the base.

Figures: Dresser, Birds of Europe, vi. pl. 425.

The Shoveller is a winter visitor to the Japanese Islands. It is a rare bird in Yezzo (Whitely, Ibis, 1867, p. 207), and probably only occurs on migration (Swinhoe, Ibis, 1875, p. 457); but in Nagasaki and Yokohama it is common. There is an example in the Swinhoe collection from Hakodadi procured by Captain Blakiston in October, and there are six examples in the Pryer collection from the Yokohama winter-market. The examples obtained by the Siebold Expedition were doubtless procured near Nagasaki (Temminck and Schlegel, Fauna Japonica, Aves, p. 128).

The Shoveller is a circumpolar species, breeding in the arctic and Subarctic Regions of both continents.

234. ANAS BOSCHAS.
(MALLARD.)

Anas boschas, Linnæus, Syst. Nat. i. p. 205 (1766).

The Mallard is one of the larger Ducks (wing from carpal joint $10\frac{1}{2}$ to 11 inches). Its axillaries and under wing-coverts are white. Its greater wing-coverts are grey with black tips, emphasized by a subterminal white band.

Figures: Dresser, Birds of Europe, vi. pl. 422.

The Mallard has long been known to occur on the Kurile Islands and in Kamtschatka (Pallas, Zoogr. Rosso-Asiat. ii. p. 256). It breeds sparingly on the Kurile Islands, and more abundantly on Yezzo (Blakiston and Pryer, Trans. As. Soc. Japan, 1882, p. 96); but it is only a winter visitor to Southern Japan. There is an example in the Swinhoe collection from Hakodadi, obtained by Captain Blakiston in March (Swinhoe, Ibis, 1877, p. 146); and there are two in the Pryer collection from the Yokohama winter-market. Mr. Ringer has obtained it at Nagasaki, where the examples procured by the Siebold Expedition were doubtless also obtained (Temminck and Schlegel, Fauna Japonica, Aves, p. 126).

The Mallard is a circumpolar species, but its breeding-range rarely reaches as far north as the Arctic Circle.

235. ANAS ZONORHYNCHA.
(DUSKY MALLARD.)

Anas zonorhyncha, Swinhoe, Ibis, 1866, p. 394.

The Dusky Mallard is one of the larger Ducks (wing from carpal joint about 11 inches). Its axillaries and under wing-coverts are white. Its greater wing-coverts are brown, broadly tipped with black. Its bill is black, broadly tipped with yellow.

Figures: Temminck and Schlegel, Fauna Japonica, Aves, pl. 82.

The Dusky Mallard is a resident in the Japanese Islands, and is common both in Yezzo and the more southerly islands (Blakiston and Pryer, Ibis, 1878, p. 213), breeding on the inland lakes. There is an example in the Swinhoe collection from Hakodadi (Swinhoe, Ibis, 1874, p. 164), and there is an example in the Pryer collection from

Yokohama. It has also occurred on the Kuriles, where it is probably a summer visitor (Blakiston and Pryer, Trans. As. Soc. Japan, 1882, p. 96). Temminck and Schlegel erroneously regarded it as a cross between *Anas pœcilorhyncha* and *Anas boschas*, but it is unquestionably a good species, whose range extends across China to Mongolia and Eastern Siberia.

236. ANAS CRECCA.
(COMMON TEAL.)

Anas crecca, Linneus, Syst. Nat. i. p. 204 (1766).

The Teal and the Garganey are the only Japanese Ducks in which the wing from carpal joint measures less than $7\frac{1}{2}$ inches. In the Teal the outer webs of the outer secondaries are velvet-black, those of the three inner ones metallic emerald-green.

Figures: Dresser, Birds of Europe, vi. pl. 426.

The Teal has long been known to occur in the Kurile Islands and in Kamtschatka (Pallas, Zoogr. Rosso-Asiat. ii. p. 263). It breeds in the Kurile Islands (Blakiston and Pryer, Trans. As. Soc. Japan, 1882, p. 97); winters sparingly in Yezzo (Whitely, Ibis, 1867, p. 207), and abundantly in Southern Japan (Blakiston and Pryer, Ibis, 1878, p. 213). There is a male (Blakiston, April) and a female (Blakiston, November) in the Swinhoe collection from Hakodadi (Swinhoe, Ibis, 1877, p. 147), and there are two examples in the Pryer collection from Yokohama. Mr. Ringer has obtained it at Nagasaki, where the examples procured by the Siebold Expedition were doubtless also obtained (Temminck and Schlegel, Fauna Japonica, Aves, p. 127). It has also been recorded from the most southerly group of the Loo-Choo Islands (Stejneger, Zeitschr. ges. Orn. 1887, p. 169).

The breeding-range of the Teal extends from the British Islands across arctic and subarctic Eurasia to the Kurile Islands.

237. ANAS FORMOSA.
(SPECTACLED TEAL.)

Anas formosa, Georgi, Reis. Russ. Reichs, i. p. 168 (1775).

The Spectacled Teal is not one of the larger Ducks (wing from carpal joint $7\frac{1}{4}$ to 8 inches). It has a small bill (from frontal feathers about 1½ inches). Its greater wing-coverts (like those of the male Pintail) are tipped with chestnut-buff.

Figures: Temminck and Schlegel, Fauna Japonica, Aves, pl. 82 B (male), pl. 82 c (female); Dresser, Birds of Europe, vi. pl. 428.

The Spectacled Teal has not yet been recorded from Yezzo or the Kurile Islands, but is a common winter visitor to Southern Japan. There is an example in the Swinhoe collection from Awomori Bay, opposite Hakodadi (Swinhoe, Ibis, 1877, p. 147), and there are three in the Pryer collection from Yokohama. There are two examples in the Norwich Museum sent by Mr. Ringer from Nagasaki.

Under the various names of *Anas formosa*, *Querquedula formosa*, *Eunetta formosa*, *Anas glocitans*, *Querquedula glocitans*, or *Eunetta glocitans*, the Spectacled Teal has been recorded as breeding throughout Eastern Siberia, and wintering in China, occasionally wandering as far as France and Calcutta. It is extremely improbable that it reaches Southern Japan without passing along the coasts of Yezzo.

238. ANAS FALCATA.
(FALCATED TEAL.)

Anas falcata, Georgi, Reis. Russ. Reichs, i. p. 167 (1775).

The Falcated Teal is not a small bird (wing from carpal joint 9 to 9¼ inches). Its axillaries and under wing-coverts are nearly white. Its greater wing-coverts are grey, shading into white at the tips.

Figures: Dresser, Birds of Europe, vi. pl. 429.

The Falcated Teal is a winter visitor to all the Japanese Islands. The Perry Expedition found it to be one of the most abundant of the water-birds of Japan, and noticed it at various points during the voyage, obtaining specimens at Hakodadi (Cassin, Exp. Am. Squad. China Seas and Japan, ii. p. 231); and there is also an example in the Swinhoe collection from Hakodadi obtained by Captain Blakiston in April (Swinhoe, Ibis, 1874, p. 164). There are eight examples in the Pryer collection from Yokohama, and it is recorded from Nagasaki (Blakiston and Pryer, Trans. As. Soc. Japan, 1882, p. 98), where the examples obtained by the Siebold Expedition were doubtless also procured (Temminck and Schlegel, Fauna Japonica, Aves, p. 127).

The Falcated Teal breeds in Eastern Siberia and winters in China. It is a rare accidental visitor to Europe.

239. ANAS CIRCIA.
(GARGANEY.)

Anas circia, Linneus, Syst. Nat. i. p. 204 (1766).

The Garganey and the Teal are the only Japanese Ducks in which the wing from carpal joint measures less than $7\frac{1}{2}$ inches. The secondaries of the Garganey are much paler than those of the Teal, and have very little metallic gloss.

Figures: Dresser, Birds of Europe, vi. pl. 427.

The Garganey is a winter visitor to all the Japanese Islands, but appears to be nowhere common. It was first recorded as a Japanese bird from an example procured in the Yokohama market (Blakiston and Pryer, Ibis, 1878, p. 214). Captain Blakiston obtained an example in Yezzo; Mr. Owston procured several in the Yokohama market (Blakiston and Pryer, Trans. As. Soc. Japan, 1882, p. 98), one of which is in the Pryer collection; and Mr. Ringer has sent several examples from Nagasaki (Seebohm, Ibis, 1884, p. 175).

The breeding-range of the Garganey extends across Europe and Southern Siberia from the British Islands to the Pacific.

240. ANAS ACUTA.
(PINTAIL.)

Anas acuta, Linneus, Syst. Nat. i. p. 202 (1766).

The Pintail is one of the larger Ducks (wing from carpal joint 10 to 11 inches). Very few of the under wing-coverts are white. It has a long bill (from frontal feathers $1\frac{3}{4}$ to $2\frac{1}{4}$ inches).

Figures: Dresser, Birds of Europe, vi. pl. 431 (male), pl. 430 (female).

The Pintail has long been known to visit the Kurile Islands (Pallas, Zoogr. Rosso-Asiat. ii. p. 280). It passes the coasts of Yezzo in spring and autumn on migration, to winter in considerable numbers in Southern Japan (Blakiston and Pryer, Ibis, 1878, p. 213). The Perry Expedition obtained examples from Hakodadi (Cassin, Exp. Am. Squad. China Seas and Japan, ii. p. 231). In the Swinhoe collection there are two examples from Awomori, on the main island opposite

Hakodadi (Swinhoe, Ibis, 1877, p. 147), and there are seven examples in the Pryer collection from Yokohama. Mr. Ringer procured it at Nagasaki, where the examples obtained by the Siebold Expedition were probably also procured (Temminck and Schlegel, Fauna Japonica, Aves, p. 128).

The Pintail is a circumpolar bird, breeding in the Arctic Regions of both continents.

241. ANAS PENELOPE.
(WIGEON.)

Anas penelope, Linneus, Syst. Nat. i. p. 202 (1766).

The Wigeon is not one of the smaller Ducks (wing from carpal joint $9\frac{1}{2}$ to $10\frac{1}{2}$ inches), but it has a small bill (from frontal feathers $1\frac{1}{4}$ to $1\frac{3}{4}$ inches). Its axillaries and under wing-coverts are mottled with brown.

Figures: Dresser, Birds of Europe, vi. pls. 532, 533.

The Wigeon passes the coasts of Yezzo in spring and autumn, and winters in great numbers in Southern Japan (Blakiston and Pryer, Ibis, 1878, p. 213). The Perry Expedition obtained an example from Hakodadi in May 1854 (Cassin, Exp. Am. Squad. China Seas and Japan, ii. p. 231); and there is an example in the Swinhoe collection, also from Hakodadi, obtained by Captain Blakiston in November (Swinhoe, Ibis, 1875, p. 457). There are four examples in the Pryer collection from Yokohama, and it has been obtained near Nagasaki (Blakiston and Pryer, Trans. As. Soc. Japan, 1882, p. 97), where the examples procured by the Siebold Expedition were doubtless also obtained (Temminck and Schlegel, Fauna Japonica, Aves, p. 128).

The Wigeon winters on the shores of Great Britain as well as on those of Japan, breeding throughout the intervening Arctic Regions. It is represented on the American continent by a nearly allied species, *Anas americana*, which is said occasionally to wander across the Atlantic as far as the British Islands.

242. ANAS GALERICULATA.
(MANDARIN DUCK.)

Anas galericulata, Linneus, Syst. Nat. i. p. 539 (1766).

The Mandarin Duck may always be recognized by the silver-grey on the outer webs of its primaries.

Figures: Gould, Birds of Asia, vii. pl. 69.

The Mandarin Duck is probably only a summer visitor to Yezzo, but a resident in the more southerly Japanese Islands, where it is very common on the small streams (Blakiston and Pryer, Ibis, 1878, p. 213). It frequents the large lakes near Hakodadi (Whitely, Ibis, 1867, p. 207). There is an example in the Swinhoe collection from Hakodadi, obtained by Captain Blakiston in September; and there are eight examples in the Pryer collection from Yokohama. Mr. Ringer has sent examples to the Norwich Museum obtained at Nagasaki, where those procured by the Siebold Expedition were doubtless also obtained (Temminck and Schlegel, Fauna Japonica, Aves, pl. 127). In autumn it frequents the lakes in the interior in large flocks, and feeds on the paddy-fields in company with the Teal (Jouy, Proc. United States Nat. Mus. 1883, p. 318).

The Mandarin Duck is a summer visitor to the valley of the Amoor, but in Central and Southern China it is a resident.

243. FULIGULA AMERICANA.
(AMERICAN BLACK SCOTER.)

Oidemia americana, Swainson and Richardson, Faun. Bor.-Amer. ii. p. 450 (1831).

The Black Scoter has dark axillaries and no white on the wing. It has a large bill (width in front of the nostrils about 1 inch).

Figures: Baird, Brewer, and Ridgway, Water-Birds N. Amer. ii. pp. 89, 90, 91 (woodcut and coloured woodcuts of head).

The Diving Ducks form a very natural group possessing two characters, either of which will distinguish them from the genus *Anas* or from the genus *Tadorna*: the hind toe is furnished with a well-developed lobe; and the nostrils are placed very far forward. The distance from the gape to the front of the nostril is more than that from the tip of the bill to the front of the nostril.

The existence of a species of Black Scoter on the Japanese coasts was recorded in 1810 (Temminck, Man. d'Orn. iv. p. 543), but probably in error, as no mention is made of it in the 'Fauna Japonica.' Captain Blakiston was undoubtedly the first to discover that it was

Head of *Fuligula americana*. ¾ natural size.

the Nearctic and not the Palæarctic species which occurs in Yezzo (Blakiston and Pryer, Ibis, 1878, p. 215, no. 52), and was probably the first discoverer of its occurrence in Japan.

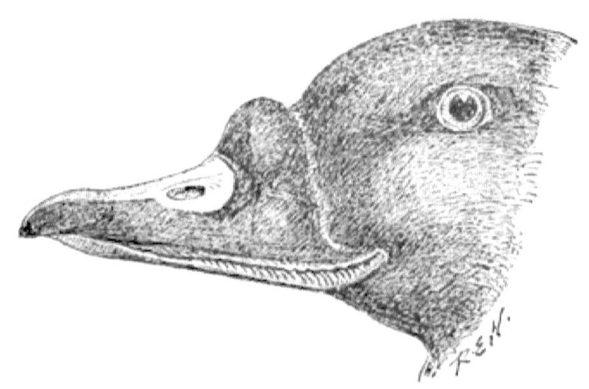

Head of *Fuligula nigra*. ½ natural size.

The American Black Scoter breeds on the Kurile Islands (Blakiston and Pryer, Trans. As. Soc. Japan, 1882, p. 100), and winters on the shores of the Japanese Seas. Captain Blakiston and Mr. Henson have sent examples from Hakodadi obtained in February and March

(Seebohm, Ibis, 1879, p. 23), and there are two in the Pryer collection from Yokohama.

The American Black Scoter breeds on the shores of the Arctic Ocean from Labrador to Alaska, and across Bering Straits to Kamtschatka.

244. FULIGULA FUSCA.
(VELVET SCOTER.)

Anas fusca, Linnæus, Syst. Nat. i. p. 196 (1766).

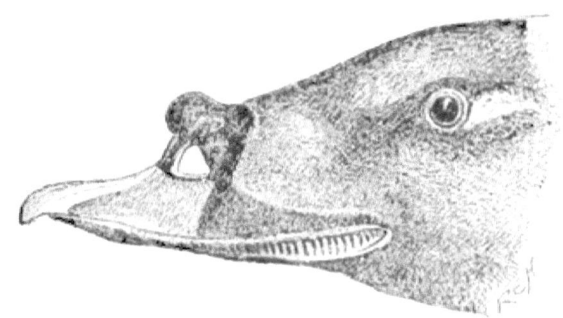

Head of *Fuligula fusca stejnegeri*. ⅔ natural size.

The Velvet Scoter has dark axillaries; but the central secondaries are white. The feathering on the side of the upper mandible approaches within a quarter of an inch of the nostrils.

The Velvet Scoter was recorded from Japan in 1840 (Temminck, Man. d'Orn. iv. p. 543), but it is not mentioned in the 'Fauna Japonica,' and we may consequently assume that Temminck discovered good reasons for doubting the accuracy of the statement in the earlier work, before the later one was published.

It has, however, been since found in various localities as a winter visitor. There is a skin in the Pryer collection obtained by Mr. Snow on the Kurile Islands.

The first record of its occurrence in Yezzo is that of two specimens obtained at Hakodadi, one on the 24th of December, 1864, and the other shot by Captain Blakiston on the 28th of January, 1865 (Whitely, Ibis, 1867, p. 209). I have a very fine male from the same locality, obtained by Mr. Henson on the 28th of February,

There are three examples in the Pryer collection from Yokohama, and one in the Norwich Museum sent by Mr. Ringer from Nagasaki. These examples all agree with others from East Siberia and China in the shape of the bill and in the extent of the frontal feathering.

The Velvet Scoter is a winter visitor to the British Islands as well as to Japan, and breeds throughout the Arctic Regions of both continents; but there is a slight difference between European, Asiatic, and American examples.

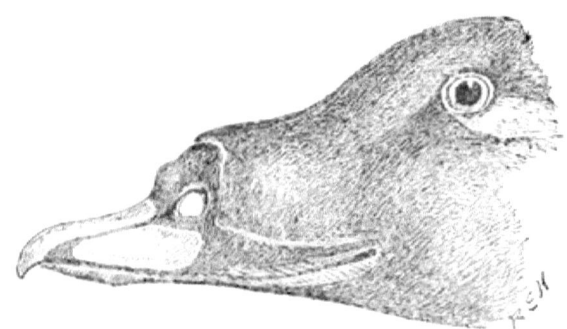

Head of *Fuligula fusca velvetina*. ⅔ natural size.

The Velvet Scoter is subject to some local variation in the shape of its bill and in the extent to which the frontal feathering is carried.

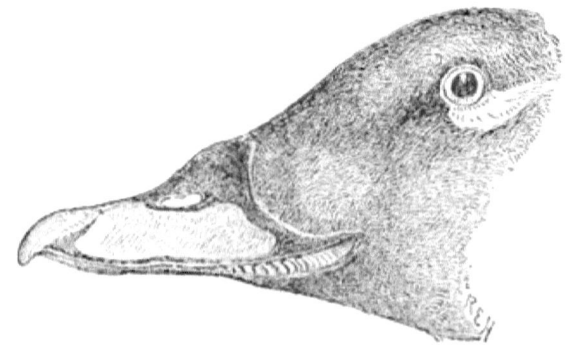

Head of *Fuligula fusca*. ⅔ natural size.

In the European or typical form the rostral knob is comparatively small; in the American form, *Fuligula fusca velvetina*, it is well developed; and in the Asiatic form, *Fuligula fusca stejnegeri*, it is enormously developed and overhangs the bill. The frontal feathering is least developed in the European form, more so in the Asiatic form, and most so in the American form. In the European form the nostrils are separated from the frontal feathering on the side of the bill by more than their width; in the American form the frontal feathering extends on the top of the bill as far as above the beginning of the nostrils; the Asiatic form agrees with the American form in the former point, and with the European in the latter.

The Japanese form was described as a new species in 1887 under the name of *Oidemia stejnegeri* (Ridgway, Man. North Amer. Birds, p. 112); but the American form appears to be so intermediate between it and the European form that it can scarcely be regarded as more than subspecifically distinct unless a much larger series than is at present attainable should hereafter prove that the apparent intergradation is not complete.

245. FULIGULA GLACIALIS.
(LONG-TAILED DUCK.)

Anas glacialis, Linneus, Syst. Nat. i. p. 203 (1766).

The Long-tailed Duck has dark axillaries and no white on the wing. It has a small bill (width in front of the nostrils about ⅜ inch). The feathering on the side of the upper mandible approaches within ¼ inch of the nostril.

Figures: Dresser, Birds of Europe, vi. pl. 444.

The Long-tailed Duck has long been known to occur on the Kurile Islands and on the coast of Kamtschatka (Pallas, Zoogr. Rosso-Asiat. ii. p. 276), and more recently Mr. Snow has observed that it is the earliest Duck to pass the Kurile Islands on its spring migration to its arctic breeding-grounds. It was first recorded as a Japanese bird from examples procured in January 1865 at Hakodadi, where it is said to be common in winter (Whitely, Ibis, 1867, p. 208). There is an example in the Swinhoe collection obtained by Captain Blakiston at Hakodadi in February (Swinhoe, Ibis, 1877, p. 147), and there is another in the Pryer collection obtained at the same time. I have a third example collected by Mr. Henson in the same locality on the

27th of March, but I can find no evidence of its occurrence in Southern Japan.

The Long-tailed Duck is a circumpolar species, breeding in the Arctic Regions of both continents.

246. FULIGULA CLANGULA.
(GOLDEN-EYE.)

Anas clangula, Linneus, Syst. Nat. i. p. 201 (1766).

The Golden-eye has dark axillaries, but the central secondaries are white. The feathering on the side of the upper mandible does not approach within half an inch of the nostrils.

Figures: Dresser, Birds of Europe, vi. pl. 410.

The Golden-eye appears to have been found by Steller on the Kurile Islands (Pallas, Zoogr. Rosso-Asiat. ii. p. 272). It is a common winter visitor to Japan, and is especially numerous on the coast of Yezzo (Blakiston and Pryer, Trans. As. Soc. Japan, 1882, p. 99). There are two examples in the Swinhoe collection from Hakodadi, one of them obtained by Captain Blakiston in November (Swinhoe, Ibis, 1877, p. 147), and there are eight examples in the Pryer collection from Yokohama. Mr. Ringer procured it at Nagasaki, where the examples obtained by the Siebold Expedition were doubtless also procured (Temminck and Schlegel, Fauna Japonica, Aves, pl. 128).

The Golden-eye is a circumpolar species, breeding in the Arctic Regions of both continents.

247. FULIGULA HISTRIONICA.
(HARLEQUIN DUCK.)

Anas histrionicus, Linneus, Syst. Nat. i. p. 204 (1766).

The Harlequin Duck has dark axillaries. It has a small bill (width in front of the nostrils about $\frac{1}{2}$ inch). The feathering on the side of the upper mandible recedes from the nostrils, curving backwards in a semicircle.

Figures: Dresser, Birds of Europe, vi. pl. 412.

The Harlequin Duck is a winter visitor to the Japanese Islands; extremely abundant on the Kuriles, less so in Yezzo, and least so in the more southerly islands (Blakiston and Pryer, Trans. As. Soc. Japan, 1882, p. 99). I have an example collected by Mr. Snow on the Kurile Islands; there is an example in the Swinhoe collection from Hakodadi, obtained by Captain Blakiston in June (Swinhoe, Ibis, 1877, p. 147), which appears to be the first recorded from Japan; and there are two in the Pryer collection from Yokohama.

The Harlequin Duck breeds in the Arctic Regions of both continents.

248. FULIGULA BAERI.
(SIBERIAN WHITE-EYED DUCK.)

Anas (Fuligula) baeri, Radde, Reis. Süd. Ost-Sibirien, ii. p. 376 (1863).

The Siberian White-eyed Duck has white axillaries and nearly white under tail-coverts. It has no white on the lores, and no white vermiculations on the back or scapulars.

Figures: David and Oustalet, Oiseaux de la Chine, pl. 124.

The Siberian White-eyed Duck has not previously been recorded from Japan, but all the notices of the occurrence of the Ferruginous Duck in those islands (Seebohm, Ibis, 1879, p. 22) probably refer to the Eastern species. It has occurred both in Yezzo and on the main island, but probably only in winter (Blakiston and Pryer, Ibis, 1878, p. 215). There are four examples in the Pryer collection from Yokohama.

The Siberian White-eyed Duck breeds in the valley of the Amoor and winters in China. It is the Eastern representative of the White-eyed Pochard, *Fuligula nyroca*, a species which ranges eastwards as far as the valley of the Obb.

249. FULIGULA FERINA.
(POCHARD.)

Anas ferina, Linneus, Syst. Nat. i. p. 203 (1766).

The Pochard has white axillaries and nearly uniform grey secondaries.

Figures: Dresser, Birds of Europe, vi. pl. 434.

The Pochard occurs both in Yezzo and in the more southerly

Japanese Islands; but whether it be a resident or only a winter visitor there seems to be no evidence to determine. Captain Blakiston shot a single example at Hakodadi (Seebohm, Ibis, 1884, p. 176), apparently the first recorded from Japan; and there are four examples in the Pryer collection from Yokohama.

The Pochard winters in the British Islands as well as in Japan, and breeds in subarctic Europe and Southern Siberia.

250. FULIGULA CRISTATA.
(TUFTED DUCK.)

Anas cristata, Leach, Syst. Cat. Mamm. &c. Brit. Mus. p. 39 (1816).

The Tufted Duck has white axillaries. Its secondaries are white, broadly tipped with black. Its under tail-coverts are black (adult male) or brown. It has very little white on the lores, and seldom any trace of white vermiculations on the back or scapulars.

Figures: Dresser, Birds of Europe, vi. pl. 437.

The Tufted Duck was probably found by Steller in the Kurile Islands (Pallas, Zoogr. Rosso-Asiat. ii. p. 266). It is a spring and autumn visitor to Yezzo, and may possibly breed there, but in Southern Japan it is only known as a winter visitor (Blakiston and Pryer, Ibis, 1878, p. 214). There is an example in the Swinhoe collection from Hakodadi, obtained by Captain Blakiston in September; I have an example sent me by Captain Blakiston from the same locality procured in May (Seebohm, Ibis, 1879, p. 22); and there are five in the Pryer collection from Yokohama. The examples obtained by the Siebold Expedition were probably procured at Nagasaki (Temminck and Schlegel, Fauna Japonica, Aves, p. 128).

The breeding-range of the Tufted Duck extends from the Atlantic to the Pacific.

It is represented on the American Continent by a nearly allied species, *Fuligula collaris*, which has a chestnut collar round the neck, and further differs from its Eurasian ally in having a shorter crest and a pale slate-grey wing-speculum.

251. FULIGULA MARILA.
(SCAUP.)

Anas marila, Linneus, Syst. Nat. i. p. 196 (1766).
Aythya affinis mariloides, Stejneger, Orn. Expl. Comm. Isl. & Kamtschatka, p. 161 (1885).

The Scaup has white axillaries, but brown under tail-coverts. There is much white on the lores, and always some white vermiculations on the back and scapulars.

Figures: Dresser, Birds of Europe, vi. pl. 436.

The Scaup is a winter visitor to the shores of the Japanese Seas. There are two examples in the Swinhoe collection from Hakodadi (Swinhoe, Ibis, 1875, p. 457), which appear to be the first recorded from Japan; and there are three in the Pryer collection from Yokohama. Mr. Ringer has obtained it at Nagasaki (Blakiston and Pryer, Trans. As. Soc. Japan, 1882, p. 98); and it has been recorded from the Loo-Choo Islands (Cassin, Proc. Acad. Nat. Sc. Philad. 1862, p. 322).

The Scaup is a circumpolar bird breeding in the Arctic Regions of both continents.

Fully adult males from Japan are precisely similar to those from the British Islands. In both the black of the sides of the head is glossed with green and not with purple. The absence of pure white on the primaries and the presence of vermiculations on the flanks are indications of immaturity.

252. SOMATERIA SPECTABILIS.
(KING EIDER.)

Anas spectabilis, Linneus, Syst. Nat. i. p. 195 (1766).

The male King Eider differs from the male Common Eider in having the crown lavender-grey instead of black, and the lower back and scapulars black instead of white. Both sexes may be distinguished by the feathering on the base of the bill. In the King Eider the feathering on the base of the mandible extends further forward in the centre than at the sides; in the Common Eider exactly the contrary is the case.

Figures: Dresser, Birds of Europe, vi. pl. 446.

The King Eider is said to be common on the Kurile Islands (Pallas, Zoogr. Rosso-Asiat. ii. p. 237), but no examples have been obtained there by recent travellers. It is a circumpolar species, breeding on the shores of the Arctic Ocean and wintering further south. It is a common winter visitor to the Aleutian Islands, and occasionally strays as far south as the British Islands.

253. SOMATERIA STELLERI.
(STELLER'S EIDER.)

Anas stelleri, Pallas, Spicilegia Zoologica, pt. vi. p. 35 (1780).

The male Steller's Eider has a blue-black ring round the neck and a black back. It has a green patch on the lores and on the nape. The female has a purple-blue speculum between two white alar bars.

Figures: Dresser, Birds of Europe, vi. pl. 447.

Steller's Eider is a winter visitor to the Kurile Islands, but has not been recorded from Japan (Blakiston and Pryer, Trans. As. Soc. Japan, 1882, p. 100, no. 51). Dr. Stejneger found it on the Commander Islands, and it visits Kamtschatka (Blakiston and Pryer, Ibis, 1878, p. 215).

Steller's Eider breeds on the shores of the Arctic Ocean in North Russia and Siberia, and on the islands in Bering Sea. It is an accidental visitor to the British coast.

254. MERGUS MERGANSER.
(GOOSANDER.)

Mergus merganser, Linneus, Syst. Nat. i. p. 208 (1766).

The Goosander is the largest of the Mergansers and differs from its allies in all plumages in the colour of its central secondary quills, which are white with completely concealed dark bases on the outer webs.

Figures: Dresser, Birds of Europe, vi. pl. 452.

The Goosander occurs on the Kurile Islands (Pallas, Zoogr. Rosso-Asiat. ii. p. 287), and is a winter visitor to the Japanese Islands.

In the Swinhoe collection there is an example from Hakodadi obtained by Captain Blakiston in April (Swinhoe, Ibis, 1875, p. 456), and in the Pryer collection there are five examples from Yokohama. The examples obtained by the Siebold Expedition were doubtless procured at Nagasaki (Temminck and Schlegel, Fauna Japonica, Aves, p. 129).

The Goosander breeds in the Arctic Regions of Russia and Siberia, and winters on the coasts of the British Islands, as well as on those of Japan.

The Goosander is also found on the American continent, but examples from the New World differ from those found in the Old World in having a black bar across the wing-coverts, for which reason they are regarded as subspecifically distinct under the name of *Mergus merganser americanus*. In both forms the basal portion of the greater wing-coverts is black, but in examples from Europe and Asia (including those from Japan) the median wing-coverts extend beyond and entirely conceal the black bases, whilst in the American form they fall short of them and thus leave exposed the narrow black bar to which allusion has been made.

255. MERGUS SERRATOR.

(RED-BREASTED MERGANSER.)

Mergus serrator, Linnæus, Syst. Nat. i. p. 208 (1766).

The Red-breasted Merganser is slightly smaller than the Goosander, and the terminal half of the central secondaries is white, but the dark bases on the outer webs are never concealed by the greater wing-coverts.

Figures: Dresser, Birds of Europe, vi. pl. 453.

The Red-breasted Merganser breeds on the Kurile Islands and winters in Japan (Blakiston and Pryer, Trans. As. Soc. Japan, 1882, p. 101). There are two examples in the Swinhoe collection from Hakodadi (Swinhoe, Ibis, 1875, p. 456), and six in the Pryer collection from Yokohama. Mr. Ringer has sent an example to the Norwich Museum from Nagasaki, where those procured by the Siebold Expedition were doubtless also obtained (Temminck and Schlegel, Fauna Japonica, Aves, p. 129).

The Red-breasted Merganser is a circumpolar bird, breeding in the subarctic districts of both continents.

256. MERGUS ALBELLUS.
(SMEW.)

Mergus albellus, Linneus, Syst. Nat. i. p. 209 (1766).

The Smew is the smallest of the Mergansers, and the central secondaries are dark with narrow white terminal bands.

Figures: Dresser, Birds of Europe, vi. pls. 454, 455.

The Smew is a winter visitor to the Japanese Islands, and is sent both to the Yezzo and the Yokohama game-markets (Blakiston and Pryer, Ibis, 1878, p. 215). There are three examples in the Pryer collection from the latter locality. The examples procured by the Siebold Expedition were probably obtained at Nagasaki (Temminck and Schlegel, Fauna Japonica, Aves, p. 129).

The Smew winters on the British coasts, as well as on those of Japan. It breeds in the intervening Arctic Regions.

The Smew is represented on the American Continent by the Hooded Merganser, *Mergus cucullatus*, a perfectly distinct species with a much longer bill and with many important differences in its colour.

Suborder XXIII. *PALAMEDEÆ*.

The Screamers are supposed to be absolutely unique amongst existing birds in having lost the uncinate processes of their ribs. In this respect they appear to be in advance of all other birds. They may otherwise be diagnosed as follows:—Plumage of upper parts with no spinal bare tract. Palate desmognathous. Young born covered with down, and able to run in a few hours. Only three species are known, which are confined to the Neotropical Region.

Subclass GALLIFORMES.

The Galliformes are believed to be the only birds which combine the following characters:—

Young born completely covered with down or feathers; maxillo-palatines not united across the middle line; coracoid articulating with the scapula at an angle more acute than $120°$; if a deep plantar tendon reach the hallux it proceeds from the *flexor longus hallucis*, and not from the *flexor perforans digitorum*.

The Subclass Galliformes contains three orders, two of which are represented in Japan.

Order TUBINARES.

The Tubinares differ so little amongst themselves that they cannot be divided into suborders of sufficient importance to claim more than family rank.

The diagnosis of the Order is therefore the same as that of the Suborder.

Suborder XXIV. *TUBINARES*.

Hallux absent or reduced to one bone; other three digits directed forwards and webbed; spinal feather-tract well defined on neck by lateral bare tracts; dorsal vertebræ heterocœlous; nasals holorhinal; external nostrils produced into tubes.

There are three families which can be well defined comprised in the Tubinares, each of which is represented in the Japanese Seas.

Family DIOMEDEIDÆ.

The Albatrosses differ from all the other families of Tubinares in having the nasal tubes separated from each other; they further differ from the Puffinidæ in having no basipterygoid processes.

They all possess the ambiens muscle.

The Albatrosses belong essentially to the southern hemisphere, but out of ten or a dozen species, two belong to the North Pacific and frequent the Japanese Seas.

257. DIOMEDEA ALBATRUS.
(STELLER'S ALBATROSS.)

Diomedea albatrus, Pallas, Spicilegia Zoologica, pt. v. p. 28 (1780).

Steller's Albatross is a very large bird (wing from carpal joint 22 to 19 inches). The pale form is nearly white, with the wings, the tail, most of the scapulars, and most of the wing-coverts brown. The dark form is entirely brown both above and below. Bill, legs, and feet pale.

Figures: (light form) Temminck, Planches Coloriées, no. 554; (both forms) Gould, Birds of Australia, vii. pl. 39; (dark form) Temminck and Schlegel, Fauna Japonica, Aves, pl. 66.

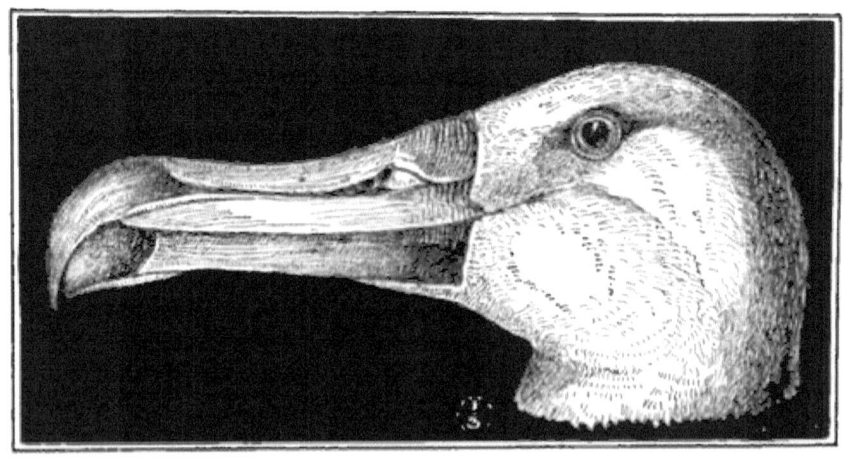

Head of *Diomedea albatrus*. ⅓ natural size.

Steller's Albatross was discovered by the illustrious traveller whose name it bears on the coast of Kamtschatka, and was described by

Pallas in the last century. It is a common species in the Japanese Seas. There is an example in the British Museum collected by Mr. Snow on Eturop, the most southerly of the Kurile Islands, and several examples have been sent by Mr. Henson from Hakodadi (Seebohm, Ibis, 1884, p. 176). There are two examples in the Pryer collection from Tokio Bay, and Mr. Ringer has obtained it at Nagasaki.

It is generally found in company with an entirely dark form, which is, on an average, slightly smaller, and has therefore been regarded by many ornithologists as the young of Steller's Albatross. Other ornithologists regard the dark form as specifically distinct. In the 'Planches Enluminées' of D'Aubenton, plate 963 represents a brown Albatross with pale bill and feet, under the title of L'Albatros de la Chine. Upon this plate the name of *Diomedea chinensis* was founded in 1820 (Temminck, Man. d'Orn. i. preface, p. cx); but its author appeared to be dissatisfied with it, and in 1828 altered it to *Diomedea brachiura* (Temminck, Planches Coloriées, Genus *Diomedea*, 75th livraison). It was afterwards rediscovered and redescribed under the name of *Diomedea derogata* (Swinhoe, Proc. Zool. Soc. 1873, p. 786). This dark form, which may, for the sake of distinction, be called Swinhoe's Albatross, is described by Messrs. Blakiston and Pryer as the commoner bird on the coasts of Yezzo, whilst Steller's Albatross is represented as the most abundant farther south. In the Swinhoe collection there is an example of the dark form obtained by Captain Blakiston at Hakodadi in July (Swinhoe, Ibis, 1874, p. 165), and there are two examples in the Pryer collection from Tokio Bay.

It seems probable that these two forms represent a dimorphic species like the Fulmar Petrel (*Fulmarus glacialis*), the Pomarine Skua (*Stercorarius pomarinus*), Richardson's Skua (*Stercorarius richardsoni*), or the Reef-Heron (*Ardea jugularis*). I have never seen any intermediate forms.

Eggs in the Pryer collection from the Bonin Islands vary in size from 4·7 by 2·9 inches to 4·3 by 3 inches; they are creamy white, profusely speckled with russet at the large end, some of the spots occasionally being larger.

Although it is figured by Gould in his 'Birds of Australia,' it is not known to have occurred in the Southern Hemisphere. It has been recorded from Bering Sea, and appears to be confined to the North Pacific Ocean, where it is common both on the Asiatic and American coasts.

258. DIOMEDEA NIGRIPES.
(AUDUBON'S ALBATROSS.)

Diomedea nigripes, Audubon, Orn. Biogr. v. p. 327 (1839).

Audubon's Albatross is slightly smaller than Steller's Albatross (wing from carpal joint $18\frac{1}{2}$ inches). It is dark brown above and below, shading into pale brown round the base of the bill. Bill, legs, and feet nearly black.

Figures: Cassin, Illustr. Birds of California, Texas, &c. pl. 35.

Head of *Diomedea nigripes*. $\frac{1}{2}$ natural size.

The claim of Audubon's Albatross to be regarded as a Japanese bird rests upon a fine adult female shot by Mr. H. Henson on the 17th of May, 1883, in the Strait of Tsugaru, between Yezzo and the main island of Japan (Seebohm, Ibis, 1884, p. 176), and a male obtained on the 27th of February on the coast of the province of Sagami near Yokohama (Seebohm, Ibis, 1885, p. 363).

Audubon's Albatross is a North-Pacific species, but its range does not extend so far north as Bering Sea.

Family PUFFINIDÆ.

The Shearwaters differ from all the other families of Tubinares in having basipterygoid processes.

They further differ from the Albatrosses in having the nasal tubes side by side in the centre of the mandible. They all possess the ambiens muscle except the genus *Pelecanoides*.

There may be about 70 species of Shearwaters of various genera, most of which belong to the Southern Hemisphere, and six of which have been recorded from the Japanese seas.

259. PUFFINUS LEUCOMELAS.
(SIEBOLD'S SHEARWATER.)

Procellaria leucomelas, Temminck, Planches Coloriées, no. 587 (1836).

Siebold's Shearwater is the largest of the Japanese Shearwaters (wing from carpal joint 13 to $12\frac{1}{2}$ inches). It is brown above and white below, but there is much white on the forehead and crown, and some brown on the under wing-coverts. The bill, legs, and feet are pale.

Figures: Temminck and Schlegel, Fauna Japonica, Aves, pl. 85.

Head of *Puffinus leucomelas*. Natural size.

Siebold's Shearwater is found in all the Japanese seas. There are several examples in the British Museum collected by Mr. Henson at Hakodadi, and there are two examples in the Pryer collection from Yokohama (Seebohm, Ibis, 1884, p. 176).

Siebold's Shearwater ranges as far south as the Malay Archipelago, where it has been recorded from the Philippines, Borneo, Celebes, the Moluccas, New Guinea, and the Duke of York Island.

260. PUFFINUS CARNEIPES.
(PINK-FOOTED SHEARWATER.)

Puffinus carneipes, Gould, Proc. Zool. Soc. 1844, p. 57.

The Pink-footed Shearwater is very slightly less than Siebold's Shearwater (wing from carpal joint 12½ inches). It is brown above and below, including the under wing-coverts, but the bill, legs, and feet are pale.

Figures: Gould, Birds of Australia, vii. pl. 57.

Head of *Puffinus carneipes*. Natural size.

The Pink-footed Shearwater is probably the species described by Pennant in his 'Arctic Zoology' as the Kurile Petrel, and very appropriately named about 1780 *Procellaria nigra* (Pallas, Spicilegia Zoologica, pt. v. p. 28). Three examples collected by Mr. Henson at Hakodadi in May have passed through my hands (Seebohm, Ibis, 1884, p. 176) and are now in the British Museum.

This species is said to breed on some islands off the coast of South-western Australia, and is probably a non-breeding summer visitor to the North Pacific.

261. PUFFINUS GRISEUS.
(SOOTY SHEARWATER.)

Procellaria grisea, Gmelin, Syst. Nat. i. p. 564 (1788).

The Sooty Shearwater is slightly smaller than the Pink-footed Shearwater (wing from carpal joint 12 to 11 inches). It is almost uniform brown, except the under wing-coverts, which are mostly white. Its bill, legs, and feet are very dark.

Figures: Dresser, Birds of Europe, viii. pl. 616.

Head of *Puffinus griseus*. Natural size.

The Sooty Shearwater was obtained by Mr. Snow on the Kurile Islands (Seebohm, Ibis, 1884, p. 33). It was procured by him as far north as Urup Island (Blakiston and Pryer, Trans. As. Soc. Japan, 1882, p. 106), whence there are several examples in the Hakodadi Museum. An example formerly in my collection is now in the British Museum.

The Sooty Shearwater breeds in the Southern Hemisphere in the Pacific, and probably in the Indian Ocean and the South Atlantic. It has frequently occurred on the British coasts, and may be regarded as a non-breeding summer visitor to the North Atlantic and the North Pacific. Its only known breeding-place is on the Chatham Islands east of New Zealand, but there can be little doubt that other breeding-grounds remain to be discovered.

262. PUFFINUS TENUIROSTRIS.
(SLENDER-BILLED SHEARWATER.)

Procellaria tenuirostris, Temminck, Planches Coloriées, text to no. 587 (1835).

The Slender-billed Shearwater is the smallest Japanese Shearwater (wing from carpal joint 10¾ to 10 inches). It is almost uniform brown, with pale grey under wing-coverts, but dark bill and feet.

Figures: Temminck and Schlegel, Fauna Japonica, Aves, pl. 86; Gould, Birds of Australia, vii. pl. 56, under the name of *Puffinus brevicaudus*.

Head of *Puffinus tenuirostris*. Natural size.

The Slender-billed Shearwater does not appear to have been procured by the Siebold Expedition, but was originally described by Temminck from an example obtained by Mons. Burger in the Sea of Japan. Capt. Rodgers procured it on the east coast of the main island of Japan a few miles north of the latitude of Yokohama (Cassin, Proc. Acad. Nat. Sc. Philad. 1862, p. 327). An example was picked up after a storm at Yoshino, Yamato, an inland town in the south of the main island forty miles from the coast (Bakiston and Pryer, Ibis, 1878, p. 218); another specimen was picked up much decayed on the beach at Kamakara in Tokio Bay (Blakiston and Pryer, Trans. As. Soc. Japan, 1882, p. 106); and there is an example in the British Museum, collected by Captain St. John at Nagasaki in May.

This Shearwater is said to breed in millions on some of the islands off the coasts of Van Diemen's Land and New Zealand, but is probably a non-breeding summer visitor to the North Pacific.

263. FULMARUS GLACIALIS.
(FULMAR.)

Procellaria glacialis, Linnæus, Syst. Nat. i. p. 213 (1766).

The Fulmar is one of the larger species (wing from carpal joint about 12 inches). It varies very much in colour; but is easily distinguished from any of the Shearwaters by the length of its nasal tubes, which nearly reach the nail of the bill; and from all its Japanese allies by the under outline of its bill, which turns suddenly upwards at the angle of the mandible, instead of following the same direction throughout.

Figures: Dresser, Birds of Europe, viii. pl. 617.

Head of *Fulmarus glacialis*. Natural size.

Steller observed the Fulmar breeding in great numbers on the Kurile Islands (Pallas, Zoogr. Rosso-Asiat. ii. p. 313), and Mr. Snow has recently observed the same fact (Blakiston and Pryer, Trans. As. Soc. Japan, 1882, p. 106). There are examples from the Kurile Islands in the Hakodadi Museum (Blakiston and Pryer, Ibis, 1878, p. 218), and there is a skin in the British Museum collected by Mr. Snow on the same islands (Seebohm, Ibis, 1879, p. 25). The latter belongs to the dark form, but there is a skin in the Pryer collection, also collected by Mr. Snow on the Kurile Islands, which is as typical of the light form.

The Fulmar is a circumpolar species and breeds in great numbers on St. Kilda. It appears to be dimorphic, but, as is common with other dimorphic species, the proportion of dark and light forms varies greatly in different localities.

264. ŒSTRELATA HYPOLEUCA.
(BONIN-ISLAND SHEARWATER.)

Œstrelata hypoleuca, Salvin, Ibis, 1888, p. 359.

The Bonin-Island Shearwater is blackish brown above, with the feathers of the back and rump margined with grey; the forehead, lores, and underparts are white; the axillaries are white, but most of the under wing-coverts are brown; quills brown, rectrices white at base. Wing $8\frac{3}{4}$, tail $4\frac{3}{4}$ inches.

The Bonin-Island Shearwater was originally described from an example sent to me by Mr. Snow from Kruzenstern Island in the North Pacific, where it had been obtained in the spring of 1883. The type remained unique in the British Museum until Mr. Holst sent me adult and young from Nakondo-Shima, one of the Parry Islands (Seebohm, Ibis, 1890, p. 105).

Family PROCELLARIIDÆ.

The Petrels appear to be intermediate between the Shearwaters, which they resemble in the position of their nasal tubes, and the Albatrosses, which, like the Petrels, have lost their basipterygoid processes.

They all possess the ambiens muscle except the genus *Fregetta*.

There may be about a score species of true Petrels in the two subfamilies, Procellariinæ and Oceanitinæ, into which the group may be divided. The family is represented in both hemispheres, and three species are recorded from Japan.

265. PROCELLARIA LEACHI.
(LEACH'S FORK-TAILED PETREL.)

Procellaria leachii, Temminck, Man. d'Orn. ii. p. 812 (1820).

Leach's Petrel is a little bird (wing from carpal joint about 6 inches). It is very dark brown above and below, except the upper tail-coverts, which are white.

Figures: Dresser, Birds of Europe, viii. pl. 613. fig. 2.

Head of *Procellaria leachi*. Natural size.

Leach's Fork-tailed Petrel breeds in the Kurile Islands (Blakiston and Pryer, Ibis, 1878, p. 218). There is an example in the Pryer collection obtained by Mr. Snow in this locality, and Captain Blakiston has sent me others (Seebohm, Ibis, 1884, p. 33).

Leach's Fork-tailed Petrel breeds on some of the western islands of Scotland and Ireland, as well as in the Bay of Fundy on the American coast. These colonies appear to be completely isolated from the colony in the North Pacific, which extends to the Commander Islands and the Aleutian Islands.

266. PROCELLARIA MELANIA.
(BLACK PETREL.)

Procellaria melania, Bonaparte, Compt. Rend. xxviii. p. 662 (1854).

The Black Petrel is slightly larger than Leach's Petrel (wing from carpal joint $6\frac{1}{4}$ inches), the fork of its tail is somewhat deeper ($1\frac{1}{4}$ inches), and the rump and upper tail-coverts are sooty brown like the rest of the plumage; but in both species the margins of the tertials and scapulars are paler.

Figures: Baird, Brewer, and Ridgway, Water-Birds N. Amer. ii. p. 107 (woodcut of tail and foot), p. 111 (woodcut of head).

The Black Petrel is only known from the type in the Paris Museum, which was collected by Mons. Delattre during his voyage from Nicaragua to California; and an example in the collection of Canon Tristram, which was obtained twenty years afterwards by Lieut. Gunn during July in Sendai Bay on the east coast of Hondo.

I have carefully compared the two examples and have no doubt that they belong to the same species.

267. PROCELLARIA FURCATA.
(GREY FORK-TAILED PETREL.)

Procellaria furcata, Gmelin, Syst. Nat. i. p. 561 (1788).

The Grey Fork-tailed Petrel is slightly larger than Leach's Petrel (wing from carpal joint about 6¼ inches). It is pale slate-grey, shading into white on the under tail-coverts and the tips of the scapulars and tertials, and into dark brown on the axillaries, under wing-coverts, lesser wing-coverts, and ear-coverts.

Figures: Cassin, Birds of California &c. pl. 47; Baird, Brewer, and Ridgway, Water-Birds N. Amer. ii. p. 413 (woodcut of head, foot, and tail).

Head of *Procellaria furcata*. Natural size.

The Grey Fork-tailed Petrel was recorded from the Kurile Islands from examples obtained by Mons. Merck (Pallas, Zoogr. Rosso-Asiat. ii. p. 315); and it has recently been found breeding on Rashua, one of the central islands of the chain, by Mr. Snow (Blakiston and Pryer, Ibis, 1878, p. 218). There are three examples in the Pryer collection obtained by Mr. Snow from this locality in June (Seebohm, Ibis, 1884, p. 33).

The breeding-range of the Grey Fork-tailed Petrel extends eastwards from the Kurile Islands and the Commander Islands, across the Aleutian chain and the islands on the south coast of Alaska, as far east as Sitka. It has occurred as far north as Bering Straits.

Order IMPENNES.

The order Impennes contains only one suborder.

Suborder XXV. *IMPENNES*.

The Penguins possess several characters, each of which is diagnostic; and are the only birds which combine the following characters:—

Young born helpless, but covered with down; spinal feather-tract not defined on neck; palate schizognathous.

There are about a score species of Penguins, which are principally confined to the Antarctic Region. On the eastern shores of the Pacific they range as far north as the equator, but on the western shores they do not approach Japan nearer than the southern coasts of Australia.

Order GALLO-GRALLÆ.

Young born covered with down or feathers; maxillo-palatines not coalesced with each other across the middle line; angle formed by the lines drawn from the junction of the scapula and coracoid to the other ends of those bones generally less and never much more than a right angle (extreme limit 110°); quill-feathers well differentiated; external nostrils not produced into tubes.

The order Gallo-Grallæ contains seven suborders, six of which are represented in Japan.

Suborder XXVI. *GAVIÆ*.

Palate schizognathous; dorsal vertebræ more or less opisthocœlous; no basipterygoid processes; spinal feather-tract well defined on the neck by lateral bare tracts.

The Gaviæ consist of four families—the Laridæ (containing about 140 species), which are cosmopolitan; the Alcidæ (containing about 30 species), confined to the Nearctic and Palæarctic Regions; the Cursoriidæ (containing about 30 species), found in the Neotropical, Ethiopian, Oriental, Australian, and the southern portion of the Palæarctic Regions; and the Œdicnemidæ (about 10 species), which are found in the tropical and subtropical countries of the world. The Laridæ and Alcidæ are the only families of this suborder which are represented in Japan.

268. ALCA TROILE.
(GUILLEMOT.)

Colymbus troile, Linnæus, Syst. Nat. i. p. 220 (1766).

The Guillemot has white tips to the secondaries at all ages and seasons, and white under wing-coverts; the bill from frontal feathers is longer than 1 inch. The combination of the first and last of these characters is found in no other Japanese species of the genus.

Figures: Baird, Brewer, and Ridgway, Water-Birds N. Amer. ii. p. 486 (woodcut of bill).

A form of the Common Guillemot known as Pallas's Guillemot is a resident on the coast of Yezzo and the Kurile Islands, and probably strays in winter along the coasts of the more southerly Japanese Islands. I have six examples collected by Mr. Snow on the Kurile Islands, and two obtained by Mr. Henson at Hakodadi. There is a third example from the latter locality, procured by Captain Blakiston, in the Swinhoe collection. The variation in the shape of the bill in even this small series is very remarkable. The length from the frontal feathers varies from 1·8 to 1·4 inches, and the height at the extremity of the nostrils from ·7 to ·45 inch.

Five of the examples are evidently identical with the birds described as *Cepphus lomvia* (Pallas, Zoogr. Rosso-Asiat. ii. p. 345); they are large (wing $8\frac{1}{2}$ to $9\frac{1}{2}$ inches), and very dark coloured, they have large thick bills with a short gonys, and the upper mandible is pale and denuded of feathers almost to the gape. Three of the examples agree with the description of *Cepphus arra* (Pallas, Zoogr. Rosso-Asiat. ii. p. 347); they are small (wing 8 inches) and of a much paler grey in colour, they have small thin bills with a long gonys, and the base of the upper mandible is feathered to the nostrils. It is possible that the slender-billed examples may be birds of the year of the thick-billed species, but, be that as it may, it is scarcely possible that either form is specifically distinct from the Common Guillemot.

Eggs in the Pryer collection resemble well-known varieties of the common form.

The Guillemot is a circumpolar species, breeding in great numbers on the rocky coasts of the British Islands, and varying much in different parts of its range, the thick-billed birds being commoner in the high north.

269. ALCA CARBO.
(SOOTY GUILLEMOT.)

Cepphus carbo, Pallas, Zoogr. Rosso-Asiat. ii. p. 350 (1826).

The Sooty Guillemot may be distinguished from all its Japanese congeners by its combination of the two characters—bill from frontal feathers longer than $1\frac{1}{4}$ inch, and secondaries never tipped with white.

Figures: Gould, Birds of Asia, vii. pl. 71; Baird, Cassin, and Lawrence, Birds of North America, pl. 97.

The Sooty Guillemot has long been known to visit the Kurile Islands (Pallas, Zoogr. Rosso-Asiat. ii. p. 350), but it was not recorded from Japan until Captain Rodgers procured it in June in the Bay of Sendai, on the east coast of Hondo, about halfway between Yokohama and Hakodadi (Cassin, Proc. Acad. Nat. Sc. Philad. 1862, p. 323). There is an example in the Swinhoe collection obtained by Captain Blakiston at Hakodadi on the 31st of May (Swinhoe, Ibis, 1875, p. 458), and I have two examples collected by Mr. Henson in the same locality, one on the 27th of March and the other on the 2nd of April. I have also an example collected by Mr. Owston at Uraga near Yokohama on the 27th of February.

The Sooty Guillemot breeds in great numbers on the rocky islands in the Gulf of Tartary (Schrenck, Reisen und Forsch. im Amur-Lande, i. p. 498) and on the south shores of the Sea of Okhotsk

(Middendorff, Sibirische Reise, ii. p. 239). It is not uncommon during summer on the coasts of Yezzo (Blakiston and Pryer, Ibis, 1878, p. 211). It is an accidental visitor to the Commander Islands (Stejneger, Orn. Expl. Comm. Isl. and Kamtsch. p. 22), but there is no record of its occurrence on the coast of Alaska or the Aleutian Islands since the time of Pallas.

The Sooty Guillemot is slightly larger than the Pigeon-Guillemot, and may always be distinguished from it by its longer bill. The upper parts are dark brown at all ages and seasons, except that the frontal feathers, the chin, and a space round the eye are pale grey. It never has any white on the upper wing-coverts, but young birds have much white on the under wing-coverts.

A. columba.	*A. carbo.*
Bill from frontal feathers 1·2 to 1·3 inch.	Bill from frontal feathers 1·4 to 1·7 inch.
Wing 6·9 to 7·2 inches.	Wing 7·1 to 7·9 inches.
More or less white on wing-coverts, except in adult summer plumage.	No white on the upper surface of the wing at any season or age.
Frontal feathers always dark brown; no pale grey round the eye, except in very young birds.	More or less pale grey on frontal feathers and round the eye.

The egg of the Sooty Guillemot (Schrenck, Reisen u. Forsch. im Amur-Lande, i. pl. xvi. fig. 1) is larger than that of the Black Guillemot, but not so large as even small examples of the egg of the Razorbill; otherwise it closely resembles them.

270. ALCA COLUMBA.
(PIGEON-GUILLEMOT.)

Cepphus columba, Pallas, Zoogr. Rosso-Asiat. ii. p. 348 (1826).

The Pigeon-Guillemot may be most easily distinguished from its Japanese congeners by its combination of the two characters—secondaries never tipped with white, and bill from frontal feathers more than an inch, but less than 1¼ inch.

Figures: Baird, Cassin, and Lawrence, Birds of North America, pl. 96. fig. 1.

The fact that the Pigeon-Guillemot occurs on the Kurile Islands was probably known to Pallas, inasmuch as he gives the local name of the bird in various languages, amongst which is that bestowed on it by the inhabitants of those islands. Of late years this bird has been found by Mr. Snow breeding in some numbers on the Kurile Islands (Blakiston and Pryer, Trans. As. Soc. Japan, 1882, p. 91); and Captain Blakiston has obtained it at Hakodadi (Seebohm, Ibis, 1884, p. 174). There are three examples in the Pryer collection obtained in June by Mr. Snow on Ketoi Island, one of the central islands of the Kurile range.

The breeding-range of the Pigeon-Guillemot extends from Bering Straits, southwards as far west as the Kurile Islands, and as far east as the Santa Barbara Islands, about 250 miles south of San Francisco.

It is a smaller bird than the Common Guillemot, and further differs from that species in always having some brown on the under wing-coverts, and in never having any white tips to the secondaries. It is very closely allied to the Sooty Guillemot, *Alca carbo*, and the only constant difference between them appears to be the shorter bill of the Pigeon-Guillemot. Immature birds are easily distinguished by the more or less obscure white bars across the wing; these bars appear entirely to disappear with age, leaving only the following differences in colour between the two species, viz.: the darker and more sooty hue of the brown, especially of the underparts, and the absence of the pale region round the eye, on the frontal feathers, and on the chin. It is said that the Pigeon-Guillemots from Kamtschatka and California always have white on the wing-coverts, but that a species occurs in Greenland (*Alca mansfeldi*) which is black all over.

An egg of this Guillemot in my collection, obtained by Mr. Snow on the Kurile Islands, does not differ in size, colour, or markings from eggs of the Black Guillemot.

271. ALCA ANTIQUA.
(BERING'S GUILLEMOT.)

Alca antiqua, Gmelin, Syst. Nat. i. p. 554 (1788).

In summer plumage Bering's Guillemot is conspicuous by its black flanks, and by the white streaks on the sides of the crown and the sides of the lower neck. In its winter plumage it resembles the Marbled Guillemot, but is easily distinguished from that species by its shorter bill (·6 instead of ·8 inch from frontal feathers), its longer tarsus (1·0 instead of ·7 inch), and by the absence of white on the scapulars. The wing from the carpal joint varies from 5·2 to 5·6 inches. No other Japanese Guillemot combines all these measurements, except Temminck's Guillemot, which has a crest in summer plumage. In winter the two species are very difficult to distinguish.

Figures : Temminck and Schlegel, Fauna Japonica, Aves, pl. 80.

Bering's Guillemot has long been known to be a visitor to the Kurile Islands (Pallas, Zoogr. Rosso-Asiat. ii. p. 368); and many examples were obtained by the Siebold Expedition at Nagasaki, whence I have a male in winter plumage procured by Mr. Collingwood on the 24th of February. I have four examples collected in June by Mr. Snow on the Kurile Islands, where he found it breeding (Blakiston and Pryer, Trans. As. Soc. Japan, 1882, p. 90). There is an example in the Swinhoe collection from Hakodadi, procured by Captain Blakiston in April (Swinhoe, Ibis, 1874, p. 166); and there are eight examples in the Pryer collection from the Yokohama market, where it is abundant during winter.

The breeding-range of Bering's Guillemot extends eastwards from the Kurile Islands and the Commander Islands, across the Aleutian chain to the islands on the south coast of Alaska as far east as Sitka. It is not known to occur further north.

272. ALCA WUMIZUSUME.
(TEMMINCK'S GUILLEMOT.)

Uria wumizusume, Temminck, Planches Coloriées, no. 579 (1835).

Temminck's Guillemot almost exactly resembles *Alca antiqua*, except that in summer plumage it is furnished with a black occipital crest.

Figures : Temminck and Schlegel, Fauna Japonica, Aves, pl. 79.

Temminck's Guillemot was originally described from Japan, whence it was probably procured near Nagasaki. It has since been recorded from Simoda (Cassin, Exp. Am. Squad. China Seas and Japan, ii. p. 233), but is probably only an occasional visitor to the main islands, as no specimens have reached this country. It is said to breed in some numbers on the cliffs of Kodushima, one of the Seven Islands (Stejneger, Proc. United States Nat. Mus. 1887, p. 482).

It is a very rare bird, and has only occurred elsewhere on the opposite shore of the Pacific south of Vancouver Island.

273. ALCA MARMORATA.
(MARBLED GUILLEMOT.)

Colymbus marmoratus, Gmelin, Syst. Nat. i. p. 583 (1788).

The Marbled Guillemot may be distinguished from all its Japanese congeners by its combination of the two characters—*scapulars streaked with white* (winter) or *buff* (summer), and *no white on the upper surface of the wings* (wing-coverts, or tips of secondaries).

Figures: Latham, Gen. Syn. iii. pt. ii. pl. 96; Audubon, Ornithological Biography, v. pl. 430; Audubon, Birds of America, vii. pl. 475.

Although the Marbled Guillemot was known both to Pallas and Latham, I can find no record of its occurrence either in the Kurile Islands or in Japan earlier than that of the female in the Swinhoe collection, which was procured in May by Captain Blakiston at Hakodadi (Swinhoe, Ibis, 1874, p. 166). There is a male in the Swinhoe collection apparently procured at the same time, but it is in full summer plumage, whilst the female is only beginning to lose its winter dress. I have another male collected by Mr. Henson at Hakodadi on the 23rd of March, which has half completed its spring moult, and a female collected by Captain Blakiston at Hakodadi in November in full winter plumage. I have a female collected by Mr. Snow on the Kurile Islands in which the feathers of the upper parts are tipped with greyish white instead of dark buff, and the under wing-coverts are mottled with grey instead of being all grey. Mr. Owston has collected this species at Yokohama (Blakiston and Pryer, Trans. As. Soc. Japan, 1882, p. 90).

The breeding-range of the Marbled Guillemot extends eastwards from the Kurile Islands across the Aleutian chain and the islands on the south coast of Alaska as far south as Vancouver Island. It is not known to occur north of the Aleutian Islands.

Dr. Stejneger divides the Marbled Guillemot into two species, which he calls *Brachyrhamphus marmoratus* and *B. perdix*; the former is probably the male and the latter the female or immature male.

The Marbled Guillemot can never be confounded with its shorter-billed ally. These two very distinct species differ in the following particulars :—

Alca marmorata.	*Alca brevirostris.*
Bill from frontal feathers ·8 to ·7 inch.	Bill from frontal feathers ·5 to ·4 inch.
Tail-feathers all brown.	Outer tail-feathers white.
No white on any of the secondaries.	White at tips of many outer secondaries.
Lores always brown.	Lores white in winter plumage.
Under wing-coverts principally white in winter plumage.	Under wing-coverts always grey.

The changes of plumage are very similar in both species, but the larger-billed form often has much white on the under wing-coverts in winter, which is not the case in any winter examples of the smaller-billed form that I have seen.

274. ALCA BREVIROSTRIS.
(KITTLITZ'S GUILLEMOT.)

Uria brevirostris, Vigors, Zool. Journ. 1828, p. 357.

Kittlitz's Guillemot may always be known by its very small bill, which measures less than half an inch from the frontal feathers.

Figures: Turner, Nat. Hist. Alaska, Birds, pl. 2 (winter plumage); Henshaw, Nat. Hist. Coll. Alaska, Birds, pl. 1 (summer plumage).

No authentic instance of the occurrence of Kittlitz's Guillemot, otherwise known by the name of the Short-billed Marbled Guillemot, in any of the Japanese Islands has been recorded. All the examples recorded as *Brachyrhamphus kittlitzi* from the Kurile Islands or Japan have proved on examination to have belonged to the longer-billed species, *Alca marmorata*. There are, however, in the Pryer collection two examples in nearly full summer plumage, obtained by Mr. Snow on the Kurile Islands, which belong to the short-billed species.

Kittlitz's Guillemot appears to be a very rare bird. Besides the two examples collected by Mr. Snow on the Kurile Islands, two examples have been obtained on the Aleutian Islands (Nelson, Cruise of the 'Corwin,' p. 117), and I have an example in winter plumage from Kamtschatka which was sent to me by Mons. Taczanowski. The Smithsonian Institution at Washington has received it from Cape Lisburne, north of Bering Straits. Nothing further is known of its range.

Kittlitz's Guillemot is remarkable for the shortness of its tarsus and the smallness of the exposed portion of its bill. Though the wing measures from 5 to 6 inches from the carpal joint, the tarsus only measures ·7, and the bill from the frontal feathers only ·4 inch. The winter plumage is grey above and white below and on the innermost scapulars. In summer the upper parts are marbled with buff, and the underparts with grey. The under wing-coverts are always grey, and the outer tail-feathers and the tips of the outermost secondaries are always white.

Genus FRATERCULA.—The Puffins are a very unfortunate group of birds, for in spite of the fact that they form a compact and well-defined genus, the variable character of their bills has caused them to be split up into numerous pseudogenera, and that to such an extent that in some cases one genus has been provided for the summer plumage and a second for the winter dress. In the genus *Fratercula* the bill is provided with one or more sheath-like structures of an orange-red colour, which appear in spring and are shed in autumn.

275. FRATERCULA CORNICULATA.
(HORN-EYED PUFFIN.)

Mormon corniculata, Naumann, Isis, 1821, p. 782.

The Horn-eyed Puffin in breeding-dress is readily diagnosed by the horn-shaped wattles above its eyes, but as these disappear before winter, a more complicated diagnosis is necessary. No other Japanese Puffin combines the characters, *bill ·9 inch or more high*, and *breast and belly white*.

Figures: Stejneger, Orn. Expl. Comm. Isl. and Kamtschatka, pl. 3; Gray, Genera of Birds, iii. pl. 174; Baird, Brewer, and Ridgway, Water-Birds N. Amer. ii. p. 529 (coloured woodcut of head).

The Horn-eyed Puffin has long been known to inhabit the Kurile Islands (Pallas, Zoogr. Rosso-Asiat. ii. p. 365), but it was for a long time confused with its Atlantic representative. It has not been recorded from Japan, but probably breeds on the Kurile Islands (Blakiston and Pryer, Trans. As. Soc. Japan, 1882, p. 89), as I have

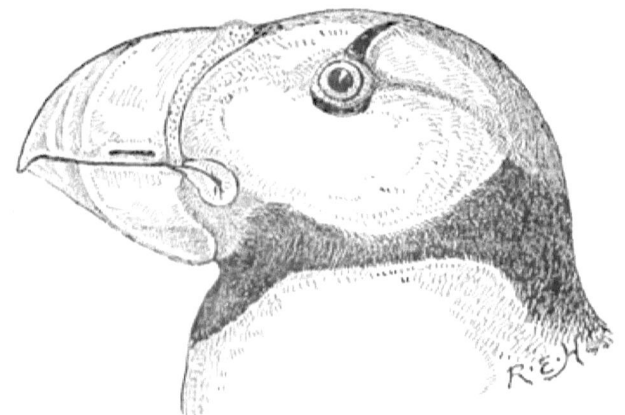

Head of *Fratercula corniculata*. ⅔ natural size.

an example obtained in June by Mr. Snow on Shiashkotan, one of the more northerly islands of the group (Seebohm, Ibis, 1884, p. 174). It breeds in great numbers on the islands of the southern shores of the Sea of Okhotsk (Middendorff, Sibirische Reise, ii. p. 240); whence its range extends eastwards to the Commander Islands, the Aleutian Islands, and the islands on the southern coast of Alaska as far south as Sitka; and northwards through Bering Straits to Cape Lisburne.

276. FRATERCULA CIRRHATA.
(TUFTED PUFFIN.)

Alca cirrhata, Pallas, Spicilegia Zoologica, pt. v. p. 7 (1780).

The Tufted Puffin may be distinguished from its Japanese allies by its combination of the two characters—*underparts entirely brown*, and *wing from carpal joint between 7 and 8 inches long*.

Figures: Daubenton, Planches Enluminées, no. 761; Bonaparte, Proc. Zool. Soc. 1851, pl. 44; Stejneger, Orn. Expl. Comm. Isl. and Kamtschatka, pl. 1; Baird, Brewer, and Ridgway, Water-Birds N. Amer. ii. p. 532 (coloured woodcut of head).

Head of *Fratercula cirrhata*. ¼ natural size.

The Tufted Puffin has long been known to inhabit the Kurile Islands (Pallas, Zoogr. Rosso-Asiat. ii. p. 364), where it breeds (Blakiston and Pryer, Ibis, 1878, p. 210). Its claim to be regarded as a Japanese bird rests upon examples from Nemoro in north-east Yezzo (Blakiston and Pryer, Trans. As. Soc. Japan, 1882, p. 88).

I have an example collected by Dr. Schrenck in Kamtschatka, and there are four examples in the Pryer collection obtained by Mr. Snow on the Kurile Islands in July.

It also breeds on many of the islands in Bering Sea, and on both shores of the North Pacific Ocean, where its range extends a little further north than Bering Straits, and as far south as the Farallon Islands, outside the harbour of San Francisco.

An egg collected by Mr. Snow on the Kurile Islands measures 2·8 by 2·0 inches; it resembles eggs of the Common Puffin, but the colour is browner, and it is streaked rather than spotted.

277. FRATERCULA MONOCERATA.
(HORN-BILLED PUFFIN.)

Alca monocerata, Pallas, Zoogr. Rosso-Asiat. ii. p. 361 (1826).

The Horn-billed Puffin in breeding-dress is easily diagnosed by the remarkable horny projection above its nostrils, but as this is cast every autumn, other characters must be found to distinguish it in winter plumage. The combination of a *wing from the carpal joint between 6·4 and 7·8 inches long*, and a *bill, exclusive of the horny projection, ·8 inch or less in height* will exclude every other Japanese species.

Figures: Gray, Genera of Birds, iii. pl. 174 (woodcut of head); Eschscholtz, Zool. Atlas, pl. 12.

Head of *Fratercula monocerata*. ½ natural size.

The Horn-billed Puffin was included by Temminck and Schlegel in the list of Japanese birds at the end of the 'Fauna Japonica' on the authority of a Japanese drawing; but it was first legitimately introduced into the Japanese list from examples obtained at Hakodadi in 1854 by Mr. Heine, the artist of the Perry Expedition (Cassin, Exp. Am. Squad. China Seas and Japan, ii. p. 233); and has since been found to be by no means a scarce bird in that locality, breeding in great numbers on the islands off the coast of Yezzo and wintering in Southern Japan (Blakiston and Pryer, Trans. As. Soc. Japan, 1882, p. 92). It is possibly to an immature or winter example of this species that the bird obtained by the Siebold Expedition at Nagasaki, and recorded as *Alca torda*, must be referred (Temminck and Schlegel, Fauna Japonica, Aves, p. 125).

I have an example from Kamtschatka collected by Dybowski; one from the Kurile Islands collected by Wossnesensky; and a third from Hakodadi collected by Mr. Henson on the 12th of May. There are two examples in the Swinhoe collection obtained by Captain Blakiston at Hakodadi (Swinhoe, Ibis, 1874, p. 166), and five in the Pryer collection from Yokohama.

The Horn-billed Puffin is not known to have occurred as far north as Bering Sea, or even on the Aleutian Islands; but it is common on the American shores of the Pacific, breeding on the islands near the coast, as far north as Sitka on the southern coast of Alaska, and as far south as the Farallon Islands near San Francisco.

278. FRATERCULA PSITTACULA.
(PARROT-BILLED PUFFIN.)

Alca psittacula, Pallas, Spicilegia Zoologica, pt. v. p. 13 (1780).

The Parrot-billed Puffin is easily recognized by the shape of its bill, *the line of the gape curves upwards to such an extent that in profile both outlines of the upper mandible appear equally convex.*

Figures: Eschscholtz, Zool. Atlas, pl. 16; Stejneger, Orn. Expl. Comm. Isl. and Kamtschatka, pl. 4. fig. 6 (head).

Head of *Fratercula psittacula*. Natural size.

The Parrot-billed Puffin has long been known to inhabit the Kurile Islands (Pallas, Zoogr. Rosso-Asiat. ii. p. 366), and was found there in 1881 by Mr. Snow (Blakiston and Pryer, Trans. As. Soc. Japan, 1882, p. 89). I have an example collected by Mr. Snow on the Kurile Islands in June (Seebohm, Ibis, 1884, p. 174), and there are two examples in the Pryer collection, also obtained by Mr. Snow in the same locality. It is said to be a comparatively rare bird, not more than half a dozen pairs being met with in a season's sea-otter hunting.

The Parrot-billed Puffin is extremely abundant in Bering Straits and for some distance north, especially on the Siberian coast. It has not been recorded from Japan, nor from the coast of Alaska east of the Aleutian Islands.

279. FRATERCULA CRISTATELLA.
(CRESTED PUFFIN.)

Alca cristatella, Pallas, Spicilegia Zoologica, pt. v. p. 18 (1780).

The Crested Puffin may be distinguished from its Japanese congeners by its combination of the two characters—*underparts entirely brown*, and *wing from carpal joint between 5 and 6 inches long*.

Head of *Fratercula cristatella*. Natural size.

Figures: Gray, Genera of Birds, iii. pl. 174. fig. 1 (woodcut of head); Stejneger, Orn. Expl. Comm. Isl. and Kamtschatka, pl. 4. fig. 4 (head in summer), fig. 5 (head in winter).

The Crested Puffin has long been known as an inhabitant of the

Kurile Islands (Pallas, Zoogr. Rosso-Asiat. ii. p. 370), and two examples were procured in 1865 on the east coast of Japan, one of them in the latitude of Yokohama, and the other about 150 miles due south of Yezzo (Whitely, Ibis, 1867, p. 209). I have two examples collected by Wossnesensky on the Kurile Islands, where it was found breeding by Mr. Snow (Blakiston and Pryer, Trans. As. Soc. Japan, 1882, p. 89), and whence I have an example obtained by Mr. Snow in June (Seebohm, Ibis, 1882, p. 168). There are two examples in the Pryer collection procured at the same time.

The breeding-range of the Crested Puffin extends from the islands in Bering Straits, south-west to the Kurile Islands, and south-east to the Aleutian Islands and Kodiak.

An egg obtained by Mr. Snow on the Kurile Islands measures 2·15 by 1·4 inches; it is dirty white outside and greenish inside.

280. FRATERCULA PYGMÆA.
(WHISKERED PUFFIN.)

Alca pygmæa, Gmelin, Syst. Nat. i. p. 555 (1788).

The Whiskered Puffin may be distinguished from its Japanese congeners by its combination of two characters—*wing less than five inches from carpal joint*, and *throat dark brown*.

Figures: Temminck, Planches Coloriées, no. 200 (erroneously named *Phalaris cristatella*); Turner, Nat. Hist. Alaska, Birds, pl. i. (breeding-plumage); Stejneger, Orn. Expl. Comm. Isl. and Kamtschatka, pl. 4. fig. 1 (head in winter), fig. 2 (head in summer).

The Whiskered Puffin is doubtless identical with the Pigmy Auk of the early writers (Pennant, Arctic Zoology, ii. p. 513), who described it in winter plumage (Latham, Gen. Syn. iii. pt. i. p. 328) from the islands in Bering Sea. It was afterwards described in summer plumage under the name of *Alca kamtschatica* (Lepechin, Nov. Act. Petropol. 1801, p. 369), to be renamed some years later as *Uria mystacea* (Pallas, Zoogr. Rosso-Asiat. ii. p. 372). Its synonymy was further complicated by Lichtenstein, who named it *Mormon superciliosa*, and by Dr. Coues, who called the young *Simorhynchus cassini*.

It is a mistake to call this Puffin the Pigmy Auk, firstly, because it is a Puffin and not an Auk, and, secondly, because there is a still smaller species belonging to the same genus.

The Whiskered Puffin has long been known to occur both on the Kurile Islands and the Japanese coasts (Pallas, Zoogr. Rosso-Asiat. ii. p. 372). Examples were taken by Mr. Heine of the Perry Expedition at Simoda and in the Bay of Yedo in April 1854 (Cassin, Exp. Am. Squad. China Seas and Japan, ii. p. 234). There are nine

Head of *Fratercula pygmæa*. Natural size.

examples in the Pryer collection obtained by Mr. Snow in June on Ushisir, one of the central islands of the Kurile range.

The breeding-range of the Whiskered Puffin extends from the Kurile Islands eastwards to the Commander Islands and the Aleutian Islands as far east as Unalaska.

281. FRATERCULA PUSILLA.
(LEAST PUFFIN.)

Uria pusilla, Pallas, Zoogr. Rosso-Asiat. ii. p. 373 (1826).

The Least Puffin may always be distinguished from its congeners by its diminutive size. The wing from the carpal joint measures from $3\frac{1}{2}$ to 4 inches.

Figures: Gray, Genera of Birds, iii. pl. 175 (summer plumage);
Nelson, Cruise of the 'Corvin,' Birds, pl. 1 (summer plumage).

Head of *Fratercula pusilla*. Natural size.

The Least Puffin was not recorded from Japan until it was described from two examples, one of which, in the Tokio Museum, was procured on the Kaga coast of Hondo, and the other, in the Hakodadi Museum, was caught in Hakodadi Bay in Yezzo (Blakiston and Pryer, Ibis, 1878, p. 210). On the 2nd of March, 1883, several examples were collected in Hakodadi Bay by Mr. Henson, two of which were sent me by Captain Blakiston (Seebohm, Ibis, 1884, p. 31), and two more came into my possession afterwards.

The Least Puffin breeds in great abundance on the islands in the North Pacific Ocean as far north as Bering Straits, and as far south as the Aleutian Islands.

282. STERCORARIUS RICHARDSONI.
(RICHARDSON'S SKUA.)

Lestris richardsoni, Swainson, Faun. Bor.-Amer. p. 433 (1831).

Richardson's Skua is a sooty-brown bird, with or without a great deal of white on the underparts. The narrow, pointed central tail-feathers are 3 or 4 inches longer than the outer ones. The shafts of most of the primaries are white.

Figures: Gould, Birds of Great Britain, v. pl. 80; Dresser, Birds of Europe, viii. pl. 611.

Richardson's Skua probably breeds on the Kurile Islands, whence examples have been brought by Mr. Snow (Blakiston and Pryer, Trans. As. Soc. Japan, 1882, p. 105). There are three examples in the Pryer collection from the Kurile Islands, all of them belonging to the dark form with no white on the underparts.

Richardson's Skua is a circumpolar species, and frequently visits the British Islands on its migrations, a few breeding in Scotland.

283. STERCORARIUS BUFFONI.
(BUFFON'S SKUA.)

Lestris buffoni, Boie, Isis, 1822, p. 562.

Buffon's Skua closely resembles Richardson's Skua, but it is not known ever to have dark underparts. The central tail-feathers are 4 to 8 inches longer than the others, and only the first and second primaries have white shafts.

Figures: Gould, Birds of Great Britain, v. pl. 81; Dresser, Birds of Europe, iii. pl. 612. fig. 1.

Buffon's Skua probably breeds on the Kurile Islands, whence examples have been brought by Mr. Snow (Blakiston and Pryer, Trans. As. Soc. Japan, 1882, p. 105).

Buffon's Skua is a circumpolar species, and occasionally visits the British Islands on its migrations.

284. STERCORARIUS POMARINUS.
(POMARINE SKUA.)

Lestris pomarinus, Temminck, Man. d'Orn. p. 514 (1815).

The Pomarine Skua differs from its allies in having broad, rounded and twisted central tail-feathers.

Figures: Gould, Birds of Great Britain, v. pl. 79; Dresser, Birds of Europe, viii. p. 610.

The Pomarine Skua is an occasional winter visitor to Japan (Seebohm, Ibis, 1884, p. 32). There is an example in the Pryer collection from Tokio Bay.

The Pomarine Skua is a circumpolar species, and frequently visits the British Islands in winter.

285. LARUS GLAUCUS.
(GLAUCOUS GULL.)

Larus glaucus, Brünnich, Orn. Bor. p. 44 (1764).

The Glaucous Gull is the largest species of the genus (wing from carpal joint 21 to 18 inches). Its legs are flesh-coloured, its orbits vermilion (in the adult), and its primaries white.

Figures: Gould, Birds of Great Britain, v. pl. 57; Dresser, Birds of Europe, viii. pl. 605 (legs and feet wrongly coloured).

The Glaucous Gull is a winter visitor to Japan; immature examples are more common than adults, but the latter are occasionally met with. There is an example in the Swinhoe collection from Hakodadi (Swinhoe, Ibis, 1874, p. 165), which was apparently the first obtained in Japan; and there is another in the Pryer collection from Yokohama.

The Glaucous Gull is a circumpolar species, and is a winter visitor to the British Islands.

286. LARUS GLAUCESCENS.
(GLAUCOUS-WINGED GULL.)

Larus glaucescens, Naumann, Naturg. Vög. Deutschl. x. p. 351 (1840).

The Glaucous-winged Gull is quite as large as a Herring-Gull, and resembles that species in the colour of its mantle. It may at once be recognized by its primaries, the pattern of which resembles that of *L. cachinnans*, but the dark part of the markings, instead of being nearly black, are lavender-grey like the mantle.

The Glaucous-winged Gull is an occasional visitor to Japan from the American coast, where it breeds in Alaska and winters in California. Examples from Hakodadi are in the Swinhoe collection, which appear to be the first obtained in Japan, and in that of Mr. Howard Saunders (Swinhoe, Ibis, 1874, p. 165). It has also occurred in Tokio Bay (Blakiston and Pryer, Trans. As. Soc. Japan, 1882, p. 104).

The Glaucous-winged Gull is a Pacific-Ocean species, breeding on the Commander Islands and in Alaska.

287. LARUS MARINUS.
(GREAT BLACK-BACKED GULL.)

Larus marinus, Linnæus, Syst. Nat. i. p. 225 (1766).

The Great Black-backed Gulls of Yezzo and the Kuriles are apparently identical with those found on the Commander Islands, and named *Larus schistisagus* (Stejneger, Auk, 1884, p. 231). They seem to intergrade with *Larus marinus* (of which they are doubtless an Eastern race), and some examples from Japan are so absolutely intermediate between the two forms that they may be referred to either, and can only be recognized as *Larus marinus schistisagus*.

The Great Black-backed Gull is one of the largest species (wing from carpal joint 20 to 18 inches). Its legs are flesh-coloured, its mantle very dark slate-grey, and the light pattern on the inner webs of its primaries is obscure and seldom much cuneated. Its orbits are vermilion.

Mr. Snow obtained a series of Gulls from the Kurile Islands, and Mr. Whitely and Captain Blakiston have both procured specimens near Hakodadi of a large size, and differing only from *L. affinis* in having flesh-coloured instead of yellow legs and feet (Whitely, Ibis, 1867, p. 210). There can be little doubt that they are identical with the Gulls described by Dr. Stejneger as *Larus schistisagus*, which I regard as the Eastern form of *Larus marinus*, and with which it appears to intergrade. The Western race differs from the Eastern one in being on an average a slightly larger bird, in having a somewhat darker mantle, and in possessing a white subterminal spot or bar on the second primary. In the Swinhoe collection is a Gull from Hakodadi (Swinhoe, Ibis, 1874, p. 165), which has a dark mantle, but no subterminal spot on the second primary; and in the Pryer collection is an example from the Kurile Islands which has the paler mantle, but possesses the subterminal spot on the second primary.

288. LARUS CACHINNANS.
(PALLAS'S HERRING-GULL.)

Larus cachinnans, Pallas, Zoogr. Rosso-Asiat. ii. p. 318 (1826).

Pallas's Herring-Gull is smaller than the Great Black-backed Gull (wing from carpal joint 19 to 18 inches). Its legs are yellow, its mantle pale slate-grey, and the light pattern on the inner webs of its primaries is very distinct and wedge-shaped. Its orbits are vermilion.

Figures: Dresser, Birds of Europe, viii. pl. 602. fig. 2.

Pallas's Herring-Gull appears to be a regular winter visitor to Japan. There are several examples in the Pryer collection from Yokohama, and many have been procured at Hakodadi (Whitely, Ibis, 1867, p. 210). It is probably to this species that the immature Gulls obtained by the Perry Expedition in Yedo Bay, and supposed to be the young of *Larus ichthyaetus*, are to be referred (Cassin, Exp. Am. Squad. China Seas and Japan, ii. p. 232). It is a winter visitor to the Bonin Islands (Seebohm, Ibis, 1890, p. 105), and there is an example in the Norwich Museum obtained by Mr. Ringer near Nagasaki.

This species has a very wide range, from Teneriffe (where I found it very abundant in May), through the Mediterranean, the Black Sea, the Caspian, the Aral Sea, and Lake Baikal, to the valley of the Amoor, but it has not occurred in the British Islands, where its place is taken by a very closely-allied species, *Larus argentatus*.

289. LARUS LEUCOPTERUS.
(ICELAND GULL.)

Larus leucopterus, Faber, Prodr. islandischen Orn. p. 98 (1822).

The Iceland Gull is a miniature Glaucous Gull, with very pale mantle and white primaries (wing from carpal joint 17 to 16 inches).

Figures: Dresser, Birds of Europe, viii. pl. 606 (orbits coloured vermilion instead of flesh-colour).

The occurrence of the Iceland Gull in Japan is not very satisfactorily proved. It is said that Captain Blakiston obtained an example (Saunders, Proc. Zool. Soc. 1878, p. 166). There is no such example in the Swinhoe collection, but it is very probable that there may have been one, as it is said to be the commonest Gull in the Bering Sea (Nelson, Nat. Hist. Coll. Alaska, p. 53).

The Iceland Gull breeds in Arctic America, wandering in winter as far west as Japan and as far east as the British Islands.

290. LARUS CRASSIROSTRIS.
(TEMMINCK'S GULL.)

Larus crassirostris, Vieillot, Nouv. Dict. d'Hist. Nat. xxi. p. 508 (1818).

Temminck's Gull is exactly the same size as the Common Gull, but is slightly darker in the colour of its mantle, which scarcely differs from that of the Eastern form of the Great Black-backed Gull.

The black band across the end of the tail of the adult of Temminck's Gull is an excellent mark of distinction from the adult of the Common Gull, in which the tail is white throughout. Immature examples of the former have three fourths of the tail nearly black, whilst those of the latter have only one third dark brown.

Figures: Temminck and Schlegel, Fauna Japonica, Aves, pl. 88 (adult and young), as *Larus melanurus*.

Temminck's Gull is the common Gull of China and Japan, and breeds in Yezzo (Blakiston, Ibis, 1862, p. 311), and probably in all the Japanese Islands. There are several examples in breeding-plumage in the Pryer collection obtained near Yokohama; and Mr. Heine, the artist of the Perry Expedition, says that it abounded in the Bay of Hakodadi in May 1854 (Cassin, Exp. Am. Squad. China Seas and Japan, ii. p. 232).

Temminck's Gull breeds in Eastern Siberia, as well as in China and Japan. Eggs in the Pryer collection resemble large eggs of *Larus canus*.

291. LARUS CANUS.
(COMMON GULL.)

Larus canus, Linneus, Syst. Nat. i. p. 224 (1766).

The Common Gull is one of the smaller species. It scarcely differs from Pallas's Herring-Gull, except in being smaller and in having greener legs. The entirely white tail distinguishes adult examples from Temmiuck's Gull. In immature examples one third of the tail is dark brown (instead of three fourths being nearly black, as in the young of Temminck's Gull). The pattern of its primaries is quite different from that of either the Black-headed Gull or the Kittiwake.

It has been stated that the Common Gulls of East Asia were a larger race than those of Western Europe. A series in the Pryer collection vary in length of wing from $13\frac{3}{4}$ to $14\frac{3}{4}$ inches, whilst a series from Europe vary from $12\frac{3}{4}$ to $14\frac{3}{4}$ inches.

Figures: Gould, Birds of Great Britain, v. pl. 60; Dresser, Birds of Europe, viii. pl. 600.

The Common Gull probably breeds on the Kurile Islands and on the coasts of Yezzo, whence many examples have been procured by Mr. Snow, Mr. Henson, and Captain Blakiston. The first identified example recorded from Japan was shot at Hakodadi on the 13th of November, 1864 (Whitely, Ibis, 1867, p. 210). There are several examples in the Pryer collection from Yokohama, but most of them are in immature plumage.

Larus californicus has been recorded from Japan (Saunders, Proc. Zool. Soc. 1878, p. 175), and also *Larus delawarensis* (Saunders, Proc. Zool. Soc. 1878, p. 177). Mr. Saunders informs me certainly that the former and probably the latter were wrongly identified, and must be referred to *Larus canus*.

The range of the Common Gull, like that of the Black-headed Gull, extends from the British Islands across Siberia to Japan, but it is an Arctic not a Temperate species, and rarely breeds where the mean temperature for the month of July is above 60°.

292. LARUS TRIDACTYLUS.
(KITTIWAKE.)

Larus tridactylus, Linnæus, Syst. Nat. i. p. 224 (1766).

The Kittiwake is one of the smaller Gulls (wing from carpal joint 13 to 14 inches). Its legs and feet are nearly black, and the hind toe is absent or very small. The colour of its mantle is like that of the Common Gull. The wing beyond the tip of the sixth primary is black, with small white tips to the fourth and fifth primaries; the rest of the wing is pearl-grey, with a black outer web to the first primary.

Figures: Gould, Birds of Great Britain, v. pl. 61; Dresser, Birds of Europe, viii. pl. 608.

The Kittiwake breeds on the Kurile Islands, whence it was probably obtained by Steller (Pallas, Zoogr. Rosso-Asiat. ii. p. 324), and whence there is an example in the Pryer collection obtained by Mr. Snow in June on the island of Rashua. In the Hakodadi Museum there are examples obtained at Nemoro, the eastern extremity of Yezzo (Blakiston and Pryer, Trans. As. Soc. Japan, 1882, p. 105).

The Kittiwake is a circumpolar species, and breeds on the coasts of the British Islands.

293. LARUS RIDIBUNDUS.
(BLACK-HEADED GULL.)

Larus ridibundus, Linneus, Syst. Nat. i. p. 225 (1766).

The Black-headed Gull is one of the smaller species (wing from carpal joint 12½ to 11¼ inches). When adult the bill, legs, and feet are coral-red, and the mantle pale grey. In summer the head is dark brown.

A white band runs down the centre of the second primary to within an inch of the tip; and the outer primary-coverts are for the most part white.

Figures: Gould, Birds of Great Britain, v. pl. 64; Dresser, Birds of Europe, viii. pl. 597. fig. 1.

The Black-headed Gull probably breeds in Yezzo (Swinhoe, Ibis, 1874, p. 165), and migrates in autumn as far south as Yokohama, whence it was procured by the Perry Expedition (Cassin, Exp. Am. Squad. China Seas and Japan, ii. p. 232), and whence there are several skins in winter plumage in the Pryer collection. There is an example in the Norwich Museum obtained at Nagasaki by Mr. Ringer. It is a Palæarctic species, breeding in the British Islands and in various inland localities across Temperate Europe and Southern Siberia to Northern Japan.

294. STERNA DOUGALLI.
(ROSEATE TERN.)

Sterna dougalli, Montagu, Orn. Dict. Suppl. (1813).

The Roseate Tern agrees with the Common and Daurian Terns in having a black forehead; the rump and upper tail-coverts are suffused with grey, the white margins of the inner webs of the primaries extend to the tips of the feathers; the bill is black, and the feet are red.

Figures: Gould, Birds of Great Britain, v. pl. 71; Dresser, Birds of Europe, viii. pl. 581.

There is an example of the Roseate Tern from the Loo-Choo Islands in the Pryer collection (Seebohm, Ibis, 1887, p. 181).

This widely spread species is known to breed in many localities in the Atlantic and Indian Oceans; and on the western shores of the

Head of *Sterna dougalli*. Natural size

Pacific is recorded from New Guinea, Australia, and New Caledonia. It formerly bred in several localities on the British coasts.

295. STERNA LONGIPENNIS.
(DAURIAN TERN.)

Sterna longipennis, Nordmann, Erman's Verz. von Thieren und Pflanzen, p. 17 (1835).

The Daurian Tern has a black forehead; the upper tail-coverts are pure white; the primaries are not margined with white towards the tip; the bill is black and the feet are brown.

Head of *Sterna longipennis*. Natural size.

Figures: Middendorff, Sibirische Reise, ii. pl. 25. fig. 4.

The Daurian Tern is the eastern representative of the Common Tern, and breeds in Eastern Siberia, wintering in Japan and China. I have an example collected by Mr. Snow on Eturop, the most southerly of the Kurile Islands, where it probably breeds (Seebohm, Ibis, 1879, p. 23). In the Tweeddale collection is a skin from Yezzo (Saunders, Proc. Zool. Soc. 1876, p. 650); and I have an example collected by Mr. Pryer in Tokio Bay.

296. STERNA MELANAUCHEN.
(BLACK-NAPED TERN.)

Sterna melanauchen, Temminck, Planches Coloriées, no. 427 (1827).

The Black-naped Tern has a black patch covering the nape and passing through each eye to the lores, in strong contrast to the white forehead and crown. The bill and the feet are black.

Figures : Gould, Birds of Australia, vii. pl. 28.

Head of *Sterna melanauchen*. Natural size.

There is an example of the Black-naped Tern from the Loo-Choo Islands in the Pryer collection (Seebohm, Ibis, 1887, p. 181). It breeds on the rocks outside Amoy Harbour (Swinhoe, Proc. Zool. Soc. 1871, p. 422); and is found in the Indian Ocean as far west as the Andaman Islands (Walden, Proc. Zool. Soc. 1866, p. 556), and in the Pacific as far south as New Caledonia (Layard, Ibis, 1879, p. 365), and as far east as the Samoa Group.

297. STERNA SINENSIS.
(ORIENTAL LESSER TERN.)

Sterna sinensis, Gmelin, Syst. Nat. i. p. 608 (1788).

The Oriental Lesser Tern is easily distinguished from the other Japanese Terns by its small size (wing from carpal joint 7 to 6 inches). It differs from our Lesser Tern in having the shafts of the three outer primaries white instead of brown. The forehead is white, and the bill and the feet are yellow.

Figures: Gould, Birds of New Guinea, v. pl. 72, as *Sternula placens*.

Head of *Sterna sinensis*. Natural size.

The Oriental Lesser Tern is the eastern representative of our Lesser Tern, and breeds in Southern Japan. There are several examples from the neighbourhood of Yokohama, and one from the Loo-Choo Islands in the Pryer collection (Seebohm, Ibis, 1887, p. 181). There can be little doubt that the examples of the Little Tern obtained by the Perry expedition in the last-named locality in August 1854 are referable to this species (Cassin, Exp. Am. Squad. China Seas and Japan, ii. p. 248). This is probably the northern limit of its range, which extends south to Northern Australia and west to Ceylon.

Eggs in the Pryer collection do not differ in any way from common varieties of those of *Sterna minuta*.

298. STERNA ALEUTICA.
(ALEUTIAN TERN.)

Sterna aleutica, Baird, Trans. Chicago Acad. Nat. Sc. i. p. 321 (1869).

The Aleutian Tern is about the size of the Common Tern (wing from carpal joint 11 to 9½ inches). Its mantle and rump are slate-grey, but its forehead and upper tail-coverts are white.

Figures: Baird, Trans. Chicago Acad. 1869, pl. 31. fig. 1.

The Aleutian Tern appears to be a more or less accidental visitor to the coasts of Japan, as there is an example in the Pryer collection labelled " Innoboye," which is probably the same as Cape Inoboga due east of Tokio.

It is a North Pacific species, breeding on the coasts of Alaska from Kadiak to Norton Sound.

299. STERNA BERGII.
(RÜPPELL'S TERN.)

Sterna bergii, Lichtenstein, Verz. Doubl. Zool. Mus. Berlin, p. 80 (1823).

Rüppell's Tern is one of the larger species (wing from carpal joint 15 to 13 inches). The black on the crown does not cover the fore-

Head of *Sterna bergii*. Natural size.

head, and ends in a point on the nape, which is white, gradually shading into slate-grey on the back.

Figures: Gould, Birds of Australia, vii. pl. 23 (sub nom. *Thalasseus pelecanoides*), pl. 24 (sub nom. *Thalasseus poliocercus*).

Rüppell's Tern has been found in the southern group of the Loo-Choo Islands (Stejneger, Proc. United States Nat. Mus. 1887, p. 393).

It is a tropical species, ranging from the Red Sea and both coasts of South Africa across the Indian Ocean to the Western Pacific. It breeds on Kelung Island, off the north coast of Formosa, and ranges southwards to Tasmania and the Fiji Islands.

300. STERNA STOLIDA.
(NODDY TERN.)

Sterna stolida, Linneus, Syst. Nat. i. p. 227 (1766).

The Noddy Tern belongs to the group in which the whole of the upper parts below the head are of a nearly uniform brown. It may easily be recognized by its pale forehead and crown, or by its wedge-shaped tail.

Figures: Gould, Birds of Australia, vii. pl. 31.

The Noddy Tern is found in all tropical seas, and occasionally strays into the more temperate regions. There is an example in the Pryer collection obtained by Mr. Harrison near the Gulf of

Head of *Sterna stolida*. Natural size.

Yedo, and there is a second example in the Tokio Museum from the same locality (Blakiston and Pryer, Trans. As. Soc. Japan, 1882,

p. 103); Mr. Cuthbert Collingwood procured a third example near the Loo-Choo Islands (Swinhoe, Proc. Zool. Soc. 1870, p. 603) : these are, so far as I know, the only recorded occurrences of this species in Japan. It breeds on the cliffs of Formosa (Swinhoe, Ibis, 1863, p. 430). It is said that two examples of this species were once shot in Ireland.

301. STERNA ANÆSTHETA.
(BRIDLED TERN.)

Sterna anæstheta, Scopoli, Deliciæ Floræ et Faunæ Insubricæ, i. p. 92 (1786).

Head of *Sterna anæstheta*. Natural size.

The Bridled Tern is smaller than the Sooty Tern, the upper parts are paler, the white nuchal collar is more distinct, the black on the lores extends as far forwards as the white, and the web on both sides of the middle toe is narrower.

Figures: Gould, Birds of Australia, vii. pl. 33.

The Bridled Tern is said to be an oceanic species, resident throughout the tropics, and only occasionally wandering into the temperate regions. An example was obtained by Mr. Cuthbert Collingwood at Hakodadi on the 4th of November (Swinhoe, Proc. Zool. Soc. 1870, p. 603), and a second example in the Pryer collection was obtained near Yokohama. It is common on the Pescadore Islands between Formosa and China (Swinhoe, Proc. Zool. Soc. 1871, p. 422).

The Bridled Tern is a tropical species, inhabiting the Indian Ocean, the Pacific, and the Atlantic. It is said to have once occurred in England.

302. STERNA FULIGINOSA.
(SOOTY TERN.)

Sterna fuliginosa, Gmelin, Syst. Nat. i. p. 605 (1788).

Head of *Sterna fuliginosa*. Natural size.

The Sooty Tern belongs to the group in which the back, rump, and upper tail-coverts are dark brown, but the underparts are white in the adult. It is larger than the Bridled Tern. The white nuchal collar does not extend across the nape, and the black on the lores does not extend as far forward as the white on the forehead.

Figures : Temminck and Schlegel, Fauna Japonica, Aves, pl. 89 ; Dresser, Birds of Europe, viii. pl. 587.

The Sooty Tern is a resident throughout the tropical seas, and is only accidentally found in more temperate regions. Its only claim to be regarded as a Japanese species is the record of the examples, doubtless from Southern Japan, obtained by the Siebold Expedition, and an example obtained on the Yaye-yama Islands, the most southerly group of the Loo-Choo Islands, by Mr. Nishi (Stejneger, Proc. United States Nat. Mus. 1887, p. 392). There are two records of its occurrence on the British Islands.

Suborder XXVII. LIMICOLÆ.

Palate schizognathous; nasals schizorhinal; dorsal vertebræ opisthocœlous; basipterygoid processes present.

The suborder Limicolæ consists of two families. The Charadriidæ, containing rather more than 150 species, are cosmopolitan. The Parridæ do not number a dozen species, and are confined to the tropical regions of the world.

Fifty species belonging to the family Charadriidæ are recorded from Japan.

303. CHARADRIUS FULVUS.
(ASIATIC GOLDEN PLOVER.)

Charadrius fulvus, Gmelin, Syst. Nat. i. p. 687 (1788).

The Asiatic Golden Plover belongs to a small group of Plovers which have the tail-feathers transversely barred. It is the only species in this group which has grey axillaries.

Figures: Temminck and Schlegel, Fauna Japonica, Aves, pl. 62. Dresser, Birds of Europe, vii. pl. 516 (summer plumage), pl. 517. figs. 2 & 3 (winter plumage).

The Asiatic Golden Plover passes the Kuriles, the Japanese Islands, and the Loo-Choo Islands in great numbers, both on the spring and on the autumn migrations. Dr. Henderson procured it at Hakodadi in October 1857 (Cassin, Proc. Acad. Nat. Sc. Philad. 1858, p. 195). There are two examples from Hakodadi in the Swinhoe collection (Swinhoe, Ibis, 1874, p. 162); and there are four from Yokohama in the Pryer collection. Captain Rodgers obtained it both on the Bonin Islands and on the Loo-Choo Islands (Cassin, Proc. Acad. Nat. Sc. Philad. 1862, p. 321), and there are four examples from the latter locality in the Pryer collection (Seebohm, Ibis, 1887, p. 180).

Although the only known breeding-grounds of the Asiatic Golden Plover are on the tundras of Eastern Siberia, it is an accidental straggler to Western Europe, and is recorded as having once been obtained on the British Islands.

304. CHARADRIUS HELVETICUS.
(GREY PLOVER.)

Tringa helvetica, Linneus, Syst. Nat. i. p. 250 (1766).

The Grey Plover belongs to a small group of Plovers which have the tail-feathers transversely barred. In this small group it may be distinguished either by its foot, which is furnished with a hind toe, or by its axillaries, which are black.

Head of *Charadrius helveticus*. Natural size.

Figures: Dresser, Birds of Europe, vii. pl. 517. fig. 1 (winter plumage), pl. 515. fig. 2 (summer plumage).

The Grey Plover passes the Kuriles, Japan, and the Loo-Choo Islands in some numbers on the spring and autumn migration. It was obtained by Dr. Henderson at Hakodadi in October 1857 (Cassin, Proc. Acad. Nat. Sc. Philad. 1858, p. 195), and there are several examples in the Swinhoe collection from the same locality (Swinhoe, Ibis, 1875, p. 452). There are five examples in the Pryer collection from Yokohama, and one from the Loo-Choo Islands. It was obtained, doubtless near Nagasaki, by the Siebold Expedition

(Temminck and Schlegel, Fauna Japonica, Aves, pl. 106). Examples from Yokohama are dated from the 24th of September to the 16th of October.

The Grey Plover is a common winter visitor to the British Islands as well as to Japan, and breeds in various localities in the intervening Arctic Regions, as well as in Arctic America.

305. CHARADRIUS MORINELLUS.
(COMMON DOTTEREL.)

Charadrius morinellus, Linneus, Syst. Nat. i. p. 254 (1766).

The Common Dotterel has grey axillaries like the Oriental Dotterel and the Asiatic Golden Plover. In summer plumage its black belly and chestnut flanks are very distinctive, and at all seasons its short bill is remarkable. The bill from the frontal feathers is shorter than the middle toe without the claw.

Charadrius morinellus.

Figures: Dresser, Birds of Europe, vii. pl. 526 (summer and winter plumage).

It is quite possible that the Common Dotterel may be a rare

visitor on migration to the Japanese Islands. In October 1857 Dr. Henderson visited Hakodadi, during the cruise of the 'Portsmouth,' and procured on the shores of a salt-water creek a bird which is recorded as being indistinguishable from the European Dotterel (Cassin, Proc. Acad. Nat. Sc. Philad. 1858, p. 195).

The breeding-range of the Common Dotterel extends from the British Islands, across Northern Europe and Siberia.

306. CHARADRIUS MINOR.
(LITTLE RINGED PLOVER.)

Charadrius minor, Wolf and Meyer, Vög. Deutschland, i. p. 1-2 (1805).

The Ringed Plovers are a widely spread and numerous section of the genus. They combine the following characters :—white axillaries, no hind toe, white belly, and a dark subterminal band across the tail. There are two Ringed Plovers in Japan. The Little Ringed Plover is a small species (length of wing from carpal joint

Head of *Charadrius minor*. Natural size.

about 4½ inches), with a slightly graduated tail (outer feathers less than a quarter of an inch shorter than the central ones).

Figures: Dresser, Birds of Europe, vii. pl. 524.

The Little Ringed Plover breeds in all the Japanese Islands. It is probably only a summer visitor to Yezzo, but a resident in Southern Japan. There are several examples in the Swinhoe collection from Hakodadi (Swinhoe, Ibis, 1875, p. 452); three in the Pryer collection from Yokohama; and there are three from the Loo-Choo Islands.

The Little Ringed Plover breeds in various localities from Western Europe to East Siberia, but is only a rare visitor to the British Islands.

307. CHARADRIUS PLACIDUS.
(HODGSON'S RINGED PLOVER.)

Charadrius placidus, Gray, Cat. Mamm. &c. Nepal &c. Hodgson, p. 70 (1863).

Hodgson's Ringed Plover scarcely differs in colour from the Little Ringed Plover, but it is a much larger bird (wing from carpal joint about $5\frac{1}{2}$ inches), with a more graduated tail (outer feathers half an inch shorter than the central ones).

Figures: Swinhoe, Proc. Zool. Soc. 1870, pl. 12.

Head of *Charadrius placidus*. Natural size.

Hodgson's Ringed Plover is a winter visitor to Southern Japan, but probably breeds in Yezzo. There are several examples in the Swinhoe collection obtained by Captain Blakiston at Hakodadi

(Swinhoe, Ibis, 1874, p. 162) ; and there are five in the Pryer collection from Yokohama. I have also four examples collected by Mr. Snow from the latter locality.

It is occasionally seen in autumn on the dry river-courses, and on the lake shores of Central Hondo (Jouy, Proc. United States Nat. Mus. 1883, p. 316).

The range of Hodgson's Ringed Plover extends from Japan westwards across Central China to the Himalayas, at least as far west as Nepal.

308. CHARADRIUS MONGOLICUS.
(MONGOLIAN SAND-PLOVER.)

Charadrius mongolicus, Pallas, Zoogr. Rosso-Asiat. ii. p. 136 (1826).

The Sand-Plovers are a large and widely distributed section of the genus which combine the two characters of having the base of the outer web of the innermost primaries white, and of having no dark subterminal band across the tail. The Mongolian Sand-Plover is one of the larger species (wing from carpal joint 5·4 to 4·9 inches), but it has a short bill (terminal vault ·3 inch or less).

Figures : Middendorff, Sibir. Reise, ii. pl. 19. figs. 2, 3.

Head of *Charadrius mongolicus*. Natural size.

The Mongolian Sand-Plover was originally described by Pallas from examples obtained in Siberia, Mongolia, and the Kurile Islands. It passes in considerable numbers along the shores of the Japanese seas both on spring and on autumn migration. I have an example obtained by Mr. Snow on the Kurile Islands (Blakiston and Pryer, Trans. As. Soc. Japan, 1882, p. 108), and there is an example in the

Swinhoe collection obtained by Captain Blakiston at Hakodadi in September (Seebohm, Ibis, 1879, p. 25). There are five examples in the Pryer collection from Yokohama, and I have three examples procured by Mr. Owston in the Yokohama market on the 21st of April.

The breeding-range of the Mongolian Sand-Plover extends from Eastern Turkestan to the valley of the Amoor.

309. CHARADRIUS CANTIANUS.
(KENTISH PLOVER.)

Charadrius cantianus, Latham, Index Orn. Suppl. p. lxvi (1801).

The Kentish Plover belongs to the section of the genus which I have called Sand-Plovers, of which it is one of the smaller species (wing from carpal joint 4·6 to 4·1 inches). The Chinese form of

Charadrius cantianus.

the Kentish Plover is found in Japan, and only differs from the typical form in having pale legs.

Figures: Dresser, Birds of Europe, vii. pl. 523 (typical form with dark legs).

The Kentish Plover appears to be only a summer visitor to Yezzo, but to be a resident in Southern Japan, whence its range extends southwards to Formosa, South China, and Hainan. There is no example from Yezzo in the Swinhoe collection, but it is probable that the typical dark-legged form is found there (Swinhoe, Ibis, 1875, p. 452). In the Pryer collection there are nine examples of the pale-legged form from Yokohama, and one of the dark-legged form unfortunately without any locality affixed.

The pale-legged race has been described as specifically distinct under the name of *Ægialites dealbatus* (Swinhoe, Proc. Zool. Soc. 1870, p. 138), but there can be little doubt that the two races intergrade, and that the Chinese race ought to be called *Charadrius cantianus dealbatus*.

310. CHARADRIUS GEOFFROYI.
(GEOFFROY'S SAND-PLOVER.)

Charadrius geoffroyi, Wagler, Syst. Av. p. 61 (1827).

Geoffroy's Sand-Plover is one of the larger species of the section (wing from carpal joint 5¾ to 5¼ inches), with a long thick bill (length from frontal feathers ·95 to ·8 inch, of which the terminal vault occupies about half).

Figures: Harting, Ibis, 1870, pl. 11; Dresser, Birds of Europe, vii. pl. 521.

Head of *Charadrius geoffroyi*. Natural size.

Geoffroy's Sand-Plover is a very rare accidental straggler to Japan, and the only authority for its occurrence in any of the Japanese islands (that I can find) is a statement (Harting, Ibis, 1870, p. 379) founded upon an example obtained by Mr. Whitely from Japan.

It is a tropical species, breeding in Formosa and Hainan, and possibly in the Red Sea.

311. LOBIVANELLUS CINEREUS.
(GREY-HEADED WATTLED LAPWING.)

Pluvianus cinereus, Blyth, Journ. As. Soc. Beng. xi. p. 587 (1842).

This species differs from every other Japanese bird in having a small wattle between the bill and the eye. No other Wattled Lapwing has both white secondaries and a hind toe.

Figures: Temminck and Schlegel, Fauna Japonica, Aves, pl. 63.

The Grey-headed Wattled Lapwing is not known to have occurred in Yezzo (Blakiston and Pryer, Ibis, 1878, p. 219), but is a not uncommon resident in Southern Japan. There are five examples in the Pryer collection from Yokohama.

Eggs in the Pryer collection measure 1·95 by 1·35 inch, and exactly resemble eggs of the Common Lapwing. They are said to be laid in April, on the grass-ridges between the paddy-fields. The cock guards the sitting hen vigilantly, driving off the birds of prey with loud laughing cries.

CHARADRIUS VEREDUS.
(ORIENTAL DOTTEREL.)

Charadrius veredus, Gould, Proc. Zool. Soc. 1848, p. 38.

The Oriental Dotterel has grey axillaries like the Common Dotterel and the Asiatic Golden Plover. Its white belly distinguishes it from either of these species in summer plumage, but at all ages and seasons it is remarkable for its small foot (middle toe without the claw shorter than the bill from the frontal feathers, and less than half the length of the tarsus).

Figures: Gould, Birds of Australia, vi. pl. 14; Harting, Ibis, 1870, pl. 6.

The only claim of the Oriental Dotterel to be regarded as a Japanese bird is a skin in Dresser's collection obtained by Capt. Conrad at Saigon, and a skin in the British Museum labelled Japan. As Saigon is in Cochin China, and as the skin in the British Museum is also marked "Celebes, Wallace," this species must be struck out of the Japanese List for the present.

The Oriental Dotterel breeds in Mongolia.

The breeding-range of the Grey-headed Wattled Lapwing extends westwards from Japan across North China to South-east Mongolia. As it is not known to have occurred either in Siberia or Yezzo it is fair to assume that it found its way to Southern Japan *viâ* Formosa or the Corea.

312. VANELLUS CRISTATUS.
(COMMON LAPWING.)

Vanellus cristatus, Wolf and Meyer, Vög. Deutschland, ii. p. 110 (1805).

The Common Lapwing has both the upper and under tail-coverts of a chestnut-buff colour.

Vanellus cristatus.

Figures: Dresser, Birds of Europe, vii. pl. 531.

The Common Lapwing is found in all the Japanese Islands, but it is not known whether any remain to breed. There are four examples in the Pryer collection from Yokohama. It is said to be very rare in Yezzo (Blakiston and Pryer, Trans. As. Soc. Japan, 1882, p. 108), but Dr. Siebold found it very common at Nagasaki (Temminck and Schlegel, Fauna Japonica, Aves, p. 106).

The range of the Common Lapwing extends from the British Islands across Europe and Southern Siberia to Japan.

313. HÆMATOPUS OSCULANS.
(JAPANESE OYSTERCATCHER.)

Hæmatopus osculans, Swinhoe, Proc. Zool. Soc. 1871, p. 405.

The Japanese Oystercatcher is an intermediate form between the European and the Australian Pied Oystercatchers, but it is more nearly allied to the former than to the latter species. The white on the outside web of the primaries appears on the third quill of the European species, not until the sixth quill of the Japanese species, and on none of the quills of the Australian species.

The occurrence of a Pied Oystercatcher on the Kurile Islands has long been known (Pallas, Zoogr. Rosso-Asiat. ii. p. 129), and the Japanese Oystercatcher has recently been found to be a resident on the Japanese coasts. I have an example obtained by Mr. Snow on the Kurile Islands, where it is possibly only a summer visitor. It is found on the coasts of Yezzo, but not in great abundance (Blakiston and Pryer, Ibis, 1878, p. 219). There are eight examples in the Pryer collection from Yokohama.

The range of the Japanese Oystercatcher extends northwards to Kamtschatka, westwards to the lower valley of the Amoor, and southwards to the east coast of China.

314. HÆMATOPUS NIGER.
(NORTH-AMERICAN BLACK OYSTERCATCHER.)

Hæmatopus niger, Pallas, Zoogr. Rosso-Asiat. ii. p. 131 (1826).

This species of Black Oystercatcher is of a uniform blackish-brown colour, with a vermilion-red bill and flesh-coloured legs.

Figures: Audubon, Orn. Biogr. v. pl. 427; Audubon, Birds of America, v. pl. 325.

The North-American Black Oystercatcher is said (Pallas, Zoogr. Rosso-Asiat. ii. p. 131) to be a common bird on the Kurile Islands, but I have never seen any examples from the Asiatic continent. I

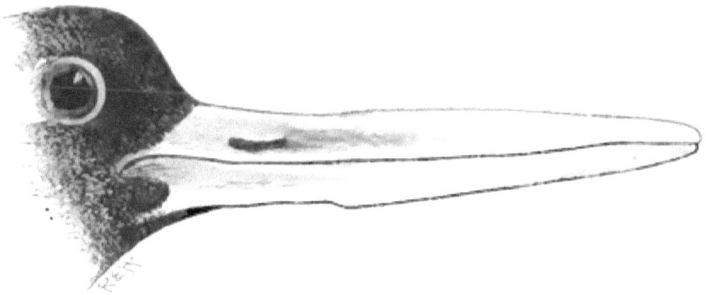

Bill of *Hæmatopus niger*. Natural size.

have a skin procured by Wossnesensky on the Aleutian Islands, where it is known as a summer visitor. Its breeding-range extends along the southern coast of Alaska as far south as the coast of Upper California. It winters on the coast of Lower California.

315. NUMENIUS ARQUATUS.
(COMMON CURLEW.)

Scolopax arquata, Linnæus, Syst. Nat. i. p. 242 (1766).

The Eastern race of the Common Curlew completely intergrades with the Western or typical race, but it was described as distinct as long ago as 1829 under the name of *Numenius lineatus* (Cuvier, Règ. An. i. p. 521). It can, however, only be regarded as subspecifically distinct under the name of *Numenius arquatus lineatus*. It is a large bird (tarsus more than three inches long), and it has a white lower back and rump.

It differs from the European form in having the lower back and the axillaries without any dark markings, in having whiter margins to the scapulars and the feathers of the upper back, and in having, on an average, a longer bill (varying from $5\frac{1}{2}$ to 8 instead of from $4\frac{1}{2}$ to 7 inches).

The Oriental race of the Common Curlew was probably found by

Steller in the Kurile Islands (Pallas, Zoogr. Rosso-Asiat. ii. p. 168). It appears to be a resident on the coasts of Southern Japan and a summer visitor to Yezzo. There are no examples in the Swinhoe

Numenius arquatus.

collection from Hakodadi; but there are six in the Pryer collection from Yokohama. I have also an example obtained by Mr. Owston in the Yokohama market in April; and it has been recorded from the Loo-Choo Islands (Cassin, Proc. Acad. Nat. Sc. Philad. 1862, p. 321).

The range of the Curlew extends from the British Islands across Europe and Southern Siberia to Japan, but, as already explained, eastern examples differ slightly from western ones.

316. NUMENIUS CYANOPUS.
(AUSTRALIAN CURLEW.)

Numenius cyanopus, Vieillot, N. Dict. d'Hist. Nat. viii. p. 306 (1817).

The Australian Curlew is a large bird (tarsus more than three inches long); but the lower back and rump scarcely differ in colour from the upper back.

Figures: Temminck and Schlegel, Fauna Japonica, Aves, pl. 66 (the description applies to the preceding species).

The Australian Curlew is a summer visitor to the Japanese Islands. There are three examples in the Swinhoe collection from Hakodadi (Swinhoe, Ibis, 1876, p. 334), and five in the Pryer collection from Yokohama.

The Australian Curlew breeds in Eastern Siberia and winters in Australia.

317. NUMENIUS PHÆOPUS.
(COMMON WHIMBREL.)

Scolopax phæopus, Linnæus, Syst. Nat. i. p. 243 (1766).

The Eastern form of the Common Whimbrel completely intergrades with the Western or typical form, and was described as distinct as long ago as 1786, under the name of *Tantalus variegatus* (Scopoli, Del. Fl. Faun. Ins. ii. p. 92).

Numenius phæopus.

The Eastern form of the Common Whimbrel is a small bird (tarsus about 2¼ inches), with the lower back much paler in colour than the upper back. It differs from the European form in having the lower back much more profusely streaked with brown, but can only claim subspecific rank under the name of *Numenius phæopus variegatus*.

Figures: Gould, Birds of Australia, vi. pl. 43.

The Eastern form of the Whimbrel is a migratory species which passes along the Japanese coasts in some numbers both in spring and in autumn. I have an example procured by Mr. Henson at Hakodadi on the 5th of October, and there is an example in the Swinhoe collection, also from Hakodadi, procured by Captain Blakiston on the 24th of May (Swinhoe, Ibis, 1877, p. 146). The Perry Expedition obtained two examples from the same locality in 1854 (Cassin, Exp. Am. Squad. China Seas and Japan, ii. p. 228). There are three examples in the Pryer collection from Yokohama; and I have an example procured by Mr. Owston in the Yokohama market on the 22nd of May. Mr. Ringer has obtained it at Nagasaki (Blakiston, Amended List of the Birds of Japan, p. 40).

The Whimbrel breeds in the Arctic Regions of the Old World, and visits the British Islands as well as Japan in spring and autumn, but the partial isolation caused by these widely separated winter-quarters appears to have given rise to a partial differentiation of the western from the eastern birds.

318. NUMENIUS MINUTUS.
(LEAST WHIMBREL.)

Numenius minutus, Gould, Proc. Zool. Soc. 1840, p. 176.

The Least Whimbrel has scarcely any trace of pale bars on the inner webs of its primaries and secondaries, and the back of the tarsus is scutellated as distinctly as the front.

Figures: Temminck and Schlegel, Fauna Japonica, Aves, pl. 67.

The Least Whimbrel is a very rare visitor on migration to Japan. Its sole claim to be regarded as a Japanese bird is the single example figured in the 'Fauna Japonica,' and a second obtained on the 3rd of October, 1883, at Giotoku, near Yokohama (Seebohm, Ibis, 1885, p. 363).

The Least Whimbrel breeds in Eastern Siberia.

319. PHALAROPUS FULICARIUS.
(GREY PHALAROPE.)

Tringa fulicaria, Linnæus, Syst. Nat. i. p. 249 (1766).

Two species of Phalarope, or lobed-footed Sandpipers, occur within Japanese territory. The Grey Phalarope has a short wide bill and a much graduated tail, the central feathers being generally more than half an inch longer than the outermost.

Figures: Dresser, Birds of Europe, vii. pl. 538.

Phalaropus fulicarius.

The Grey Phalarope is a winter visitor to the Kurile Islands, whence it was procured by Mons. Merk (Pallas, Zoogr. Rosso-Asiat. ii. p. 205); but it has not yet been recorded from Japan proper. I have an example collected on the Kurile Islands by Mr. Snow (Seebohm, Ibis, 1884, p. 33).

The Grey Phalarope is a circumpolar species, breeding in the Arctic Regions of both continents, and occasionally visiting the British Islands in winter.

320. PHALAROPUS HYPERBOREUS.
(RED-NECKED PHALAROPE.)

Tringa hyperborea, Linnæus, Syst. Nat. i. p. 249 (1766).

The Red-necked Phalarope has a short slender bill, gradually tapering to the point.

Figures: Dresser, Birds of Europe, vii. pl. 537.

The Red-necked Phalarope is a winter visitor to the Japanese Islands. I have two examples procured by Mr. Snow on the Kuriles (Blakiston and Pryer, Trans. As. Soc. Japan, 1882, p. 113); and there

Phalaropus hyperboreus.

is an example from Hakodadi in the Swinhoe collection (Swinhoe, Ibis, 1875, p. 455), whence several examples, the first recorded from Japan, were procured by the Perry Expedition in 1854 (Cassin, Exp. Am. Squad. China Seas and Japan, ii. p. 230). There are four examples in the Pryer collection from Yokohama, and it has been obtained on the Loo-Choo Islands (Stejneger, Proc. United States Nat. Mus. 1887, p. 394).

The Red-necked Phalarope is a circumpolar bird, breeding in the Arctic Regions of both continents and in Scotland.

321. TOTANUS FUSCUS.
(DUSKY REDSHANK.)

Scolopax fusca, Linneus, Syst. Nat. i. p. 243 (1766).

The Dusky Redshank belongs to a small group of partially web-footed Sandpipers, which have the lower back and rump unstreaked white. It has white secondaries transversely barred with grey.

Figures: Dresser, Birds of Europe, viii. pl. 568. fig. 2 (summer plumage), pl. 569. fig. 1 (winter plumage), pl. 567. fig. 2 (young in first plumage).

The Dusky Redshank, sometimes called the Spotted Redshank, is a winter visitor to the Japanese coasts. There are several examples in the Swinhoe collection from Hakodadi (Swinhoe, Ibis, 1875,

Totanus fuscus.

p. 453), and nine in the Pryor collection from Yokohama. It is recorded by Captain Blakiston in the 'Chrysanthemum' as having been obtained by Mr. Snow on the Kurile Islands.

The Dusky Redshank breeds on the tundras of the Old World from Lapland to Bering Straits. It visits the British Islands in winter as well as Japan.

322. TOTANUS CALIDRIS.
(COMMON REDSHANK.)

Scolopax calidris, Linneus, Syst. Nat. i. p. 245 (1766).

The Common Redshank has a white lower back and rump, but its secondaries are nearly uniform white.

Figures: Dresser, Birds of Europe, viii. pl. 567. fig. 1 (summer plumage), pl. 569. fig. 2 (winter plumage), pl. 568. fig. 1 (young in first plumage).

The Common Redshank has only very recently been added to the list of Japanese birds; but there can be little doubt that it is a frequent visitor on migration, since the Japanese Islands lie on the

Totanus calidris.

direct route from its most easterly breeding-grounds in Siberia to the Malay Archipelago, where it is known to winter. A young male in first plumage was obtained at Giotoku, near Yokohama, on the 4th of September, 1883 (Seebohm, Ibis, 1885, p. 363).

The breeding-range of the Common Redshank extends from the British Islands across Europe and Southern Siberia to the Sea of Okhotsk, and possibly to Yezzo.

323. TOTANUS GLOTTIS.
(GREENSHANK.)

Scolopax glottis, Linneus, Syst. Nat. i. p. 245 (1766).

The Greenshank has a white lower back and rump; but its secondaries are nearly uniform grey. It is a large bird (wing from carpal joint about 7 inches).

Figures: Dresser, Birds of Europe, viii. pl. 570.

The Greenshank is a spring and autumn visitor to the coasts of the Japanese islands. Dr. Henderson obtained it at Hakodadi in October 1857 (Cassin, Proc. Acad. Nat. Sc. Philad. 1858, p. 196).

Totanus glottis.

There are several examples in the Swinhoe collection from Hakodadi (Swinhoe, Ibis, 1875, p. 453), and there are ten in the Pryer collection from Yokohama.

The Greenshank breeds in the Arctic Regions from Scotland across Lapland and Siberia to Kamtschatka.

324. TOTANUS STAGNATILIS.
(MARSH-SANDPIPER.)

Totanus stagnatilis, Bechstein, Orn. Taschenb. ii. p. 292 (1803).

The Marsh-Sandpiper has a white lower back and rump, and nearly uniform grey secondaries, like the Greenshank; but it is a smaller bird (wing from carpal joint about $5\frac{1}{2}$ inches long).

Figures: Dresser, Birds of Europe, viii. pl. 566.

The Marsh-Sandpiper is a very rare visitor on migration to the Japanese islands. The only example that I know of from Japan is the one described from Mr. Owston's collection (Blakiston and Pryer, Trans. As. Soc. Japan, 1882, p. 109, no. $95\frac{1}{2}$). It was probably obtained in the Yokohama market.

The breeding-range of the Marsh-Sandpiper extends from the delta of the Rhone, across Europe and Southern Siberia, and it is said to have once occurred in the British Islands (Littleboy, Trans. Hertfordshire Nat. Hist. Soc. 1888, p. 78).

325. TOTANUS INCANUS.

(WANDERING TATTLER.)

Scolopax incana, Gmelin, Syst. Nat. i. p. 658 (1788).

The Wandering Tattler has grey lower back, rump, and upper tail-coverts, and unbarred dark-grey axillaries.

Figures: Temminck and Schlegel, Fauna Japonica, Aves, pl. 65.

The Asiatic race of the Wandering Tattler passes along the shores of the Japanese islands in considerable numbers, both in spring and autumn. I have an example obtained by Mr. Snow on the Kurile Islands (Blakiston and Pryer, Trans. As. Soc. Japan, 1882, p. 109). Mr. Heine, the artist of the Perry Expedition, says that it was frequently seen on the sandy beach of the bay of Hakodadi, where examples were obtained in May 1854 (Cassin, Exp. Am. Squad. China Seas and Japan, ii. p. 229). There are four examples in the Swinhoe collection from the same locality (Swinhoe, Ibis, 1874, p. 163), and eleven in the Pryer collection from Yokohama.

Captain Rodgers procured it on the Bonin Islands (Cassin, Proc. Acad. Nat. Sc. Philad. 1862, p. 321); and there are four examples in the Pryer collection from the Loo-Choo Islands (Seebohm, Ibis, 1887, p. 180).

The Asiatic race of the Wandering Tattler breeds in Eastern Siberia and winters in Australia.

Mr. Holst sent me an example from Peel Island, in the centre of the Bonin group. All these records appear to refer to the Asiatic race of this species, which may be regarded as subspecifically distinct under the name of *Totanus incanus brevipes*, which was described in 1817 under the name of *Totanus brevipes* (Vieillot, N. Dict. d'Hist. Nat. vi. p. 410).

A second example sent by Mr. Holst from Peel Island appears to belong to the typical form. The whole of the underparts are barred; the nasal groove extends for two thirds of the length of the bill; and the back of the tarsus is very imperfectly scutellated. The American Wandering Tattler must therefore be regarded as a rare wanderer on migration to the Bonin Islands (Seebohm, Ibis, 1890, p. 104).

326. TOTANUS GLAREOLA.
(WOOD-SANDPIPER.)

Tringa glareola, Linneus, Syst. Nat. i. p. 149 (1758); Gmelin, Syst. Nat. i. p. 677 (1788).

The Wood-Sandpiper is a small bird (wing from carpal joint about 5 inches); the lower back is nearly the same colour as the mantle; and the predominant colour of the upper tail-coverts, axillaries, and under wing-coverts is white.

Figures: Dresser, Birds of Europe, viii. pl. 565.

Totanus glareola.

The Wood-Sandpiper passes along the Japanese coasts both on spring and on autumn migration. It is recorded from the Kurile Islands (Blakiston and Pryer, Trans. As. Soc. Japan, 1882, p. 110), and I have an example collected at Hakodadi on the 24th of August by Mr. Henson, and a second from the same locality collected by Mr. Snow in September. There is also an example in the Swinhoe collection procured by Captain Blakiston in May at Hakodadi (Swinhoe, Ibis, 1874, p. 163), and two in the Pryer collection from Yokohama. Mr. Ringer has procured it at Nagasaki, where the examples obtained by the Siebold Expedition were probably also procured (Temminck and Schlegel, Fauna Japonica, Aves, p. 110).

The Wood-Sandpiper breeds in the Arctic and Subarctic Regions of Europe and Siberia, and passes along the British coasts as well as those of Japan on its migrations.

327. TOTANUS OCHROPUS.
(GREEN SANDPIPER.)

Tringa ochropus, Linneus, Syst. Nat. i. p. 250 (1766).

The Green Sandpiper has white upper tail-coverts and brown axillaries narrowly barred with white.

Figures: Dresser, Birds of Europe, viii. pl. 564.

Totanus ochropus.

The Green Sandpiper is a winter visitor to Japan. There is an example in the Swinhoe collection from Hakodadi, where it was procured by Captain Blakiston (Swinhoe, Ibis, 1875, p. 453); and there are five examples in the Pryer collection from Yokohama. Mr. Ringer has procured it at Nagasaki, where the examples obtained by the Siebold Expedition were probably also procured (Temminck and Schlegel, Fauna Japonica, Aves, p. 110).

The Green Sandpiper breeds in the subarctic regions of the Old World, visiting the British Islands as well as Japan on its migrations.

328. TOTANUS TEREKIUS.
(TEREK SANDPIPER.)

Scolopax terek, Latham, Index Orn. ii. p. 724 (1790).

The Terek Sandpiper has no white on the primaries or rump, a great deal of white on the secondaries, and nothing but white on the axillaries. Its beak is recurved like that of the Greenshank.

Figures: Dresser, Birds of Europe, viii. pl. 572.

The Terek Sandpiper is probably a spring and autumn visitor on migration to all the Japanese Islands, though it has not yet been recorded from Yezzo. It is occasionally found in the Yokohama market (Seebohm, Ibis, 1884, p. 33), whence I have two examples in the Pryer collection, and a third collected by Mr. Owston.

The Terek Sandpiper breeds in the Arctic Regions from Archangel to Kamtschatka, but is not known to have visited the British Islands.

329. TOTANUS HYPOLEUCUS.
(COMMON SANDPIPER.)

Tringa hypoleucos, Linnæus, Syst. Nat. i. p. 250 (1766).

The Common Sandpiper has white axillaries, large patches of white on most of the primaries and secondaries, but no white on the rump or upper tail-coverts.

Figures: Dresser, Birds of Europe, viii. pl. 563.

The Common Sandpiper is probably a summer visitor to Yezzo, and a resident in the more southerly Japanese Islands. It has been seen on Eturop, the most southerly of the Kuriles (Blakiston and Pryer, Trans. As. Soc. Japan, 1882, p. 140), and there are several examples in the Swinhoe collection from Hakodadi (Swinhoe, Ibis,

1874, p. 163). There are five examples in the Pryer collection from Yokohama, and Mr. Ringer has sent an example to the Norwich Museum procured at Nagasaki (Blakiston, Am. List Birds of Japan, p. 36), where the examples procured by the Siebold Expedition were

Totanus hypoleucus.

probably obtained (Temminck and Schlegel, Fauna Japonica, Aves, p. 108). Captain Rodgers procured it on the Loo-Choo Islands, and also on one of the Bonin Islands * (Cassin, Proc. Acad. Nat. Sc. Philad. 1862, p. 322). Mr. Holst has lately sent an example from Peel Island (Seebohm, Ibis, 1890, p. 104).

The breeding-range of the Common Sandpiper extends from the British Islands across Europe and Siberia to Japan.

330. TOTANUS PUGNAX.
(RUFF.)

Tringa pugnax, Linnæus, Syst. Nat. i. p. 247 (1766).

The Ruff has white axillaries, but no white on the primaries, secondaries, or central upper tail-coverts.

Figures: Dresser, Birds of Europe, viii. pls. 557, 558.

---

* The Bonin Island example is recorded under the name of *Tringoides empusa* of Gould.

The Ruff is a very rare visitor on migration to the Japanese islands (Blakiston and Pryer, Ibis, 1878, p. 221). A single example has occurred in Yezzo (Seebohm, Ibis, 1884, p. 33); there is a single example in the Pryer collection from the Yokohama market, and two others were obtained in the neighbourhood of Yokohama on the 13th of October (Seebohm, Ibis, 1885, p. 364).

The Ruff is a winter visitor to the British Islands, its breeding-range extending across Europe and Siberia to Kamtschatka.

331. LIMOSA RUFA.
(BAR-TAILED GODWIT.)

Limosa rufa, Brisson, Orn. v. p. 281 (1760); Leach, Syst. Cat. Mamm. &c. Brit. Mus. p. 32 (1816).

The Eastern race of the Bar-tailed Godwit has been described as distinct from the Western race, under the name of *Limosa uropygialis* (Gould, Proc. Zool. Soc. 1848, p. 38).

In the Bar-tailed Godwit the basal half of the tail-feathers does not differ much from the terminal half in colour. The Eastern form

Limosa rufa.

of the Bar-tailed Godwit differs from the European form in having the lower back and rump much more profusely streaked with brown; but as the two forms completely intergrade, it can only be regarded as subspecifically distinct, under the name of *Limosa rufa uropygialis*.

Figures: Gould, Birds of Australia, vi. pl. 29 (winter plumage).

The Siberian form of the Bar-tailed Godwit was found on the Kurile Islands by Steller (Pallas, Zoogr. Rosso-Asiat. ii. p. 181), and passes the coasts of the Japanese Islands in some numbers in spring and autumn. I have an example from the Kurile Islands collected by Mr. Snow in July.

Dr. Henderson obtained it at Hakodadi in October 1857 (Cassin, Proc. Acad. Nat. Sc. Philad. 1858, p. 196), and there are several examples in the Swinhoe collection from Hakodadi (Swinhoe, Ibis, 1875, p. 453), and eight in the Pryer collection from Yokohama. It was obtained by the Siebold Expedition probably near Nagasaki (Temminck and Schlegel, Fauna Japonica, Aves, p. 114).

The Bar-tailed Godwit breeds on the tundras of Lapland and Siberia, visiting the British coasts as well as those of Japan on its migration. Eastern examples differ slightly as described from Western ones.

332. LIMOSA MELANURA.
(BLACK-TAILED GODWIT.)

Limosa melanura, Leisler, Nachtr. Bechst. Naturg. Deutschl. ii. p. 153 (1813).

The Eastern form of the Black-tailed Godwit has been described as a distinct species, under the name of *Limosa melanuroides* (Gould, Proc. Zool. Soc. 1846, p. 84); but as it intergrades with the Western form it can only be regarded as subspecifically distinct under the name of *Limosa melanura melanuroides*.

The Black-tailed Godwit has the terminal portion of the tail-feathers black in strong contrast to the basal half, which is pure white.

The Eastern form of the Black-tailed Godwit is rather smaller than the European form (wing from carpal joint 7 to 8 instead of 8 to 9 inches; tarsus $2\frac{1}{4}$ to 3 instead of 3 to $3\frac{3}{4}$ inches).

Figures: Gould, Birds of Australia, vi. pl. 28 (winter plumage).

The Siberian form of the Black-tailed Godwit passes in spring and autumn on migration along the Japanese coasts. There is an example in the Swinhoe collection from Hakodadi (Swinhoe, Ibis, 1875, p. 453), and six in the Pryer collection from Yokohama. The examples procured by the Siebold Expedition were probably obtained near Nagasaki (Temminck and Schlegel, Fauna Japonica, Aves, p. 114).

Limosa melanura.

The Black-tailed Godwit breeds in subarctic Europe and Siberia, visiting on its migrations the British coasts as well as those of Japan. Eastern examples are slightly smaller than Western ones.

333. MACRORHAMPHUS GRISEUS.
(AMERICAN SNIPE-BILLED SANDPIPER.)

Scolopax grisea, Gmelin, Syst. Nat. i. p. 658 (1788).

The Alaskan or Western race of the American Snipe-billed Sandpiper was described in 1833 as distinct from the Canadian or Eastern race under the name of *Limosa scolopaceus* (Say, Long's Exped. ii. p. 170).

The Canadian and Alaskan Snipe-billed Sandpipers resemble small Bar-tailed Godwits, with the bills of Snipes.

The Alaskan form is on an average a slightly larger bird than the typical Canadian form, and may be regarded as subspecifically distinct under the name of *Macrorhamphus griseus scolopaceus*.

Figures: Lawrence, Ann. Lyc. New York, v. pl. 1.

The Alaskan form of the American Snipe-billed Sandpiper has occurred twice in Japan. There is one example in the Blakiston collection, shot amongst a flock of Eastern Golden Plover in Yezzo on the 13th of October (Swinhoe, Ibis, 1875, p. 454), which I have

M. scolopaceus (winter plumage). *M. griseus* (summer plumage).

examined (Seebohm, Ibis, 1884, p. 33); and I have a second example, procured by Mr. Owston in the Yokohama market on the 13th of March.

The Alaskan form of the American Snipe-billed Sandpiper breeds in the Arctic Regions of America west of the Rocky Mountains.

334. STREPSILAS INTERPRES.
(TURNSTONE.)

Tringa interpres, Linneus, Syst. Nat. i. p. 248 (1766).

The Turnstone is peculiar in having a dark rump between a white lower back and white upper tail-coverts. It combines the cleft toes of the Snipes with the position of the nasal aperture (extending beyond the basal fourth of the bill) of the Plovers.

Figures: Dresser, Birds of Europe, vii. pl. 532.

The Turnstone probably breeds on the Kuriles, whence I have two examples obtained by Mr. Snow (Blakiston and Pryer, Trans. As. Soc. Japan, 1882, p. 109), and is undoubtedly a winter visitor to the southern Japanese Islands (Swinhoe, Ibis, 1876, p. 334). There

Strepsilas interpres.

are seven examples in the Pryer collection from Yokohama. I have not seen an example from Yezzo, but it is said to be occasionally found on migration on the coasts of that island (Blakiston and Pryer, Ibis, 1878, p. 219).

The Turnstone is a circumpolar bird, and is as common in winter on the British coasts as it is on those of Japan.

335. TRINGA CRASSIROSTRIS.
(JAPANESE KNOT.)

Tringa crassirostris, Temminck and Schlegel, Fauna Japonica, Aves, p. 107 (1847).

Amongst the cleft-toed Sandpipers the Japanese Knot is the only one which has white on the upper tail-coverts, and a straight bill more than an inch and a half long.

Figures: Temminck and Schlegel, Fauna Japonica, Aves, pl. 64 (summer and winter plumage).

The Japanese Knot, so called because it was originally discovered in Japan, is only a spring and autumn visitor on migration to the islands whose name it bears. I have an example obtained by Mr. Snow on the Kurile Islands. It was first obtained in Yezzo in October 1857, by Dr. Henderson (Cassin, Proc. Acad. Nat. Sc. Philad. 1858, p. 196), and afterwards in 1861 (Blakiston, Ibis, 1862, p. 330), and there are seven examples in the Pryer collection from Yokohama.

The Japanese Knot probably breeds somewhere in Eastern Siberia.

336. TRINGA CANUTUS.
(KNOT.)

Tringa canutus, Linnæus, Syst. Nat. i. p. 251 (1766).

Four of the cleft-toed Sandpipers have the ground-colour of the upper tail-coverts white. Two of these are smaller birds (wing from carpal joint less than $5\frac{1}{4}$ inches), but the third, the Japanese Knot,

Tringa canutus.

is larger (wing 7·6 to 7 inches, instead of 6·8 to 6·2 inches; bill from frontal feathers 1·8 to 1·6, instead of 1·5 to 1·1 inch).

Figures: Dresser, Birds of Europe, viii. pls. 555, 556.

The Knot passes the Japanese coasts on its spring and autumn migrations from its arctic breeding-grounds to its tropic winter-quarters, but apparently not in very great numbers. It has not yet been recorded from Yezzo, but it is occasionally found in the Yokohama market (Seebohm, Ibis, 1884, p. 34), whence I have an example in the Pryer collection, and a second obtained from Mr. Owston.

The Knot also passes the British coasts on its migrations to its arctic breeding-grounds. It may be regarded as a circumpolar species.

337. TRINGA ALPINA.
(DUNLIN.)

Tringa alpina, Linneus, Syst. Nat. i. p. 249 (1766).

The Pacific race of the Dunlin has been described as distinct from the European race under the name of *Pelidna pacifica* (Coues, Proc. Acad. Nat. Sc. Philad. 1861, p. 189).

The Dunlin belongs to the section of cleft-toed Sandpipers which have a great deal of white on the seventh, eighth, and ninth secondaries, but little or none on the central upper tail-coverts. Its hind toe and its black legs distinguish it from the other species in the section.

Tringa alpina.

Examples from East Asia and America are on an average slightly larger birds, and may be recognized as subspecifically distinct under the name of *Tringa alpina pacifica*.

Figures: Dresser, Birds of Europe, viii. pl. 548.

The Pacific race of the Dunlin passes in some numbers along the Japanese coasts in spring and autumn on migration. I have six examples procured by Mr. Snow on the Kuriles (Blakiston and Pryer, Trans. As. Soc. Japan, 1882, p. 111), where it probably breeds. Dr. Henderson obtained it in Yezzo in October 1857 (Cassin, Proc. Acad. Nat. Sc. Philad. 1858, p. 196), and there are several examples in the Swinhoe collection from Hakodadi (Swinhoe, Ibis, 1875, p. 455). There are eleven examples in the Pryer collection from Yokohama, where a few probably remain during winter. Mr. Ringer has procured it at Nagasaki, where it was also obtained by the Siebold Expedition (Temminck and Schlegel, Fauna Japonica, Aves, p. 108).

The Dunlin is a circumpolar species, and visits the British Islands in great numbers, but European examples are, on an average, slightly smaller than those from Asia and America.

338. TRINGA MARITIMA.
(PURPLE SANDPIPER.)

Tringa maritima, Gmelin, Syst. Nat. i. p. 678 (1788).

The Purple Sandpiper has a nearly black rump and upper tail-coverts; but the seventh, eighth, and ninth secondaries are nearly all white.

Tringa maritima.

Figures: Dresser, Birds of Europe, viii. pl. 554.

The Purple Sandpiper is a rare winter visitor to the Kurile Islands, whence it was obtained by Mons. Merk (Pallas, Zoogr. Rosso-Asiat. ii. p. 190). I can find no record of its occurrence in Japan, but I have an example obtained by Wossnesensky on Urup, one of the Kurile Islands, and there are two examples in the Pryer collection, obtained by Mr. Snow on one of the islands of that group.

The Purple Sandpiper is a circumpolar species, and is a winter visitor to the British Islands.

339. TRINGA ARENARIA.
(SANDERLING.)

Tringa arenaria, Linneus, Syst. Nat. i. p. 251 (1766).

The Sanderling is the only cleft-toed Sandpiper without a hind toe. Figures: Dresser, Birds of Europe, viii. pls. 559, 560.

The Sanderling appears to be a rare visitor to Japan. There is only one example from Hakodadi in the Swinhoe collection (Swinhoe, Ibis, 1875, p. 454), and although Messrs. Blakiston and Pryer state

Tringa arenaria.

that there are examples from Yokohama in the Hakodadi Museum, there are none in the Pryer collection. I have, however, two examples obtained by Mr. Owston in the Yokohama market on the 1st and 11th of May respectively.

The Sanderling is a circumpolar bird, and is a common visitor to the coasts of the British Islands in spring and autumn.

340. TRINGA PLATYRHYNCHA.
(BROAD-BILLED SANDPIPER.)

Tringa platyrhincha, Temminck, Man. d'Orn. p. 398 (1815).

The Broad-billed Sandpiper has little or no white on the secondaries and upper tail-coverts. Its bill is very flat, slightly widened towards the middle, and more than a fourth of the length of the wing.

Figures: Dresser, Birds of Europe, viii. pl. 545.

The Broad-billed Sandpiper appears to be a rare winter visitor to the Japanese coasts. Captain Blakiston collected an example in August at Hakodadi (Seebohm, Ibis, 1884, p. 33), and there is one in the Pryer collection from the Yokohama market.

The breeding-range of the Broad-billed Sandpiper extends from the Atlantic to the Pacific; and it is a rare winter visitor to Great Britain as well as to Japan.

341. TRINGA MINUTA.
(LITTLE STINT.)

Tringa minuta, Leisler, Nachtr. Bechst. Naturg. Deutschl. i. p. 74 (1812).

The Eastern form of the Little Stint was described as a distinct species as long ago as 1776 under the name of *Trynga ruficollis* (Pallas, Reise Russ. Reichs, iii. p. 700).

The Little Stint is a small bird (wing from carpal joint less than 4 inches), with black legs and feet, and a narrow bill, broadest at the base.

The Eastern form has a chestnut chin, throat, and upper breast in summer plumage, but in winter the two forms are indistinguishable. The Eastern form may be regarded as subspecifically distinct, under the name of *Tringa minuta ruficollis*.

Figures: Seebohm, Charadriidæ, pl. 15.

The Eastern form of the Little Stint, or Red-throated Stint, is common during the spring and autumn migrations on the Japanese coasts. I have four examples procured by Mr. Snow on the Kurile

Islands (Blakiston, Am. List Birds of Japan, p. 37). Dr. Henderson obtained it (for the first time in Japan) at Hakodadi in October 1857 (Cassin, Proc. Acad. Nat. Sc. Philad. 1858, p. 196), and there are several examples in the Swinhoe collection from the same locality (Swinhoe, Ibis, 1875, p. 155). There are ten examples in the Pryer collection from Yokohama, and Mr. Ringer procured it at Nagasaki.

The Little Stint breeds on the tundras of Lapland and Siberia, passing the British coasts as well as those of Japan on its migrations. Eastern examples differ in summer plumage from Western ones.

342. TRINGA SUBMINUTA.
(MIDDENDORFF'S STINT.)

Tringa subminuta, Middendorff, Reise in Nord. u. Ost. Sibir. ii. p. 222 (1853).

Middendorff's Stint is a small bird (wing from carpal joint less than 4 inches), with pale legs and toes, and grey outer tail-feathers.

Figures: Middendorff, Reise Nord. Ost. u. Sibir. ii. pl. 19. fig. 6.

Middendorff's Stint is a visitor on migration to the Japanese coasts. Mr. Snow obtained it on the Kurile Islands (Blakiston, Am. List Birds of Japan, p. 37), and there are several examples in the Swinhoe collection from Hakodadi (Swinhoe, Ibis, 1875, p. 455), and I have six examples collected by Mr. Henson in the same locality between the 20th of August and the 19th of September. There is only one example in the Pryer collection from Yokohama.

Middendorff's Stint breeds in Eastern Siberia, and winters in the islands of the Malay Archipelago.

343. TRINGA PYGMÆA.
(SPOON-BILLED SANDPIPER.)

Platalea pygmæa, Linneus, Syst. Nat. i. p. 231 (1766).

The Spoon-billed Sandpiper resembles the Red-throated Stint in size, colour, and seasonal variations of plumage; but its bill is three times as wide near the tip as it is at the base.

Figures: Harting, Ibis, 1869, pl. 18 (summer plumage); Gray, Genera of Birds, iii. pl. 152. fig. 1 (winter plumage).

The Spoon-billed Sandpiper is an occasional winter visitor to the Japanese coasts. Captain Blakiston obtained it at Hakodadi (Swinhoe, Ibis, 1875, p. 455); I have an example obtained on the 8th of

Head of *Tringa pygmæa*. Natural size.

October by Mr. Henson, also at Hakodadi; and there are three examples in the Pryer collection from Yokohama.

The Spoon-billed Sandpiper breeds in some unknown country north of Bering Straits.

344. TRINGA ACUMINATA.

(SIBERIAN PECTORAL SANDPIPER.)

Totanus acuminatus, Horsfield, Trans. Linn. Soc. xiii. p. 192 (1820).

The Siberian Pectoral Sandpiper is not very small (wing from carpal joint more than $4\frac{3}{4}$ inches). It has pale legs and feet, dark central upper tail-coverts, and little or no white on the secondaries beyond a narrow margin.

Figures: Jardine and Selby, Ill. Orn. ii. pl. 91; Gould, Birds of Australia, vi. pl. 30.

The Siberian form of the Pectoral Sandpiper is a common visitor on spring and autumn migration to the Japanese coasts. Captain

Blakiston procured it at Hakodadi (Swinhoe, Ibis, 1875, p. 455). I have a very fine series collected by Mr. Snow at Yokohama, and there are eight skins in the Pryer collection also from Yokohama. It has occurred at Nagasaki (Blakiston and Pryer, Trans. As. Soc. Japan, 1882, p. 112).

Rectrices of *Tringa acuminata*.

The Siberian form of the Pectoral Sandpiper is confined to East Siberia and Alaska during the breeding-season, and differs very slightly from the Western form, which breeds in the arctic regions of America.

345. RHYNCHÆA CAPENSIS.
(PAINTED SNIPE.)

Scolopax capensis, Linnæus, Syst. Nat. i. p. 246 (1766).

The Painted Snipe has a decurved bill, which is not much longer than the tarsus, but is considerably longer than the difference in length between the shortest and longest primary.

Figures: Milne-Edwards and Grandidier, Hist. Madag. Ois., Atlas, iii. pl. 261; Shelley, Birds of Egypt, pl. 11.

The Painted Snipe has been only once found in Yezzo (Seebohm, Ibis, 1884, p. 178), but is a common resident in Southern Japan. There is an example in the Swinhoe collection from Yokohama (Swinhoe, Ibis, 1877, p. 146); and there are four in the Pryer collection from the same locality. Mr. Ringer has sent examples to the Norwich Museum obtained at Nagasaki (Blakiston and Pryer, Trans. As. Soc. Japan, 1882, p. 122), where it was also procured by the Siebold Expedition (Temminck and Schlegel, Fauna Japonica, Aves, p. 113).

The Painted Snipe breeds in China, India, and South Africa, and is one of the few examples of tropical species which breed in Japan.

The genus SCOLOPAX is a remarkably isolated one, and is easily diagnosed from all other genera of the Gallo-Grallinæ group of birds by well-marked osteological characters. Its affinities are not so close with *Tringa* and the other genera of Limicolæ containing species whose toes, like those of the Snipes, are cleft to the base, as with *Limosa* and *Macrorhamphus*.

In most birds the eye is more or less protected above by an overhanging orbital septum which forms an arch springing from the postfrontal to the lachrymal. In a few genera (*Dendrocygna*, and many genera of the Psittacidæ amongst desmognathous birds) the orbital septum is continued below the eye as well as above it, thus forming a complete ring. This is the case in every species of *Scolopax* which I have been able to examine, including the Woodcock and the Jack Snipe, and is not the case, so far as I know, in any other species of schizognathous birds. The Snipes further differ from *Charadrius*, *Vanellus*, *Totanus*, &c., in having a strongly marked nasal keel to strengthen the upper mandible. It might be regarded as an ossified nasal septum, but it is of a very different character to that found in the Raptores or Striges. It thickens above as it joins the nasal processes of the premaxilla, and behind as it nears the ethmoid. It is in every sense a maxillary keel, and has no connexion whatever with the maxillo-palatines. This maxillary keel is as well-developed in *Limosa* as in *Scolopax*, and nearly as well in *Macrorhamphus*. It is more or less present in *Tringa* and *Ereunetes*, but entirely absent in *Vanellus*, *Charadrius*, and *Totanus*. If it be regarded as a good character, it completely disposes of the importance hitherto attached to the presence or absence of any remains of the webs which probably once connected the toes in all the species comprised in the Limicolæ. In subdividing the Charadriidæ, it seems as if *Strepsilas* and *Rhynchæa* must be expelled from the Scolopacinæ, and *Limosa*, *Macrorhamphus*, and *Ereunetes* admitted.

346. SCOLOPAX AUSTRALIS.
(LATHAM'S SNIPE.)

Scolopax australis, Latham, Index Orn., Suppl. p. lxv (1801).

Latham's Snipe is a large bird (wing from carpal joint $6\frac{1}{2}$ to 6 inches). It has 18 tail-feathers, of which only two on each side are less than ·3 inch in width.

Figures: Gould, Birds of Australia, vi. pl. 40.

Latham's Snipe is a common visitor to the Japanese Islands, probably breeding in Yezzo, and certainly doing so on the mountains of Southern Japan. There is an example from Hakodadi in the Swinhoe collection, procured by Captain Blakiston in May (Swinhoe, Ibis, 1874, p. 163), and there are three examples in the Pryer collection from Yokohama. I have also an example obtained by Mr. Owston in the Yokohama market on the 4th of April.

Latham's Snipe is probably confined to Japan during the breeding-season, but in autumn it passes the Philippine Islands and the coasts of China on its migration to winter in Australia.

347. SCOLOPAX SOLITARIA.
(JAPANESE SOLITARY SNIPE.)

Gallinago solitaria, Hodgson, Proc. Zool. Soc. 1836, p. 8.

The Solitary Snipe is a large bird (wing from carpal joint 6·4 to 6 inches). It has many feathers on the upper parts, especially the outer margins of the scapulars, streaked with white instead of buff.

The typical form is slightly larger (wing 6·4 to 6·8 inches), has fewer bars on the lower breast, broader pale dorsal stripes, and more white marbling near the tips of the primaries.

Figures: Temminck and Schlegel, Fauna Japonica, Aves, pl. 68.

The Japanese race of the Solitary Snipe is a resident in Japan,

I have an example collected at Sapporo in Yezzo, on the 27th of January, by Captain Blakiston, who states that it frequents during winter some spring-water creeks which remain unfrozen in the severest weather (Blakiston and Pryer, Trans. As. Soc. Japan, 1882, p. 114). Mr. Heine, the artist of the Perry Expedition, says that it was not uncommon near Hakodadi, where it kept in the meadows and marshy woods (Cassin, Exp. Am. Squad. China Seas and Japan, ii. p. 227). There are six examples in the Pryer collection from Yokohama; Mr. Ringer has obtained it at Nagasaki, whence he has sent an example to the Norwich Museum.

It has been obtained in Central Hondo in winter, and frequents marshy places, the banks of streams, and soft boggy ground, but is not found in the paddy-fields. It resembles the Woodcock in its habits more than the true Snipe (Jouy, Proc. United States Nat. Mus. 1888, p. 317).

The Japanese race of the Solitary Snipe has been described as a distinct species under the name of *Gallinago japonica* (Swinhoe, Ibis, 1873, p. 364), but it appears completely to intergrade with its continental ally, and can only be regarded as subspecifically distinct under the name of *Scolopax solitaria japonica*.

348. SCOLOPAX MEGALA.
(SWINHOE'S SNIPE.)

Gallinago megala, Swinhoe, Ibis, 1861, p. 343.

Swinhoe's Snipe is a medium-sized bird (wing from carpal joint 5·6 to 5·2 inches). It has 20 tail-feathers, of which 12 (6 on each side) are narrow (varying from ·15 to ·3 inch wide).

Figures: Seebohm, Charadriidæ, p. 479 (woodcut of tail).

Swinhoe's Snipe appears to have been overlooked by Japanese collectors, but it is doubtless a frequent visitor on spring and autumn migration, as there are two examples in the Pryer collection from the Yokohama market, and I have a third example from the same locality obtained by M. Boucard's collector.

Swinhoe's Snipe breeds in South-east Siberia, and winters in the islands of the Malay Archipelago.

349. SCOLOPAX GALLINULA.
(JACK SNIPE.)

Scolopax gallinula, Linnæus, Syst. Nat. i. p. 244 (1766).

The Jack Snipe is the smallest of all the Japanese Snipes (wing from carpal joint 4 to 4·3 inches). It has a purple gloss on its mantle, and metallic green on the inside webs of its scapulars.

Scolopax gallinula.

Figures: Dresser, Birds of Europe, vii. pl. 544.

The Jack Snipe passes the Japanese coasts on its spring and autumn migrations. It has been procured at Hakodadi on the 3rd of October (Whitely, Ibis, 1867, p. 206); and there are five examples in the Pryer collection from Yokohama. I have also three examples obtained by Mr. Owston in Yokohama, one of them in November.

The Jack Snipe is a regular winter visitor to the British Islands, and doubtless breeds across the Arctic Regions of the Old World, though it has not been recorded in Siberia east of the Taimur Peninsula.

350. SCOLOPAX STENURA.
(PINTAIL SNIPE.)

Scolopax stenura, Bonaparte, Ann. Stor. Nat. Bologna, iv. fasc. xiv. p. 335 (1830).

The Pintail Snipe is about the size of the Common Snipe (wing from carpal joint 5·3 to 4·9 inches). It has 26 tail-feathers, of which 16 (8 on each side) are very narrow (not exceeding ·1 inch across).

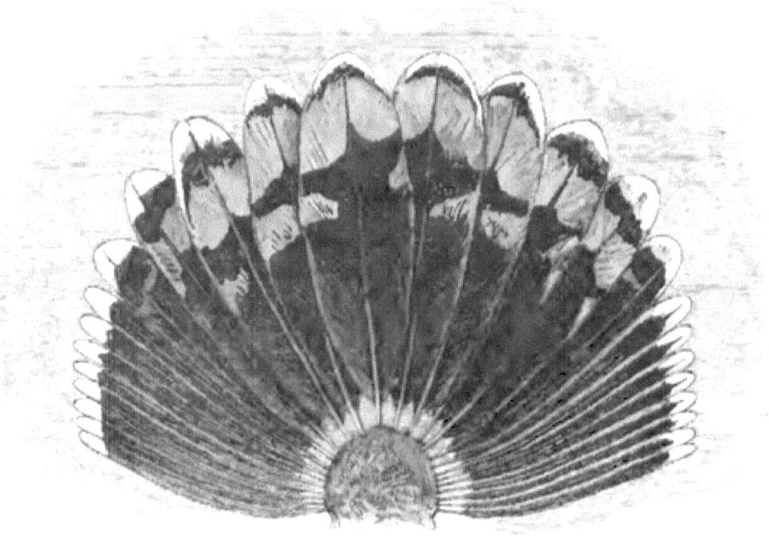

Rectrices of *Scolopax stenura*.

Figures: Radde, Reis. im Süd. von Ost-Sibir. ii. pl. 13; Hume and Marshall, Game-Birds India &c. iii. p. 339; Seebohm, Charadriidæ, p. 477 (woodcut of tail).

The sole claim of the Pintail Snipe to be regarded as a Japanese bird rests on the examples obtained by the Perry Expedition; but as Japan lies on the direct route between the arctic breeding-grounds and the tropic winter-quarters of this species, it is remarkable that its presence in these Islands has not been more often detected (Cassin, Exp. Am. Squad. China Seas and Japan, ii. p. 227).

The Pintail Snipe breeds in East Siberia, and winters in the islands of the Malay Archipelago.

351. SCOLOPAX GALLINAGO.
(COMMON SNIPE.)

Scolopax gallinago, Linneus, Syst. Nat. i. p. 244 (1766).

The Common Snipe is a medium-sized bird (wing from carpal joint about 5 inches). It has 14 tail-feathers, of which none are less than ·4 inch broad.

Figures: Dresser, Birds of Europe, vii. pls. 542, 543.

Scolopax gallinago.

The Common Snipe is abundant in Japan, but I can find no evidence that it breeds there. It has occurred on the Kurile Islands (Blakiston and Pryer, Trans. As. Soc. Japan, 1882, p. 114); there are three examples in the Swinhoe collection from Hakodadi (Swinhoe, Ibis, 1874, p. 163); and eight in the Pryer collection from Yokohama, whence I have also four examples collected by Mr. Snow in September, January, and April. The examples obtained by the Siebold Expedition were doubtless procured near Nagasaki (Temminck and Schlegel, Fauna Japonica, Aves, p. 112).

The Common Snipe is a circumpolar bird, but American examples differ from those of the Old World. It breeds in the British Islands.

352. SCOLOPAX RUSTICOLA.
(WOODCOCK.)

Scolopax rusticola, Linnæus, Syst. Nat. i. p. 243 (1766).

The Woodcock has silvery-white tips on the under surface of its tail-feathers, all the feathers of the breast are barred, and there are rudimentary bars on the margins of both webs of the primaries.

Figures: Dresser, Birds of Europe, vii. pl. 510.

Scolopax rusticola.

The Woodcock breeds in Yezzo and on the mountains of Southern Japan. To the north island it is only a summer visitor, but south of the Straits of Tsugaru it is a resident. There is an example from Hakodadi in the Swinhoe collection (Swinhoe, Ibis, 1877, p. 145); and there are six examples in the Pryer collection from Yokohama. Mr. Ringer has sent examples from Nagasaki, where those procured by the Siebold Expedition were probably also obtained (Temminck and Schlegel, Fauna Japonica, Aves, p. 112).

The Woodcock breeds in England as well as in Japan, its range extending through the Himalayas, as well as through Southern Siberia.

Suborder XXVIII. GRALLÆ.

Maxillo-palatines not coalesced with each other across the middle line, nor with the vomer; nasals schizorhinal; dorsal vertebræ heterocœlous; young born not only covered with down, but able to run in a few hours.

The Grallæ may be regarded as almost cosmopolitan, but of the six families which it contains, only two are represented in the Japanese Empire.

The Turnicidæ are a small family (about 25 species), confined to the tropical and subtropical parts of the Old World, one species extending its range in the west into Southern Europe, and another in the east reaching the Loo-Choo Islands. The Pteroclidæ are a still smaller family (about 16 species) with a similar range in the west, but in the east not extending beyond the Bay of Bengal, and consequently not reaching Japan. The Rhinochetidæ only comprises one species confined to New Caledonia; the Mesitidæ one species confined to Madagascar; and the Eurypygidæ two species confined to tropical America. The Gruidæ are a small family (about 16 species) of which no less than five visit Japan; two inhabit North America, one Australia, and the rest are either Palæarctic, Oriental, or Ethiopian.

353. GRUS CINEREA.
(COMMON CRANE.)

Grus cinerea, Bechstein, Naturg. Deutschl. iv. p. 103 (1800).

The Common Crane is a grey bird, with the nape, crown, forehead, lores, ear-coverts, chin, and throat black. No other Japanese Crane has a black nape.

Figures: Temminck and Schlegel, Fauna Japonica, Aves, pl. 72.

The Common Crane is a winter visitor to Japan. It has not been recorded from Yezzo, and the only authority that I know of for its occurrence in Japan is the example figured in the 'Fauna Japonica' under the name of *Grus cinerea longirostris*. There is an example in the Swinhoe collection from South-east China.

Grus cinerea.

The breeding-range of the Common Crane extends from Europe across Southern Siberia to Kamtschatka. It no longer breeds in the British Islands, but is occasionally seen on migration.

354. GRUS LEUCOGERANUS.
(SIBERIAN WHITE CRANE.)

Grus leucogeranus, Pallas, Reise Russ. Reichs, ii. p. 138 (1773).

The Siberian White Crane is white all over, except the primaries which are black, and the fore part of the head which is almost bare of feathers. No other Japanese Crane has a white body and a white neck.

Figures: Temminck, Planches Coloriées, no. 467 (adult); Temminck and Schlegel, Fauna Japonica, Aves, pl. 73 (young); Dresser, Birds of Europe, vii. pl. 507 (adult).

The Siberian White Crane is principally known in Japan as a spring and autumn visitor on migration, but it is probable that some remain to winter. It has not been recorded from Yezzo, nor have I seen an example from any part of Japan, except those figured in the 'Fauna

Grus leucogeranus.

Japonica,' of which there are three in the Leyden Museum (Schlegel, Mus. d'Hist. Nat. Pays-Bas, v. pt. ii. p. 5).

The Siberian White Crane breeds in Eastern Siberia, and occasionally wanders in winter as far west as Europe, and as far east as China and Japan. Its usual winter-quarters appear to be in the plains of Northern India, but even there it is said to be somewhat local and rare (Hume, Ibis, 1868, p. 28).

355. GRUS JAPONENSIS.
(SACRED CRANE.)

Ardea (Grus) japonensis, Müller, Natursyst. Suppl. p. 110 (1776).

The Sacred Crane has a white body like the Siberian White Crane, but the forehead, lores, chin, fore neck, lower hind neck, and disintegrated tertials are black. No other Japanese Crane has a white body and a black fore neck.

Figures: Wolf, Zool. Sketches, series i. pl. 46; Tegetmeier, Nat. Hist. Cranes, pp. 13, 53.

Grus japonensis.

The Sacred Crane, so called because it was formerly held sacred in Japan, and was only allowed to be hawked with great ceremony by nobles of the highest rank (otherwise known as the Manchurian Crane, because it breeds in that country), has been known as a Japanese bird from time immemorial. It is the *Ciconia grus japonensis* of Brisson; the *Ardea grus* β of Gmelin; the Japan Crane of Latham; la Grue du Japon of Buffon; and the *O-tsuri* or *Tsurisama* of the Japanese. In 1823 it received the name of *Grus viridirostris* (Vieillot, Tableau Encycl. et Méth. iii. p. 1141); in 1829 that of *Grus collaris* (Temminck, Planches Coloriées, text to no. 449); and in 1854 that of *Antigone montignesia* (Bonaparte, Compt. Rend. xxxviii. p. 661).

The Sacred Crane is found in all the Japanese Islands. It has been recorded from Yezzo as late as January (Blakiston and Pryer, Trans. As. Soc. Japan, 1882, p. 121; erroneously called *Grus leucauchen*). Mr. Ringer sent an example from Nagasaki (Seebohm, Ibis, 1884, p. 178), but for some reason or other, possibly its sacred character, it was not obtained by the Siebold Expedition.

The Sacred Crane breeds in Eastern Siberia and Japan (David and Oustalet, Ois. de la Chine, p. 436), but is supposed to be only a winter visitor to China.

356. GRUS LEUCAUCHEN.
(WHITE-NAPED CRANE.)

Grus leucauchen, Temminck, Planches Coloriées, no. 449 (1827).

The White-naped Crane is a grey bird, with the crown, nape, hind neck, and upper throat white. It is the only Crane in which the

Grus leucauchen.

white on the crown is continued along the hind neck to the mantle but not beyond.

Figures: Tegetmeier, Nat. Hist. Cranes, p. 36.

The White-naped Crane is the most abundant Crane in Japan, and is found in all the islands (Blakiston and Pryer, Ibis, 1878, p. 225, no. 137). Mr. Ringer sent two skins from Nagasaki (Seebohm, Ibis, 1884, p. 177), where the examples procured by the Siebold Expedition were probably also obtained (Temminck and Schlegel, Fauna Japonica, Aves, p. 119).

It breeds in Eastern Siberia.

The White-naped Crane is rather unfortunate in its name, inasmuch as the Sacred Crane, the Asiatic White Crane, and the White-headed Crane have also white napes.

357. GRUS MONACHUS.
(WHITE-HEADED CRANE.)

Grus monacha, Temminck, Planches Coloriées, no. 555 (1835).

The White-headed Crane is a grey bird like the Common Crane, and, like that species, the lores, forehead, and crown are black. It

Grus monachus.

differs, however, in having the rest of the head and the whole of the upper neck white. It is the only Japanese Crane that combines the two characters—body grey, upper neck entirely white.

2 A

Figures: Temminck and Schlegel, Fauna Japonica, Aves, pl. 74.

The White-headed Crane has not been recorded from Yezzo, but is not uncommon in the more southerly islands (Blakiston and Pryer, Trans. As. Soc. Japan, 1882, p. 121). There is an example in the Pryer collection from Yokohama; and Mr. Ringer sent three examples from Nagasaki (Seebohm, Ibis, 1884, p. 178).

The White-headed Crane certainly breeds in East Siberia, and probably also in Southern Japan (David and Oustalet, Ois. de la Chine, p. 134).

358. TURNIX BLAKISTONI.
(BLAKISTON'S HEMIPODE.)

Arcoturnix blakistoni, Swinhoe, Proc. Zool. Soc. 1871, p. 401.

Figures: Temminck, Planches Coloriées, no. 60. fig. 2 (female); Gould, Birds of Asia, vii. pl. 11 (male); Sykes, Trans. Zool. Soc. ii. pl. 4 (male).

Blakiston's Hemipode was described from a single example, a male, procured by Captain Blakiston near Canton, and now in the Swinhoe collection. A second male has been described from the Loo-Choo Islands (Stejneger, Proc. United States Nat. Mus. 1886, p. 635). A very adult female with black throat belongs (as Dr. Stejneger suggests) to the same species; it was sent by Mr. Pryer from the Loo-Choo Islands, and is dated July (Seebohm, Ibis, 1887, p. 179).

Blakiston's Hemipode will probably prove to be a local race of *Turnix pugnax*. It forms a connecting-link between the two Indian races known as *Turnix plumbipes* and *Turnix taigoor*. These three races differ in typical examples as follows:—

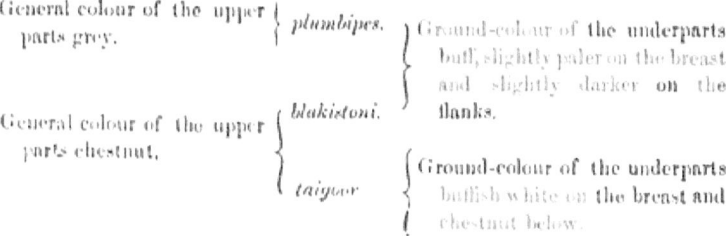

I cannot see any specific difference between *T. pugnax* from Java and *T. taigoor* from India, or between *T. plumbipes* from India and *T. rostrata* from China and Formosa. Both forms occur in the Burma peninsula. The size and thickness of the bill appear to be only individual characteristics, and vary greatly in both forms.

Suborder XXIX. *FULICARIÆ*.

Palate schizognathous; nasals holorhinal; dorsal vertebræ heterocœlous; episternal process not perforated to receive the feet of the coracoids; posterior processes of the ilia separated sufficiently to show a broad sacrum; young born able to run in a few hours.

The Fulicariæ, or Holorhinal Grallæ, comprise six families, two of which are represented in Japan. The *Otididæ* are a small family (about 25 species) distributed in most suitable localities in the Old World, one species ranging as far east as Japan. The *Opisthocomidæ* consist of only one species found in Brazil. The *Cariamidæ* consist of only two species, also found in Brazil. The *Heliornithidæ* may comprise half a dozen species—one in tropical America, four in tropical Africa, and one in Borneo. The *Psophiidæ* are also found in Brazil, and number half a dozen species. The *Rallidæ* are almost cosmopolitan, and number nearly 150 species, of which eight are recorded from the Japanese Empire.

The Grallæ (schizognathous Grallæ) and the Fulicariæ (holorhinal Grallæ) are very closely allied, and are the remains of a group which was probably once very numerous.

359. OTIS DYBOWSKII.
(EASTERN GREAT BUSTARD.)

Otis dybowskii, Taczanowski, Journ. Orn. 1874, p. 331.

The Eastern Great Bustard is found in Japan; but as the only examples procured were shot in November and December, it is uncertain whether it be a resident or only a winter visitor from

Siberia (Blakiston and Pryer, Trans. As. Soc. Japan, 1882, p. 124). Two examples have passed through my hands; one of them, shot in the presence of Captain Blakiston on the 12th of November, near the mouth of the Iskari River on the north-west coast of Yezzo, is obviously not quite adult; the other, shot in December at Jasahai, fifteen miles north of Nagasaki, after a gale of wind, appears to be adult (Seebohm, Ibis, 1884, p. 178). It is somewhat smaller than the Western species, the bill is slightly longer and more slender, the head is paler in colour, and the lesser wing-coverts are grey, like the greater and median wing-coverts, instead of being mottled with brownish buff and black, like the back.

The breeding-range of the Eastern Great Bustard probably extends across Eastern Siberia to Japan; but examples from the latter locality have not yet been procured in summer.

360. CREX PUSILLA.
(PALLAS'S CRAKE.)

Rallus pusillus, Pallas, Reise Russ. Reichs, iii. p. 700 (1776).

Pallas's Crake is a small bird (wing from carpal joint about $3\frac{1}{2}$ inches). The outer web of the first primary is white; the under tail-coverts are white, barred with black; and it has no spots on the sides of the throat or breast.

Figures: Hume and Marshall, Game-Birds of India, Burmah, and Ceylon, ii. pl. 35.

Pallas's Crake is a resident in all the Japanese Islands. Captain Blakiston sent me an example from Yezzo (Seebohm, Ibis, 1884, p. 35); there are two examples in the Pryer collection from Yokohama; and Mr. Ringer has sent an example to the Norwich Museum obtained at Nagasaki (Blakiston and Pryer, Trans. As. Soc. Japan, 1882, p. 123).

Pallas's Crake has been most unaccountably confused with Baillon's Crake, but the difference between the two species has been recently pointed out (Ogilvie Grant, Ann. & Mag. Nat. Hist. 1890, vol. v. p. 80).

The breeding-range of Baillon's Crake extends from the British Islands across Europe to Africa and Madagascar. Neither species is known to occur in Asia Minor, Persia, or South-west Siberia; but

Pallas's Crake is a regular summer visitor to South-east Siberia, and appears to be a resident in India, Burma, China, and Japan. In winter it visits the Philippines, Borneo, and the Andaman Islands.

In fully adult birds the sexes are alike. In Baillon's Crake the sides of the head are slate-grey; in Pallas's Crake this slate-grey is interrupted by a broad brown band, which begins at the base of the bill, passes over the lores, and is continued behind the eye to the nape.

361. CREX FUSCA.
(RUDDY CRAKE.)

Rallus fuscus, Linneus, Syst. Nat. i. p. 262 (1766).

The Ruddy Crake is a small bird (wing from carpal joint $4\frac{1}{2}$ to $3\frac{1}{2}$ inches). It is olive-brown above, with a vinous-chestnut breast. There is no white on any of the quills, except occasionally a spot or two on the outer web of the first primary.

Figures: Temminck and Schlegel, Fauna Japonica, Aves, pl. 78.

The Ruddy Crake is found in all the Japanese Islands, and is probably a summer visitor to Yezzo and a resident in Southern Japan. There is an example in the Swinhoe collection from Hakodadi (Swinhoe, Ibis, 1874, p. 163), whence examples were obtained by the Perry Expedition twenty years previously (Cassin, Exp. Am. Squad. China Seas and Japan, ii. p. 229). There are three examples in the Pryer collection from Yokohama; and Mr. Ringer has sent examples to the Norwich Museum obtained at Nagasaki (Blakiston and Pryer, Trans. As. Soc. Japan, 1882, p. 123). Japanese examples belong to the Siberian race of this species, and have been described as distinct from the Indian species under the name of *Gallinula erythrothorax* (Temminck and Schlegel, Fauna Japonica, Aves, p. 121), but they appear to be only entitled to subspecific distinction as *Crex fusca erythrothorax*.

The Siberian form of the Ruddy Crake breeds in the valley of the Lower Amoor and in China. It is not known to differ in any way except in size from the typical form, whose range extends from the Philippine Islands and the Malay Archipelago to the Malay peninsula, Burma, India, and Ceylon. Japanese examples vary in length of wing from 4·4 to 4·6 inches; Chinese examples from 4·1 to

4·6 inches; whilst in Siberia it is said to attain a length of 5 inches. Indian and Burmese examples vary from 3·5 to 3·8 inches, and two examples in the Swinhoe collection from Formosa measure 3·8 inches.

Young in first plumage are much darker than adults, and have no vinous chestnut on the breast. It is probably one of these immature examples that was procured by Mr. Nishi on one of the Yaye-yama Islands, the most southerly group of the Loo-Choo chain, and was described as a new species under the name of *Porzana phæopyga* (Stejneger, Proc. United States Nat. Mus. 1887, p. 394). The length of wing of this example is given as 4·1 inches.

362. CREX UNDULATA.
(SWINHOE'S CRAKE.)

Porzana undulata, Taczanowski, Journ. Orn. 1874, p. 333.

Swinhoe's Crake is a very small bird (wing from carpal joint about 3 inches). There is a great deal of white on most of the secondaries.

Figures: Swinhoe, Ibis, 1875, pl. 3 (under the name of *Porzana exquisita*).

Swinhoe's Crake is found in all the Japanese Islands, and is probably a resident. There is an example in the Swinhoe collection from Hakodadi (Swinhoe, Ibis, 1876, p. 335), and there is an example in the Pryer collection from Yokohama.

The range of Swinhoe's Crake extends westwards from Japan to South-eastern Siberia and North-eastern China.

363. CREX SEPIARIA.
(LOO-CHOO CRAKE.)

Euryzona sepiaria, Stejneger, Proc. United States Nat. Mus. 1887, p. 336.

The Loo-Choo Crake is as large as the Common Corn-Crake (wing from carpal joint 5·9 inches). It is a brown bird, with a pale throat, and the underparts below the breast black barred with white.

The Loo-Choo Crake is only known from a single example obtained by Mr. Nishi on one of the Yaye-yama Islands, the most southerly group of the Loo-Choo chain. It is represented in China by a smaller species, *C. mandarina* (wing 5 inches), with a chestnut breast (David and Oustalet, Ois. de la Chine, pl. 123); and in the Philippine Islands by an equally small species, *C. fasciata*, with pale bands across the quills and wing-coverts.

364. RALLUS AQUATICUS.
(WATER-RAIL.)

Rallus aquaticus, Linnæus, Syst. Nat. i. p. 262 (1766).

The Water-Rail is a medium-sized bird (wing from carpal joint $5\frac{1}{2}$ to 5 inches). Its bill is longer than its head, and the feathers of the upper parts are buffish brown, with nearly black centres.

Figures: Hume and Marshall, Game-Birds of India, Burmah, and Ceylon, ii. p. 257 (Eastern race).

The Indian form of the Water-Rail is a resident in all the Japanese Islands. There is an example in the Swinhoe collection from Hakodadi (Swinhoe, Ibis, 1874, p. 163), and there are ten examples in the Pryer collection from Yokohama. Mr. Ringer has sent examples to the Norwich Museum from Nagasaki, where those procured by the Siebold Expedition were probably also obtained (Temminck and Schlegel, Fauna Japonica, Aves, p. 122).

The breeding-range of the Water-Rail extends from the British Islands across Europe to Chinese Turkestan; and probably inosculates with that of the Eastern form of this species, which extends from Eastern Siberia and Japan to China, Burma, and India.

"The Eastern form of the Water-Rail is, on an average, slightly larger than the Western race: the slate-grey on the underparts is always more or less suffused with brown, the dark brown of the lores extends also below and behind the eye, and the under tail-coverts are more barred with black; but no one of these characters is always constant" (Seebohm, British Birds, ii. p. 533). It was described as a distinct species in 1849 (Blyth, Journ. As. Soc. Beng. xviii. p. 820), but it can scarcely be regarded as more than subspecifically distinct, and may be distinguished as *Rallus aquaticus indicus*.

The Rails are very closely allied to the Crakes, but differ from them in having the bill longer instead of shorter than the head. The Rails are almost cosmopolitan, but the Crakes are confined to the Old World.

365. GALLICREX CINEREUS.
(WATER-COCK.)

Fulica cinerea, Gmelin, Syst. Nat. i. p. 703 (1788).

The Water-Cock is nearly as large as the Coot (wing from carpal joint 8½ to 7 inches). The under tail-coverts are white (male) or buff (female), barred (in both sexes) with dark brown.

Figure: Gray, Fasc. Birds of China, pl. 10.

The claim of the Water-Cock to be regarded as a Japanese bird rests upon a single example procured by Mr. Ringer at Nagasaki during June (Seebohm, Ibis, 1884, p. 178).

The Water-Cock is a resident in most parts of the Oriental Region, but can only be regarded as an accidental visitor to Southern Japan.

366. FULICA ATRA.
(COMMON COOT.)

Fulica atra, Linnæus, Syst. Nat. i. p. 257 (1766).

The Coot is a large bird (wing from carpal joint 8¼ to 7¾ inches). The under tail-coverts are all black.

Figures: Temminck and Schlegel, Fauna Japonica, Aves, pl. 77; Dresser, Birds of Europe, vii. pl. 504. fig. 2.

The Common Coot is a resident in all the Japanese islands. I have examined skins from Yezzo collected by Captain Blakiston, and there are two examples in the Pryer collection from Yokohama. I have also examined an example from Nagasaki collected by Mr. Ringer (Seebohm, Ibis, 1884, p. 178), and Mr. Pryer records it from the Loo-Choo Islands (Seebohm, Ibis, 1887, p. 180).

The breeding-range of the Common Coot extends from the British Islands across Europe and South Siberia to Japan.

367. GALLINULA CHLOROPUS.
(WATER-HEN.)

Fulica chloropus, Linnæus, Syst. Nat. i. p. 258 (1766).

The Water-Hen is much less than the Coot (wing from carpal

joint 6¾ to 6 inches). The central under tail-coverts are black, but the lateral ones are white.

Figures: Dresser, Birds of Europe, vii. pl. 503.

The Water-Hen is a resident in all the Japanese Islands (Blakiston and Pryer, Ibis, 1878, p. 225). There are four examples in the Pryer collection from Yokohama, and three from the Loo-Choo Islands (Seebohm, Ibis, 1887, p. 180). From the last-named locality it was obtained by the Perry Expedition in 1854 (Cassin, Exp. Am. Squad. China Seas and Japan, ii. p. 245).

The Water-Hen is a circumpolar species, and is very common in the British Islands.

Suborder XXX. *PYGOPODES*.

The Grebes and the Divers possess the following characters, which are not combined in any other bird:—

Cnemial process of tibia produced forwards to a remarkable degree; posterior processes of the ilium approximated to such an extent that

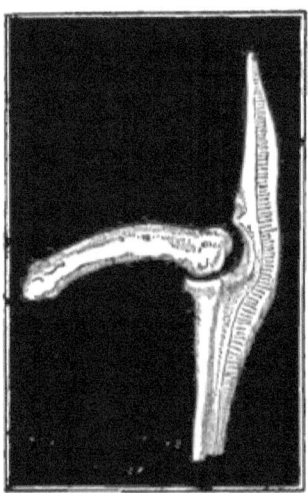

Femur and front part of tibia.

the sacrum is almost entirely concealed; spinal feather-tract not defined on the neck; palate schizognathous.

The Pygopodes consist of two families. The *Colymbidæ* contain only four species, which are confined during the breeding-season to the Arctic Regions, and three of which visit Japan in winter. The *Podicipidæ* number between twenty and thirty species, which are

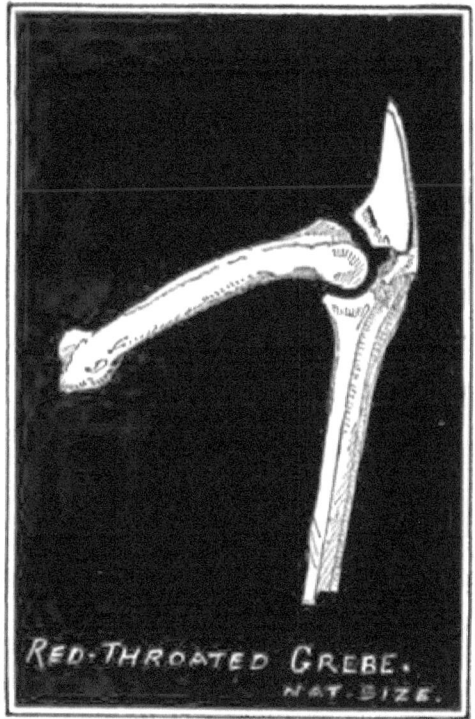

Femur and front part of tibia.

distributed over the rest of the world, including Madagascar, Australia, and New Zealand, but not the Pacific Islands. Five of the Podicipidæ breed in Japan.

368. COLYMBUS ADAMSI.
(WHITE-BILLED DIVER.)

Colymbus adamsii, Gray, Proc. Zool. Soc. 1859, p. 167.

This species may always be recognized by its bill, which is nearly white, not decurved, and very large (height at nostrils $1\frac{1}{4}$ to 1 inch, instead of, as in *C. glacialis*, from 1 to $\frac{7}{8}$ inch).

Figures: Elliot's Birds of North America, pl. 63.

There can scarcely be any doubt that the Great Northern Divers which Steller observed on the coasts of the Kurile Islands (Pallas, Zoogr. Rosso-Asiat. ii. p. 341) belonged to the species with a large white bill, and not to the perfectly distinct species with a much smaller black bill. It is to the white-billed species that the Kamtschatkan birds must be referred (Stejneger, Orn. Exp. Comm. Isl. & Kamtschatka, p. 14); and it is the White-billed Diver that has occurred on the Japanese coasts. The occurrence of this species in Japan was first ascertained from an example procured by Captain Blakiston at Hakodadi in January (Swinhoe, Ibis, 1877, p. 146); and I have a fine example collected by Mr. Henson at Hakodadi (Seebohm, Ibis, 1884, p. 32). It has also occurred as far south as Nagasaki (Saunders, Ibis, 1883, p. 318).

The White-billed Diver is probably a circumpolar species, and breeds in Siberia and Arctic America; it occasionally visits the British Islands in winter.

369. COLYMBUS ARCTICUS.
(BLACK-THROATED DIVER.)

Colymbus arcticus, Linnæus, Syst. Nat. i. p. 221 (1766).

In this species the forehead and crown are uniform unmottled grey at all ages and seasons; and the height of the bill at the nostrils never reaches ¾ of an inch.

Figures: Gould, Birds of Great Britain, v. pl. 44; Dresser, Birds of Europe, viii. pl. 627.

The Black-throated Diver is a common winter visitor to the coasts of Yezzo (Whitely, Ibis, 1867, p. 208), and probably of all the other Japanese Islands. In the Pryer collection there are five examples in winter plumage from Yokohama, and I have an example in summer dress collected by Captain Blakiston in May (Seebohm, Ibis, 1879, p. 22). Mr. Ringer has also procured it near Nagasaki, where the examples obtained by the Siebold Expedition were probably also procured (Temminck and Schlegel, Fauna Japonica, Aves, p. 123).

The range of the Black-throated Diver extends from the British Islands, across North Europe and Siberia to Japan, and across Bering Straits into Alaska and Arctic America as far east as Hudson's Bay.

It is impossible to say whether any of the examples belong to the form which American ornithologists call *Colymbus arcticus pacificus*,

An example in winter plumage measures 2·5 inches from the frontal feathers to the tip of the bill; a second, in summer plumage with a pale nape, measures 2·3 inches; whilst a third, moulting into summer plumage with a dark nape, measures only 1·9 inches.

370. COLYMBUS SEPTENTRIONALIS.
(RED-THROATED DIVER.)

Colymbus septentrionalis, Linnæus, Syst. Nat. i. p. 220 (1766).

In this species the forehead and crown are always mottled.

Figures: Gould, Birds of Great Britain, v. pl. 45; Dresser, Birds of Europe, viii. pl. 628.

It is probably to Steller that we are indebted for the local name of the Red-throated Diver on the Kurile Islands, whence we may infer its frequent occurrence in that locality (Pallas, Zoogr. Rosso-Asiat. ii. p. 343). It probably breeds there, as I have an example in breeding-plumage obtained by Mr. Snow in June on Rashua, one of the central islands of the chain. The first example recorded from Japan was obtained at Hakodadi in January 1865 (Whitely, Ibis, 1867, p. 208); and there is an example in winter plumage in the Swinhoe collection obtained by Captain Blakiston at Hakodadi (Swinhoe, Ibis, 1874, p. 163). It is said to be occasionally obtained in Tokio Bay (Blakiston and Pryer, Ibis, 1878, p. 211), but there are no examples in the Pryer collection. It is probably only a winter visitor to the Japanese Islands. Dr. Stejneger found it breeding in great abundance on the Commander Islands.

The Red-throated Diver is a circumpolar species, and breeds in Scotland and the north of Ireland.

371. PODICEPS RUBRICOLLIS.
(RED-NECKED GREBE.)

Colymbus rubricollis, Gmelin, Syst. Nat. i. p. 592 (1788).

The Red-necked Grebe has dark brown lores both in summer and winter, and its wing from carpal joint measures more than 6 inches.

Figures: Temminck and Schlegel, Fauna Japonica, Aves, pl. 78 B (under the name of *Podiceps rubricollis major,* the Eastern form).

It is not known that the Eastern form of the Red-necked Grebe differs in any way from the Western form except in size. The Western form varies in length of wing from the carpal joint from 6·3 to 7 inches, and the Eastern form from 7·3 to 8 inches; and the former varies in length of bill from the frontal feathers from 1·4 to 1·6 inches, and the latter from 1·65 to 2·2 inches.

I have an example of the Eastern form of the Red-necked Grebe, collected by Captain Blakiston at Hakodadi in January, and which I erroneously identified as *Podiceps cristatus* * (Seebohm, Ibis, 1882, p. 369), as Temminck appears to have done in 1840 (Temminck, Man. d'Orn. iv. p. 418). There is an example in winter plumage in the Pryer collection from Yokohama. The examples figured in the 'Fauna Japonica' were probably obtained near Nagasaki.

The Red-necked Grebe is almost a circumpolar bird, but European and West-Asiatic examples are not quite so large as those from East Asia and America. The Eastern form breeds in the valley of the Amoor, in Kamtschatka, and across Alaska and British North America to Greenland. It can only be regarded as subspecifically distinct from its Western ally, and must bear the name of *Podiceps rubricollis major*, given to it by Temminck and Schlegel about 1847.

PODICEPS CRISTATUS.

(GREAT CRESTED GREBE.)

Colymbus cristatus, Linneus, Syst. Nat. i. p. 222 (1766).

The Great Crested Grebe differs from its allies at all seasons in having nearly white lores.

Figures: Gould, Birds of Great Britain, v. pl. 38; Dresser, Birds of Europe, viii. pl. 629.

The occurrence of the Great Crested Grebe in Japan rests on very poor authority. It was not mentioned in Temminck and Schlegel's 'Fauna Japonica,' but was recorded from Japan in 1840 (Temminck, Man. d'Orn. iv. p. 418). The statement (Seebohm, Ibis, 1882, p. 369) that a skin from Hakodadi in winter plumage was correctly identified, was a blunder. It is a skin of *P. rubricollis holbœlli*, as were probably also the two examples previously recorded from the same locality (Whitely, Ibis, 1867, p. 208), inasmuch as a reference is given to the 'Fauna Japonica,' which relates to that species, and not to the Great Crested Grebe. There are, however, undoubted examples in the Swinhoe collection from Amoy and Ningpo, so that its occurrence in Japan is by no means improbable.

The breeding-range of the Great Crested Grebe is very extensive, reaching from the British Islands across Europe to South Africa, India, Australia, and New Zealand.

It has been described as a distinct species under the name of *Podiceps holbœlli* (Reinhardt, Vidensk. Meddel. 1853, p. 76). The fact that in 1783 the name of *Colymbus major* (Boddaert, Tabl. Planches Enluminées, p. 24, no. 404) was given to a Grebe now known as *Æchmophorus occidentalis*, can scarcely be pleaded as a reason for adopting Reinhardt's name instead of that given by Temminck and Schlegel.

372. PODICEPS NIGRICOLLIS.
(BLACK-NECKED GREBE.)

Podiceps nigricollis, Brehm, Vög. Deut-chl. p. 936 (1831).

The Black-necked Grebe is easily recognized in breeding-dress by its black throat and neck, and at all seasons by its slightly upturned bill. It may also be recognized at all times by its having not only white secondaries, but much white on many of the innermost primaries.

Figures: Gould, Birds of Great Britain, v. pl. 41; Dresser, Birds of Europe, viii. pl. 632.

The Black-necked or Eared Grebe is a common bird in Japanese collections, but I cannot find any evidence that it breeds in Japan. There are four examples in the Swinhoe collection from Hakodadi, two in winter dress and two in summer plumage, the latter dated April (Swinhoe, Ibis, 1874, p. 163). There are no fewer than thirteen in the Pryer collection from Yokohama, five of them in summer dress, but none of them dated. It has been obtained by Mr. Ringer at Nagasaki, where the examples procured by the Siebold Expedition were probably also obtained (Temminck and Schlegel, Fauna Japonica, Aves, p. 123).

The Black-necked Grebe has a wide range, but a very peculiar one. It is a resident in South Africa and in the basin of the Mediterranean. It is a summer visitor to Central Europe, Southern Russia, and Southern Siberia; but, although it winters on the Mekran coast and on the coast of Scinde on the one side, and in China and Japan on the other, it has not been recorded from any other part of India or Burma. It is a somewhat rare visitor to the British Islands.

It is represented on the American continent by a very closely allied species, *Podiceps californicus*, from which it may prove to be only subspecifically distinct.

373. PODICEPS CORNUTUS.
(SCLAVONIAN GREBE.)

Colymbus auritus, Linneus, Syst. Nat. i. p. 222 (1766, nec auctorum plurimorum).
Colymbus cornutus, Gmelin, Syst. Nat. i. p. 591 (1788).

The Sclavonian Grebe is about the same size as the Black-necked Grebe, with which it is often confounded in winter plumage. In breeding-dress it may be recognized by its combination of the two characters, ear-coverts black and fore neck chestnut; but in winter plumage a more minute examination is necessary. At all seasons it combines the two characters—length of wing from carpal joint varying between 5·8 and 5·2 inches, and white on secondaries but on none of the primaries.

Figures: Gould, Birds of Great Britain, v. pl. 40; Dresser, Birds of Europe, viii. pl. 631.

The first example of the Sclavonian Grebe recorded from Japan was shot in Hakodadi harbour on the 26th of January, 1865 (Whitely, Ibis, 1867, p. 209). There are two examples in the Swinhoe collection collected by Captain Blakiston near Hakodadi, one of them dated October (Swinhoe, Ibis, 1875, p. 456); there are three examples in winter plumage in the Pryer collection from Yokohama; and Mr. Ringer has procured it near Nagasaki (Blakiston and Pryer, Trans. As. Soc. Japan, 1882, p. 92). It is probably only a winter visitor to Japan.

The Sclavonian Grebe is a circumpolar species, breeding for the most part a little to the south of the Arctic Circle on both continents, and wandering still further southwards in winter. It is a winter visitor to the British Islands.

374. PODICEPS MINOR.
(LITTLE GREBE.)

Colymbus minor, Gmelin, Syst. Nat. i. p. 591 (1788).

The Little Grebe well deserves its name, and the fact that the length of its wing from the carpal joint is only 4 inches or less is sufficient to distinguish it from all its Old World allies, except from *P. nestor* in South Australia, and from *P. rufipectus* in New Zealand.

Figures: Gould, Birds of Great Britain, v. pl. 12; Dresser, Birds of Europe, viii. pl. 633.

The Little Grebe is a common summer visitor to Yezzo, and a still commoner resident in the more southerly Japanese Islands; but the first authentic occurrence of this species in Japan was that of an example obtained in September by Captain Blakiston in South Yezzo (Swinhoe, Ibis, 1875, p. 456). There are 12 examples in all stages of plumage in the Pryer collection from Yokohama; and Mr. Ringer has procured it at Nagasaki (Blakiston, Amended List of the Birds of Japan, p. 32).

The Little Grebe breeds in all the temperate and subtropical parts of the Old World, from the British Islands to Japan, and from South Africa to the Malay Archipelago and Australia, with the exception of Siberia and Mongolia.

Suborder XXXI. *GALLINÆ*.

Palate schizognathous; basipterygoid processes articulating with the pterygoids as far from the quadrates as possible; episternal process perforated to receive the feet of the coracoids; nasals holorhinal.

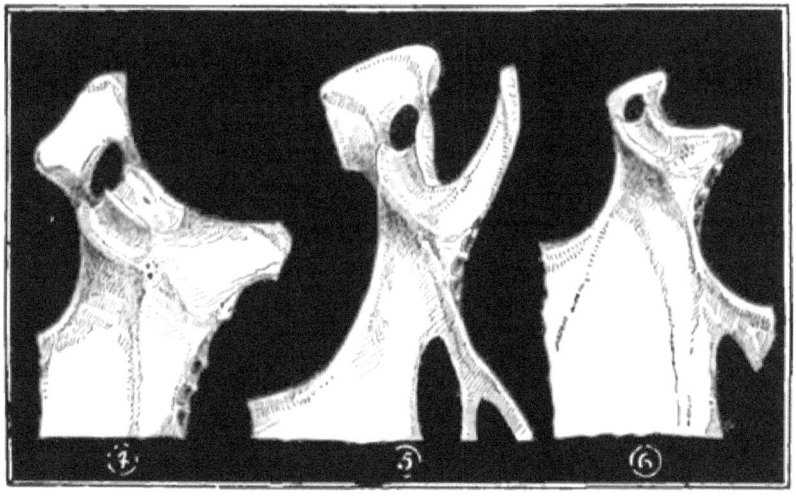

Front portion of sternum of (4) *Crax carunculata*, (5) of *Lophophorus impeyanus*, (6) of *Megapodius rubripes*.

The Gallinæ consists of three families. The *Phasianidæ* contains about 250 species, which are nearly cosmopolitan. The *Cracidæ* contains about 50 species, which are confined to the tropical regions of

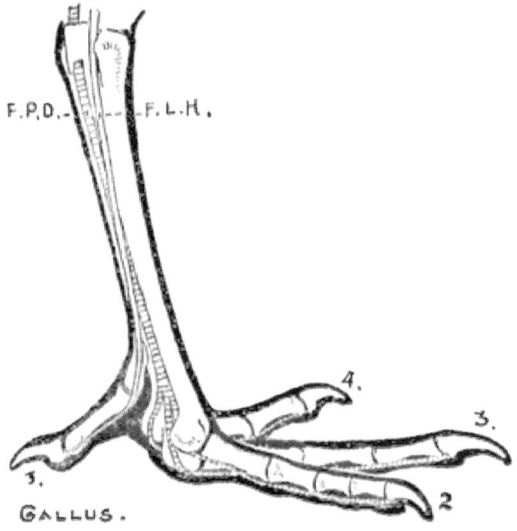

Deep plantar tendons of Domestic Fowl.

the American continent. The *Megapodidæ* numbers about a score species, which inhabit Australia and the islands of the Malay Archipelago.

Seven species of Phasianidæ are found in Japan.

375. PHASIANUS TORQUATUS.
(CHINESE RING-NECKED PHEASANT.)

Phasianus torquatus, Gmelin, Syst. Nat. i. p. 742 (1788).

The Chinese Ring-necked Pheasant has a white ring round the neck, and the colour of the rump and upper tail-coverts is for the most part lavender-grey.

Figures: Gould, Birds of Asia, vii. pl. 39.

The Chinese Ring-necked Pheasant is not found on any of the main islands of Japan, but is said to be common on Tsu-sima, an island in the Straits of Corea (Blakiston and Pryer, Trans. As. Soc. Japan, 1882, p. 127).

376. PHASIANUS VERSICOLOR.
(JAPANESE GREEN PHEASANT.)

Phasianus versicolor, Vieillot, La Galerie des Oiseaux, ii. p. 23, pl. 205 (1-29).

The Japanese Green Pheasant has the breast, belly, and flanks metallic green.

Figures: Gould, Birds of Asia, vii. pl. 40; Temminck, Planches Coloriées, no. 486 (male), no. 493 (female).

The Japanese Green Pheasant is peculiar to Japan, but its range does not extend into Yezzo. There are examples in the Paris Museum procured by l'Abbé Fauire near Aomori, in the north of Hondo, but it is not known to occur north of the Tsugaru Straits. There are a pair in the Swinhoe collection from Tokio (Swinhoe, Ibis, 1875, p. 452), and a fine series in the Pryer collection from the same district. The latter includes many hybrids between the Japanese Green Pheasant and the Chinese Ring-necked Pheasant (Blakiston and Pryer, Ibis, 1878, p. 226). A very interesting account of the habits of this Pheasant is to be found in the narrative of the Perry Expedition (Cassin, Exp. Am. Squad. China Seas and Japan, ii. p. 223). It is written by Mr. Heine, the artist of the expedition, who met with the birds at Simoda in the province of Idsu. Eggs in the Pryer collection resemble dark olive varieties of those of the Common Pheasant.

377. PHASIANUS SŒMMERINGI.
(COPPER PHEASANT.)

Phasianus sœmmeringii, Temminck, Planches Coloriées, no. 487 (male), no. 488 (female) (1830).

The Copper Pheasant is almost entirely coloured crimson and gold.

Figures: Gould, Birds of Asia, vii. pl. 37.

The Copper Pheasant is peculiar to Japan. I have several examples from Nagasaki, for which I am indebted to the kindness of Mr. Ringer. The examples procured by the Siebold Expedition were also presumably obtained near Nagasaki (Temminck and Schlegel, Fauna Japonica, Aves, p. 104). It does not occur in Yezzo (Blakiston and Pryer, Trans. As. Soc. Japan, 1882, p. 126), but an example obtained at Simoda on the coast south-west of Yokohama appears to be referable to this species (Cassin, Exp. Am. Squad. China Seas and Japan, ii. p. 226), though there is some evidence that it was a tame bird. It is not unreasonable to assume that this species was differentiated in the island of Kiu-siu.

378. PHASIANUS SCINTILLANS.
(HONDO COPPER PHEASANT.)

Phasianus scintillans, Gould, Ann. & Mag. Nat. Hist. 3rd ser. xvii. p. 150 (1864).

The Hondo Copper Pheasant is for the most part arrayed in crimson and gold, but the scapulars, wing-coverts, and upper tail-coverts are more or less edged with white.

Figures: Gould, Birds of Asia, vii. pl. 38.

In the Pryer collection there is a remarkably fine series of Pheasants, but of eight examples of the Copper Pheasant all are referable to this species. Under these circumstances it is not unreasonable to assume that it is the only species of Copper Pheasant found near Yokohama. On the other hand, there cannot be any doubt that it occurs also near Nagasaki (Elliot, Mon. Phasianidæ, text to pl. 13). All the examples in the Paris Museum procured by l'Abbé Fauire near Aomori in the north of Hondo belong to this species.

It differs from its ally in having conspicuous and more or less complete white margins to the scapulars, the wing-coverts, and the upper tail-coverts. The flank-feathers are very conspicuous in consequence of their margins being so much paler, and being separated from the body of the feather by a black line. Each black band across the tail is emphasized by a white band, above which the ground-colour of the feather is much paler, causing the black speckling to be much more conspicuous.

Mr. Elliot asserts that intermediate forms occur, and there can be no doubt that they would interbreed freely if they had the opportunity. It is not known that the females differ.

Eggs in the Pryer collection are rather longer than eggs of the Common Pheasant, and are creamy buff in colour.

The Hondo Copper Pheasant was probably differentiated in the main island of Japan, but may have been introduced into some of the other islands in a few localities.

379. TETRAO MUTUS.
(COMMON PTARMIGAN.)

Tetrao mutus, Montin, Sällsk. Handl. Lund, i. p. 155 (*circa* 1780).

The Common Ptarmigan is variously mottled with brown, buff, and grey in the breeding-season, but the wings are always white, and the whole plumage (except the lores of the male) is white in winter.

Figures: Dresser, Birds of Europe, vii. pl. 477. fig. 1 (autumn plumage), pl. 478 (summer plumage).

The Ptarmigan breeds on the main island of Japan at an elevation of nearly 10,000 feet above the level of the sea, whence I have an example of an adult female and young in down, collected by Mr. Miwa in July on Renge-yama, one of the more westerly peaks of the Tate-yama range, about a hundred miles north-west of Yokohama (Seebohm, Ibis, 1884, p. 35). The example in winter plumage, obtained by Mr. Snow on the Kuriles (Seebohm, Ibis, 1884, p. 179), may also belong to this species, the male of which has black lores in winter.

The Ptarmigan is a circumpolar species, and breeds in Scotland and various mountain-ranges in Europe and across Asia to Japan. There are several local races of this species, but it is not known that Japanese examples differ from Scotch ones.

It was long ago recorded from the Kurile Islands (Pallas, Zoogr. Rosso-Asiat. ii. p. 61); and the fact that some of the examples obtained in winter by Mr. Snow on these islands have black lores (Seebohm, Ibis, 1884, p. 179), whilst others (received since that date, and presumably females) have white lores, appears to prove that the Kurile Island species is not the Willow-Grouse, though it may still be an open question whether it be or be not the Rock-Ptarmigan, *Tetrao mutus rupestris*.

The Rock-Ptarmigan is difficult to distinguish from the allied races; but it is said that neither sex ever acquires the dark breast characteristic of the spring plumage of the cocks of the typical form.

380. TETRAO BONASIA.
(HAZEL-GROUSE.)

Tetrao bonasia, Linneus, Syst. Nat. i. p. 275 (1766).

The Hazel-Grouse is less than the Ptarmigan, the autumn plumage of which it slightly resembles, except that its wings are brown. The tarsi are feathered, but not the toes.

Figures: Dresser, Birds of Europe, vii. pl. 486.

The Hazel-Grouse is a resident on the island of Yezzo (Whitely, Ibis, 1867, p. 204), but is not known from any of the more southerly islands. I have five examples collected by Mr. Henson near Hakodadi; and there are five other examples from Yezzo in the Pryer collection.

The breeding-range of the Hazel-Grouse extends from Japan across Siberia and Europe to the Pyrenees, but does not reach the British Islands.

The Hazel-Grouse of Japan is the same as that which is found in the valley of the Amoor, in South-western Siberia, and in Europe (whence I have examples from Russia, Sweden, the Carpathians, and the Apennines). It differs from the race which I found in the lower valley of the Yenesay in being more rufous in colour and in having a longer tail (5·4 to 4·5 inches, instead of 4·6 to 4·1 inches).

381. COTURNIX COMMUNIS.
(COMMON QUAIL.)

Coturnix communis, Bonnaterre, Tabl. Encycl. et Méthod. i. p. 217 (1790).

The Quail is a small bird (wing from carpal joint $4\frac{1}{2}$ to $3\frac{1}{2}$ inches), and bears a remarkable resemblance to a Hemipode, but may be at once distinguished by its hind toe.

Figures: Temminck and Schlegel, Fauna Japonica, Aves, pl. 61, under the name of *Coturnix vulgaris japonica*.

The Quail was originally described as a Japanese bird by Temminck and Schlegel, in the 'Fauna Japonica,' from examples obtained by the Siebold Expedition, most probably near Nagasaki, whence examples have been sent to the Norwich Museum by Mr. Ringer. In May 1854 it was procured by the Perry Expedition at Hakodadi

(Cassin, Exp. Am. Squad. China Seas and Japan, ii. p. 227), whence there are several examples in the Swinhoe collection obtained by Captain Blakiston (Swinhoe, Ibis, 1875, p. 452). There are seven examples in the Pryer collection from Yokohama. It breeds on all the main islands of Japan, and a few remain during winter in Yezzo (Blakiston and Pryer, Ibis, 1878, p. 226).

The Common Quail is subject to some climatic variation. Tropical forms resident in South Africa and China are on an average slightly smaller than forms from more temperate regions. They also differ in the colour of the throat of the male in summer plumage, that of examples from China being chestnut-buff, and from South Africa rusty buff. The two forms completely intergrade, and both occur in Japan—the pale-throated form, it is said, as a winter visitor from Siberia, and the rufous-throated form as a resident. If the latter be regarded as subspecifically distinct, it may bear the name of *Coturnix communis japonica*, and its range may be given as Eastern Siberia, Japan, and North China. On the other hand, it is said that the rufous-throated form frequently occurs in South Europe, and that the Quails of India and South China belong to the pale-throated or typical race.

Suborder XXXII. *CRYPTURI*.

Keel of sternum well developed; the cartilage which connects the ilium with the ischium behind the acetabulum not ossified.

There are about 50 species of Tinamous, which are confined to the Neotropical Region.

Subclass STRUTHIONIFORMES.

In the Struthioniformes the keel of the sternum has become obsolete; the basipterygoid processes are very large, and placed on the basisphenoid rather than on its rostrum; the oil-gland is absent; and there are no lateral bare tracts on the neck.

The subclass Struthioniformes contains two orders.

Order **APTERYGES.**

The order Apteryges contains only one suborder.

Suborder XXXIII. *APTERYGES.*

Keel of sternum obsolete; basipterygoid processes present; hallux present.

There are three species of Kiwi, all of them peculiar to New Zealand.

Order **RATITÆ.**

Keel of sternum obsolete; basipterygoid processes very large, and placed on the basisphenoid rather than on its rostrum; hallux absent.

The order Ratitæ contains three suborders.

Suborder XXXIV. *RHEÆ.*

Keel of sternum obsolete; length of humerus more than the combined length of six dorsal vertebræ; hallux absent; second, third, and fourth digits present and directed forwards; basipterygoid processes present, and articulating with the pterygoids as near the quadrates as possible.

There are three species of Rhea, all of them peculiar to the Neotropical Region.

Suborder XXXV. *CASUARII.*

Keel of sternum obsolete; vomer coalesced with the maxillo-palatines in front; length of humerus less than the combined length of three dorsal vertebræ.

There are a dozen species of Emu and Cassowary, which are only known from the Australian Region.

Suborder XXXVI. *STRUTHIONES.*

Keel of sternum obsolete; first and second digits absent.

Two species of Ostrich are known. They inhabit the Ethiopian Region, the range of one of them extending to the south of the Mediterranean subregion of the Palæarctic Region.

INDEX.

Accentor alpinus, 55.
—— —— erythropygius, 55.
—— erythropygius, 55.
—— modularis rubidus, 56.
—— rubidus, 56.
Accipiter gularis, 205.
—— nisoides, 205.
—— nisus, 204.
—— palumbarius, 204.
—— stevensoni, 205.
Accipitres, 191.
Accipitrinæ, 191.
Acredula caudata, 87.
—— trivirgata, 87.
Acrocephalus bistrigiceps, 71.
—— fasciolatus, 72.
—— orientalis, 71.
Ægialites dealbatus, 310.
Ægithalus consobrinus, 88.
Alauda alpestris, 119.
—— arvensis, 118.
—— —— japonica, 118.
—— —— pekinensis, 118.
Alaudinæ, 117.
Alca antiqua, 276.
—— brevirostris, 279.
—— carbo, 274.
—— cirrhata, 281.
—— columba, 275.
—— cristatella, 285.
—— kamtschatica, 286.
—— marmorata, 278.
—— monocerata, 283.
—— psittacula, 284.

Alca pygmæa, 286.
—— torda, 284.
—— troile, 273.
—— wumizusume, 277.
Alcedo bengalensis, 176.
—— coromanda, 173.
—— —— major, 173.
—— guttatus, 174.
—— ispida, 175.
—— —— bengalensis, 175.
—— lugubris, 175.
Ampelis garrulus, 110.
—— japonicus, 110.
Anas acuta, 246.
—— baeri, 254.
—— boschas, 243.
—— circia, 246.
—— clangula, 253.
—— clypeata, 242.
—— cornuta, 241.
—— crecca, 244.
—— cristata, 255.
—— falcata, 245.
—— ferina, 254.
—— formosa, 244.
—— fusca, 250.
—— galericulata, 248.
—— glacialis, 252.
—— histrionicus, 252.
—— javanica, 240.
—— marila, 256.
—— penelope, 247.
—— rutila, 241.
—— spectabilis, 256.

Anas stelleri, 257.
—— strepera, 242.
—— zonorhyncha, 243.
Anser albifrons, 237.
—— brachyrhynchus, 236.
—— cygnoides, 235.
—— gambeli, 237.
—— hutchinsi, 236.
—— hyperboreus, 238.
—— —— nivalis, 238.
—— minutus, 238.
—— nigricans, 240.
—— segetum, 236.
—— —— serrirostris, 236.
—— vulgaris, 237.
Anseres, 233.
Anseriformes, 207.
Anthus cervinus, 117.
—— maculatus, 115.
—— pratensis japonicus, 116.
—— spinoletta, 116.
—— —— japonicus, 116.
Antigone montignesia, 352.
Apteryges, 375.
Aquila chrysaetus, 199.
—— lagopus, 200.
—— pelagica, 199.
Aquilinae, 191.
Ardea alba, 216.
—— amurensis, 225.
—— caledonica, 223.
—— cinerea, 215.
—— coromanda, 219.
—— egrettoides, 218.
—— garzetta, 218.
—— grus, 351.
—— immaculata, 219.
—— intermedia, 217.
—— japonensis, 351.
—— javanica, 224.
—— jugularis, 220.
—— —— greyi, 221.
—— malaccensis, 220.
—— modesta, 216.
—— nycticorax, 222.
—— patruelis, 225.
—— sacra, 221.
—— sinensis, 227.

Ardea stellaris, 226.
Ardeola prasinosceles, 225.
Ardetta eurhythma, 227.
—— macrorhyncha, 224.
—— stagnatilis, 224.
Arcturnx blakistoni, 354.
Arundinax blakistoni, 73.
Astur gularis, 205.
Aythya affinis mariloides, 256.

Bombycilla phoenicoptera, 110.
Bombycivora japonica, 110.
Botaurus eurhythma, 227.
—— sinensis, 227.
—— stellaris, 226.
Brachyrhamphus kittlitzi, 279.
—— marmoratus, 278.
—— perdix, 278.
Branta albifrons, 237.
Bubo blakistoni, 184.
—— maximus, 183.
Buccrotes, 180.
Buphus bacchus, 226.
Butaster indicus, 196.
Buteo hemilasius, 201.
—— japonicus, 202.
—— polyogenys, 196.
—— pyrrhogenys, 196.
—— vulgaris, 201.
—— —— japonicus, 202.
—— —— plumipes, 202.
Butorides schrenckii, 225.

Cancroma coromanda, 219.
—— leucoptera, 226.
Caprimulgus jotaka, 178.
Carbo bicristatus, 210.
—— capillatus, 209.
—— filamentosus, 209.
Carpodacus erythrinus, 123.
—— roseus, 123.
—— sanguinolentus, 124.
Carpophaga ianthina, 165.
—— jouyi, 167.
—— versicolor, 166.
Casuarii, 376.
Cepphus arra, 274.
—— carbo, 274.

Cepphus columba, 275.
—— lomvia, 273.
Certhia familiaris, 91.
—— scandulaca, 91.
Ceryle guttata, 174.
Cettia cantans, 74.
—— cantillans, 76.
—— diphone, 77.
—— squamiceps, 74.
Chætura caudacuta, 178.
Charadrius cantianus, 309.
—— fulvus, 303.
—— geoffroyi, 310.
—— helveticus, 304.
—— minor, 306.
—— mongolicus, 308.
—— morinellus, 305.
—— placidus, 307.
—— veredus, 311.
Chaunoproctus ferreirostris, 122.
Chelidon blakistoni, 144.
—— dasypus, 144.
—— namiyei, 142.
Ciconia boyciana, 228.
—— grus japonensis, 351.
Cinclus pallasi, 54.
Circus æruginosus, 203.
—— cyaneus, 202.
—— spilonotus, 203.
—— uliginosus, 203.
Cisticola cisticola, 77.
Coccothraustes ferreirostris, 122.
—— japonicus, 120.
—— personatus, 121.
—— vulgaris, 120.
Coccyges, 168.
Columba gelastis, 161.
—— humilis, 162.
—— iris, 166.
—— kitlizii, 166.
—— janthina, 165.
—— livia, 160.
—— metallica, 166.
—— orientalis, 160.
—— risoria, 162.
—— sieboldi, 163.
—— versicolor, 166.
Columbæ, 160.

Colymbus adamsi, 362.
—— arcticus, 363.
—— —— pacificus, 363.
—— auritus, 367.
—— cornutus, 367.
—— cristatus, 365.
—— marmoratus, 278.
—— minor, 367.
—— rubricollis, 364.
—— septentrionalis, 364.
—— troile, 273.
Coraciæ, 176.
Coracias orientalis, 179.
Coraciiformes, 172.
Corvinæ, 93.
Corvus caryocatactes, 99.
—— corax, 94.
—— corone, 96.
—— cyanus, 99.
—— dauricus, 97.
—— japonensis, 95.
—— macrorhynchus, 94.
—— —— japonensis, 95.
—— —— — levaillanti, 95.
—— neglectus, 97.
—— pastinator, 98.
Coturnix communis, 373.
—— —— japonica, 373.
—— vulgaris japonica, 374.
Cotyle riparia, 144.
Crateropodinæ, 64.
Crex fusca, 357.
—— —— erythrothorax, 357.
—— pusilla, 356.
—— sepiaria, 358.
—— undulata, 358.
Crypturi, 374.
Cuculi, 168.
Cuculus canorinus, 170.
—— canorus, 169.
—— horsfieldi, 170.
—— hyperythrus, 171.
—— intermedius, 169.
—— kelungensis, 170.
—— monosyllabicus, 170.
—— optatus, 170.
—— poliocephalus, 171.
—— saturatus, 170.

Cyanopolius cyanus, 93.
Cygnus bewicki, 235.
—— musicus, 234.
Cypselus pacificus, 177.

Demiegretta ringeri, 221.
Dendrocopus richardsi, 149.
Dendrocygna javanica, 240.
Diomedea albatrus, 261.
—— brachiura, 262.
—— chinensis, 262.
—— derogata, 262.
—— nigripes, 263.
Diomedeidæ, 260.
Dryobates namiyei, 153.
—— subcirris, 152.

Emberiza aureola, 138.
—— ciopsis, 131.
—— elegans, 137.
—— fucata, 134.
—— lapponica, 140.
—— nivalis, 140.
—— personata, 136.
—— rustica, 134.
—— rutila, 138.
—— schœniclus, 133.
—— —— palustris, 133.
—— spodocephala, 137.
—— sulphurata, 135.
—— variabilis, 139.
—— yessoensis, 132.
Ephialtes elegans, 188.
Erithacus akahige, 50.
—— calliope, 52.
—— cyaneus, 53.
—— komadori, 52.
—— namiyei, 51.
Eurylæmi, 145.
Eurystomus orientalis, 179.
Euryzona sepiaria, 358.

Falco æsalon, 193.
—— apivorus, 197.
—— ater, 197.
—— buteo japonicus, 202.
—— chrysaetus, 199.
—— cyaneus, 202.

Falco gyrfalco, 192.
—— haliætus, 195.
—— indicus, 196.
—— lagopus, 200.
—— nisus, 204.
—— palumbarius, 204.
—— peregrinus, 192.
—— subbuteo, 193.
—— tinnunculus, 194.
—— —— japonicus, 194.
Falconiformes, 181.
Falconinæ, 191.
Ficedula coronata, 69.
Fratercula cirrhata, 281.
—— corniculata, 280.
—— cristatella, 285.
—— monocerata, 283.
—— psittacula, 284.
—— pusilla, 287.
—— pygmæa, 286.
Fregata minor, 214.
Fringilla brunneinucha, 128.
—— kawarahiba, 127.
—— kittlitzi, 128.
—— linaria, 125.
—— montana, 130.
—— montifringilla, 126.
—— rosea, 123.
—— rutilans, 131.
—— sinica, 127.
—— spinus, 125.
Fringillinæ, 120.
Fulica atra, 360.
—— chloropus, 360.
—— cinerea, 360.
Fulicariæ, 355.
Fuligula americana, 248.
—— baeri, 254.
—— clangula, 253.
—— cristata, 255.
—— ferina, 254.
—— fusca, 256.
—— glacialis, 252.
—— histrionica, 253.
—— marila, 256.
Fulmarus glacialis, 268.

Galgulus amaurotis, 65.

Gallicrex cinereus, 360.
Galliformes, 260.
Gallinæ, 368.
Gallinago japonica, 343.
Gallinula chloropus, 360.
—— erythrothorax, 357.
Gallo-Grallæ, 272.
Garrulus brandti, 100.
—— glandarius japonicus, 101.
—— japonicus, 101.
—— sinensis, 101.
Gaviæ, 273.
Gecinus awokera, 147.
—— canus, 148.
Geocichla sibirica, 44.
—— terrestris, 44.
—— varia, 43.
Grallæ, 348.
Grus cinerea, 348.
—— —— longirostris, 349.
—— japonensis, 351.
—— leucauchen, 352.
—— leucogeranus, 349.
—— monachus, 353.
—— viridirostris, 351.

Hæmatopus niger, 313.
—— osculans, 313.
Halcyon coromanda, 173.
Halcyones, 173.
Haliaetus albicilla, 198.
—— pelagicus, 199.
Hapalopteron familiare, 66.
Herodiones, 214.
Heterornis pyrrhogenys, 109.
Hierococcyx hyperythrus, 171.
Hirundininæ, 141.
Hirundo alpestris, 142.
—— —— japonica, 142.
—— caudacuta, 178.
—— gutturalis, 141.
—— javanica, 142.
—— —— namiyei, 142.
—— pacifica, 177.
—— riparia, 144.
—— rustica, 141.
—— —— gutturalis, 141.
Hypsipetes amaurotis, 64.

Hypsipetes pryeri, 66.
—— squamiceps, 65.
—— —— pryeri, 66.
Hypurolepis domicola, 142.

Ianthœnas jouyi, 167.
—— nitens, 165.
Ibis melanocephala, 232.
—— nippon, 232.
Icoturus namiyei, 51.
Impennes, 272.
Ixos familiaris, 66.
Iyngipicus kisuki, 156.
—— —— nigrescens, 156.
—— —— seebohmi, 156.
Iynx torquilla, 157.

Lamellirostres, 233.
Lamprotornis pyrrhogenys, 108.
—— pyrrhopogon, 108.
Laniinæ, 103.
Lanius bucephalus, 106.
—— lucionensis, 105.
—— magnirostris, 104.
—— major, 103.
—— superciliosus, 104.
Larus cachinnans, 291.
—— californicus, 294.
—— canus, 293.
—— crassirostris, 293.
—— delawarensis, 294.
—— glaucescens, 290.
—— glaucus, 290.
—— ichthyaetus, 292.
—— leucopterus, 292.
—— marinus, 291.
—— —— schistisagus, 291.
—— melanurus, 293.
—— ridibundus, 295.
—— schistisagus, 291.
—— tridactylus, 294.
Lestris buffoni, 289.
—— pomarinus, 289.
—— richardsoni, 288.
Limicolæ, 303.
Limosa melanura, 329.
—— —— melanuroides, 329.
—— melanuroides, 329.

Limosa rufa, 328.
—— —— uropygialis, 329.
—— scolopaceus, 330.
—— uropygialis, 328.
Lobivanellus cinereus, 311.
Locustella fasciolata, 72.
—— lanceolata, 73.
—— ochotensis, 73.
—— subcerthiola, 73.
Loxia curvirostra, 121.
—— enucleator, 122.
Lusciniola pryeri, 79.
Lusciniopsis hendersonii, 74.
—— japonica, 73.

Macrorhamphus griseus, 330.
—— —— scolopaceus, 330.
Megalurus pryeri, 79.
Megascops elegans, 189.
Mergus albellus, 258.
—— merganser, 257.
—— serrator, 258.
Merula cardis, 45.
—— eclenops, 50.
—— chrysolaus, 48.
—— fuscata, 46.
—— naumanni, 47.
—— obscura, 49.
—— pallida, 47.
Milvus ater, 197.
—— —— melanotis, 197.
—— melanotis, 197.
Mimogypes, 180.
Monticola cyanus, 53.
—— —— solitaria, 53.
Montifringilla brunneinucha, 128.
Mormon corniculata, 280.
—— superciliosa, 286.
Motacilla alpina, 55.
—— amurensis, 111.
—— aurorea, 57.
—— blakistoni, 111.
—— boarula, 114.
—— calliope, 52.
—— cervina, 117.
—— cyane, 53.
—— cyanurus, 58.
—— flava, 111.

Motacilla grandis, 113.
—— japonica, 113.
—— lugens, 111.
—— lugubris, 113.
—— luteola, 60.
—— maura, 57.
—— melanope, 114.
Motacillinæ, 111.
Muscicapa cinereo-alba, 63.
—— cyanomelæna, 59.
—— gularis, 59.
—— hylocharis, 61.
—— latirostris, 62.
—— melanoleuca, 59.
—— narcissina, 61.
—— sibirica, 62.
Muscipeta princeps, 63.
—— principalis, 63.
Musophagi, 168.

Niltava cyanomelæna, 59.
Ninox scutulata, 187.
Nisaetus nipalensis, 200.
Nucifraga caryocatactes, 99.
—— —— leptorhynchus, 99.
Numenius arquatus, 314.
—— —— lineatus, 314.
—— cyanopus, 315.
—— minutus, 317.
—— phæopus, 316.
—— —— variegatus, 317.
Nycticorax crassirostris, 222.
—— goisagi, 223.
—— javanicus, 224.
—— —— stagnatilis, 225.
—— nycticorax, 222.
—— prasinosceles, 225.

(Estrelata hypoleuca, 269.
Oidemia americana, 248.
—— stejnegeri, 252.
Oriolus squamiceps, 65.
Otis dybowskii, 355.
Otus scops japonicus, 190.
—— semitorques, 188.

Palamedeæ, 259.
Pandion haliaetus, 195.

Parinæ, 79.
Parus ater, 82.
—— atriceps, 83.
—— —— minor, 84.
—— castaneiventris, 86.
—— caudatus, 87.
—— commixtus, 84.
—— minor, 83.
—— palustris, 81.
—— —— baikalensis, 81.
—— —— japonicus, 81.
—— sieboldi, 85.
—— trivirgatus, 87.
—— varius, 85.
—— —— castaneiventris, 86.
Passer montanus, 130.
—— russatus, 131.
—— rutilans, 131.
—— saturatus, 130.
Passeres, 42.
Passeriformes, 42.
Pelecano-Herodiones, 207.
Pelecanus carbo, 208.
—— leucogaster, 212.
—— minor, 214.
—— piscator, 213.
Pelidna pacifica, 334.
Pericrocotus cinereus, 106.
—— teginæ, 107.
Pernis apivorus, 107.
Phaeton rubricauda, 213.
Phalacrocorax bicristatus, 211.
—— capillatus, 209.
—— carbo, 208.
—— pelagicus, 210.
Phalaris cristatella, 286.
Phalaropus fulicarius, 318.
—— hyperboreus, 318.
Phasianus scintillans, 371.
—— sœmmeringi, 370.
—— torquatus, 369.
—— versicolor, 370.
Phœnicopteri, 233.
Phyllopneuste borealis, 69.
Phylloscopus borealis, 69.
—— coronatus, 69.
—— fuscatus, 79.
—— tenellipes, 70.

Phylloscopus xanthodryas, 70.
Pica caudata, 102.
—— —— kamtschatkensis, 103.
—— —— leucoptera, 103.
Picariæ, 173.
Pico-Passeres, 42.
Picus awokera, 147.
—— canus, 148.
—— —— yessoensis, 148.
—— kisuki, 156.
—— leuconotus, 152.
—— —— subcirris, 153.
—— major, 153.
—— —— japonicus, 154.
—— martius, 149.
—— minor, 155.
—— namiyei, 153.
—— noguchii, 151.
—— richardsi, 149.
Pinicola enucleator, 122.
Platalea leucorodia, 229.
—— major, 229.
—— minor, 231.
—— pygmæa, 338.
Plataleæ, 229.
Pluvianus cinereus, 311.
Podiceps cornutus, 367.
—— cristatus, 365.
—— minor, 367.
—— nigricollis, 366.
—— rubricollis, 364.
—— —— holbœlli, 365.
—— —— major, 364.
Porzana phæopyga, 358.
—— undulata, 358.
Pratincola maura, 57.
Procellaria glacialis, 268.
—— grisea, 266.
—— furcata, 271.
—— leachi, 270.
—— leucomelas, 264.
—— melania, 270.
—— nigra, 265.
—— tenuirostris, 267.
Procellariidæ, 269.
Psittaci, 181.
Puffinidæ, 263.
Puffinus brevicaudus, 267.

Puffinus carneipes, 265.
—— griseus, 266.
—— leucomelas, 264.
—— tenuirostris, 267.
Pygopodes, 361.
Pyrrhula erythrina, 123.
—— grisciventris, 129.
—— —— rosacea, 129.
—— —— kurilensis, 129.
—— orientalis, 129.
—— sanguinolentus, 124.

Rallus aquaticus, 359.
—— fuscus, 357.
—— pusillus, 356.
Raptores, 182.
Ratitæ, 375.
Regulus cristatus, 80.
—— —— orientalis, 80.
—— japonicus, 80.
Rheæ, 375.
Rhynchæa capensis, 340.
Ruticilla aurorea, 57.

Salicaria cantans, 74.
—— cantillans, 76.
—— brunneiceps, 78.
—— turdina orientalis, 71.
Saxicola rubicola, 57.
Scansores, 146.
Schœnicola pyrrhulina, 133.
—— yessoensis, 132.
Scolopax arquata, 314.
—— australis, 342.
—— calidris, 320.
—— capensis, 340.
—— fusca, 319.
—— gallinago, 346.
—— gallinula, 344.
—— glottis, 321.
—— grisea, 330.
—— incana, 323.
—— phæopus, 316.
—— rusticola, 347.
—— solitaria, 342.
—— —— japonica, 343.
—— stenura, 345.
—— terek, 326.

Scops elegans, 188.
—— japonicus, 189.
—— —— pryeri, 190.
—— —— scops, 189.
—— —— semitorques, 188.
Serpentarii, 206.
Simorhynchus cassini, 286.
Siphia luteola, 60.
Sitta albifrons, 92.
—— amurensis, 92.
—— —— clara, 93.
—— cæsia, 92.
—— —— amurensis, 93.
—— uralensis, 92.
Somateria spectabilis, 256.
—— stelleri, 257.
Spizaetus nipalensis, 200.
—— orientalis, 201.
Steganopodes, 207.
Stercorarius buffoni, 289.
—— pomarinus, 289.
—— richardsoni, 288.
Sterna aleutica, 299.
—— anæstheta, 301.
—— bergii, 299.
—— dougalli, 295.
—— fuliginosa, 302.
—— longipennis, 296.
—— melanauchen, 297.
—— sinensis, 298.
—— stolida, 300.
Sternula placens, 298.
Strepsilas interpres, 331.
Striges, 182.
Strix brachyotus, 187.
—— fuscescens, 185.
—— hirsuta japonica, 187.
—— nyctea, 185.
—— otus, 186.
—— rufescens, 186.
—— scops, 189.
—— —— japonicus, 189.
—— scutulata, 187.
—— uralensis, 185.
—— —— fuscescens, 185.
Struthiones, 376.
Struthioniformes, 375.
Sturnia pyrrhogenys, 108.

Sturninæ, 107.
Sturnus cineraceus, 107.
—— sericeus, 108.
Sula fiber, 212.
—— leucogastra, 212.
—— piscatrix, 213.
—— sinicadygna, 212.
Surnia nyctea, 185.
Sylvia akahige, 50.
—— cisticola, 77.
—— diphone, 77.
—— komadori, 52.
—— lanceolata, 73.
—— ochotensis, 73.
Sylviinæ, 68.

Tadorna cornuta, 241.
—— rutila, 241.
Tantalus melanocephalus, 232.
Tarsiger cyanurus, 58.
Terpsiphone princeps, 63.
Tetrao bonasia, 373.
—— mutus, 372.
—— —— rupestris, 372.
Thalasseus pelecanoides, 300.
—— poliocercus, 300.
Totanus acuminatus, 339.
—— brevipes, 323.
—— calidris, 320.
—— fuscus, 319.
—— glareola, 324.
—— glottis, 321.
—— hypoleucus, 326.
—— incanus, 323.
—— —— brevipes, 323.
—— ochropus, 325.
—— pugnax, 327.
—— stagnatilis, 322.
—— terekius, 326.
Treron permagna, 164.
—— sieboldi, 163.
Tribura squameiceps, 74.
Tringa acuminata, 339.
—— alpina, 334.
—— —— pacifica, 334.
—— arenaria, 336.
—— canutus, 333.
—— crassirostris, 332.

Tringa fulicaria, 318.
—— glareola, 324.
—— helvetica, 304.
—— hyperborea, 318.
—— hypoleucos, 326.
—— interpres, 331.
—— maritima, 335.
—— minuta, 337.
—— —— ruficollis, 337.
—— ochropus, 325.
—— platyrhyncha, 337.
—— pugnax, 327.
—— pygmæa, 338.
—— subminuta, 338.
Tringoides empusa, 327.
Trochili, 145.
Troglodytes fumigatus, 89.
—— —— kurilensis, 91.
Trogones, 150.
Trynga ruficollis, 337.
Tubinares, 260.
Turdinæ, 43.
Turdus amaurotis, 64.
—— cardis, 45.
—— chrysolaus, 48.
—— cœnops, 50.
—— cyanus, 53.
—— daulias, 48.
—— eunomus, 46.
—— fuscatus, 46.
—— jouyi, 49.
—— manillensis, 54.
—— naumanni, 47.
—— obscurus, 49.
—— pallidus, 47.
—— sibiricus, 44.
—— solitarius, 54.
—— terrestris, 44.
—— varius, 43.
Turnix blakistoni, 354.
—— pugnax, 354.
Turtur humilis, 162.
—— orientalis, 160.
—— risorius, 162.
—— stimpsoni, 161.

Upupa epops, 159.
Upupæ, 158.

Uria brevirostris, 279.
—— mystacea, 286.
—— pusilla, 287.
—— wumizusume, 277.

Vanellus cristatus, 312.
Vultur albicilla, 198.

Xanthopygia narcissina, 61.

Xanthopygia tricolor, 61.

Yunx japonica, 158.
—— torquilla, 157.

Zosterops japonica, 68.
—— loochooensis, 67.
—— palpebrosa, 67.
—— —— nicobarica, 67.

www.ingramcontent.com/pod-product-compliance
Lightning Source LLC
Chambersburg PA
CBHW020101020526
44112CB00032B/688